Emmanuel Stefanakis

# Geographic Databases & Information Systems

Stefanakis, E., 2014. *Geographic Databases and Information Systems*.
Edition 1.1; pp. 386; Fredericton, NB, Canada; March 2016.


Cover: "Thematic Layers in Space", by Eleni Ioannidou, 2003.

*In memory of*

*Professor Y.C. Lee*

# Preface

The management and analysis of geographic data is a major scientific domain nowadays. It is estimated that 80% of economic and political decisions worldwide engage explicitly or implicitly with some form of geographic information. The Geographic Information Systems (GIS) has been a response technology to meet the urgent demands of users and applications.

The aim of this textbook is to present GIS from a technological perspective. Emphasis is given to the core of these systems, which is the Geographic Database (GDB) along with the corresponding Database Management System (GDBMS). These two components are largely responsible for the performance and efficiency of a GIS. The fundamental methods and algorithms to analyze the geographic data are also at the heart of the debate.

The textbook is organized into fifteen (15) chapters, which are grouped into three (3) parts (Figure 1). Part I discusses the nature of geographic data (Chapter 1), the coordinate systems associated with the earth's surface (Chapter 2), the projection of the earth's surface on paper and digital maps (Chapter 3), and concludes with an introduction to geographic information systems (Chapter 4).

In Part II, the discussion focuses on the technology of database systems. After presenting the basic concepts of these systems (Chapter 5), the process of their design is elaborated (Chapter 6). Then, some advanced database system design topics are visited (Chapter 7) along with the mechanisms of structuring data to support efficient management (Chapter 8).

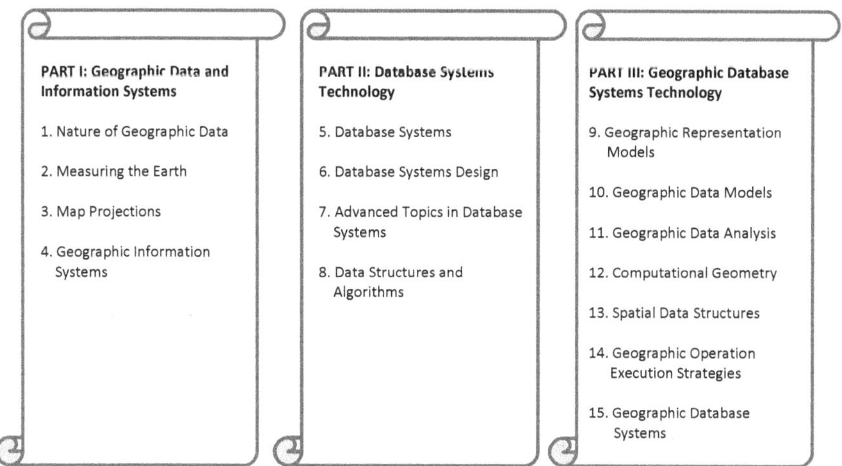


**Fig. 1.** Organization of content into three parts and fifteen chapters.

In Part III the discussion relates to geographic databases. The representation and data models for geographic data are presented (Chapters 9 and 10) followed by a classification of geographic data analysis operations (Chapter 11). The principles of computational geometry (Chapter 12), spatial data structures (Chapter 13), and optimization strategies for the execution of geographic procedures (Chapter 14) are discussed, before the state-of-the-art in geographic database systems technology (Chapter 15) is presented.

This resource aims to provide an aid and reference to both students (undergraduate and graduate level) and professionals. Prior knowledge of basic computer concepts is necessary to understand the content. Readers with discipline in software engineering and database systems can skip the Chapters in Part II.

Hoping that this textbook will fulfill its main objective, which is to contribute to knowledge transfer and promote research, readers are kindly requested to notify the author their comments and concerns, which will contribute constructively in improving the content of future editions.

*Emmanuel Stefanakis*

# Contents

# Part I: Geographic Data and Information Systems

# Part II: Database Systems Technology

# Part III: Geographic Database Systems Technology

# PART I

# GEOGRAPHIC DATA & INFORMATION SYSTEMS

# 1 Nature of Geographic Data

## 1.1 Categories of Geographic Data

*Geographic data* is a special category of data that is distributed in *space* and changes over *time*. On one hand, it is accompanied with references to the locations in space where it occurs or describes. On the other hand, it changes over time. This change can be so slow that can be ignored (e.g., change of the seashore, climate, or distribution of human ages in a country, etc.) or so fast that is of high significance to a domain (e.g., traffic load of a highway, temperature, fire front in a forest fire, etc.).

Geographic data is usually classified into four categories (Figure 1.1; according to Maguire *et al.*): (a) *physical objects* (e.g., buildings, roads, rivers, forests, etc.), (b) *administrative units* (e.g., land parcels, prefectures, national parks, etc.), (c) *geographic phenomena* (e.g., temperature, humidity, accidents, distribution of sea mammals, etc.), and (d) *derived information* (e.g., poverty level, unemployment rate, land suitability for cultivation, etc.).

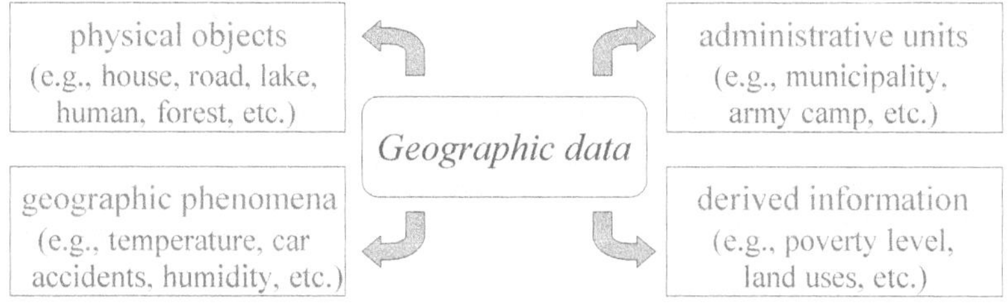


**Fig. 1.1** Categories of geographic data.

## 1.2 Geographic Entities

In order to describe the complex reality, it is assumed that the world consists of a set of discrete and interrelated units, which are named as entities. *Entity* is any unit or object with *physical* or *conceptual* existence.

A simplified geographic application in the domain of *cadastre* will be examined next, in order to make the concept of the entity clear. The scope of a cadastre is to record the estates, owners, and related ownerships.

In the simplified example of Figure 1.2, the following *physical entities* are involved in a cadastral application: the estates (two land parcels and one building), and the physical person (the human as owner of the estates).

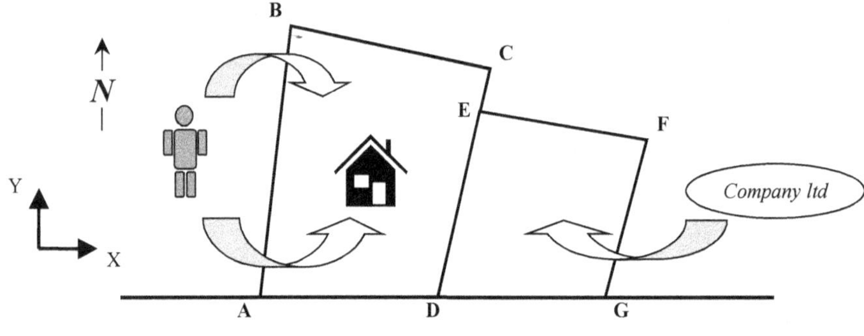



**Fig. 1.2** Physical and conceptual entities in a cadastral application (arrows represent the ownerships).

In addition, a *conceptual entity* is involved in the cadastral application: the Company as a legal entity and owner of one parcel. Conceptual entities (company, bank, municipality, etc.) have no physical existence, however, they exist conceptually and this is the reason they are treated as entities.

The individual entities (physical or conceptual) are interrelated each other in different ways depending on the application domain to reflect the semantics of reality. In the cadastral application (Figure 1.2) the ownerships of owners on estates represent the *relationships* between the entities.

## 1.3 Dimensions of Geographic Entities

Each entity is described by a number of *attributes* (properties or characteristics), which are explicitly related to the application. For example, the entity "building" has the following attributes in a cadastral application: address, area, date of construction, etc. The attributes associated with geographic entities are classified into three basic categories, which are named as *dimensions* of geographic entities. These dimensions are (Figure 1.3): (a) *identifier*, (b) *spatial* dimension, and (c) *thematic* dimension. Following, these dimensions are described in brief.

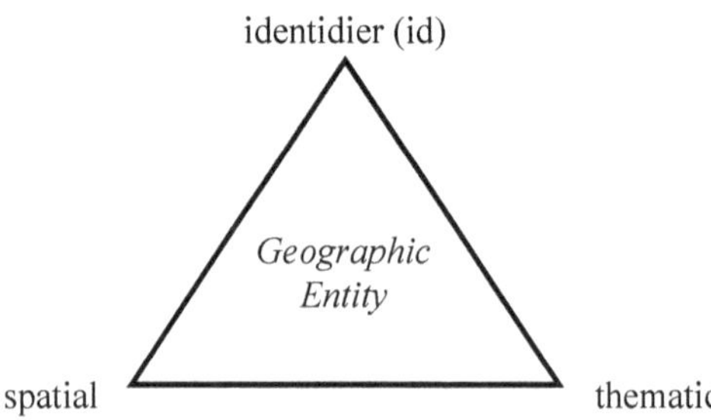



**Fig. 1.3** Dimensions of geographic entities.

## *1.3.1 Identifier*

The *identifier* is the dimension which provides a means to refer to (identify) geographic entities. Entities are usually assigned arbitrary names, geographic names, or numeric codes as identifiers. For example, in the cadastral application each individual physical or conceptual entity is assigned a unique identifier.

Physical persons (humans) are identified by their social insurance number (SIN); parcels by appropriate cadastral codes (e.g., a Cadastral Parcel Code, CPC); and legal entities by unique brand names (e.g., City of Fredericton or Walmart).

It is important for the system assigning the identifiers to be precise, consistent, and able to guarantee uniqueness. In practice, the establishment of such a system has proved a challenging task.

## *1.3.2 Spatial Dimension*

The *spatial dimension* incorporates all attributes that describe the spatial characteristics of geographic entities (Figure 1.4): (a) *position*, (b) *geometry*, (c) *graphical properties*, and (d) *spatial relations* with other entities.

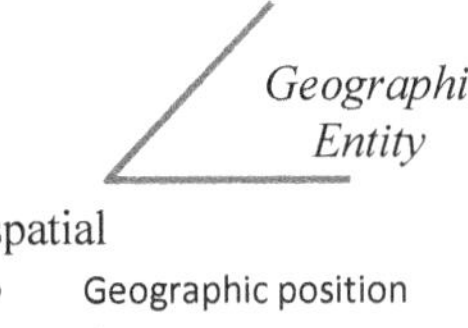


**Fig. 1.4** The spatial dimension of geographic entities.

### 1.3.2.1 Geographic Position

The *geographic position* is needed for locating or delimiting geographic entities on or around the earth's surface. The locational definitions can be rather complicated, because geographic data tends to occur in irregular and complex patterns. The geographic position can be specified using: (a) a *direct* (absolute) position (in reference to a coordinate system, e.g., Cartesian, projective, or geographic), (b) an *indirect* position (e.g., the address: street name, number, zip code, etc.), or (c) a *relative* position (e.g., north of, inside, etc.) to a reference point or entity.

In the cadastral example (Figure 1.2), the geographic position of the land parcel ABCD can be defined by: (a) a direct position (e.g., the X,Y coordinates – in the established Cartesian coordinate system – of its vertices or center point), (b) an indirect position (e.g., the parcel address in the established street network: 110 Maple street, Fredericton), or (c) a relative position to another entity (e.g., stating that the parcel lies west of parcel EFGD by sharing the arc DE).

### 1.3.2.2 Geometry

*Geometric attributes* describe the basic geometric properties of geographic entities, such as the shape, perimeter, areal extent, or volume. These attributes are derived from the direct position of geographic entities and are usually treated as non-spatial (thematic) attributes.

Example geometry attributes for the entities in Figure 1.2 are: parcel ABCD is a polygon with four edges; its area is equal to $800m^2$.

### 1.3.2.3 Graphical Properties

Geographic entities are usually visualized in graphical representations, such as maps or diagrams. The visualization (mapping) of geographic entities is realized using appropriate graphics, which are called *graphical properties*. In a map visualization, the graphical properties of an entity specify the (map) *symbols* used for the depiction of the entity on a paper or digital map.

The symbols are usually specified in advance and their parameters are described in a *symbol library*. The symbol library plays the role of the dictionary and assists the communication through diagrams and maps. Some of the typical symbols, adopted for the representation of geographic entities in topographic maps, are: the red line for the road network, black dot for the villages, green patterns for the vegetation, etc. Figure 1.2 shows the representation of geographic entities in a cadastral diagram. The black and continuous line represents the parcel outlines, icon represent the building, and so on. The *scale* is a basic parameter for the selection of the appropriate symbols (graphical attributes) in metric representations (e.g., maps and plans). A geographic entity may be assigned different graphical properties for different scales (multi-resolution or granularity; see Chapter 9).

### 1.3.2.4 Spatial Relations

*Spatial relations* describe the relationships between geographic entities as a result of their relative positions (e.g., north of, adjacent to, near to, inside, etc.). Spatial relationships (i.e., topological, direction, distance) are generally numerous and may be complex. They may be either explicitly defined or calculated as needed.

Example spatial relations in the cadastral application are the following: the building lies inside the parcel ABCD, parcels ABCD and EFGD are adjacent and share the arc DE, parcel EFGD lies north of the street and east of the parcel ABCD, etc. Spatial relations are examined in Chapter 9.

### *1.3.3 Thematic Dimension*

The *thematic dimension* includes all thematic or non-spatial attributes of geographic entities, e.g., vegetation type, owner's name, road category, etc. Obviously, the thematic attributes that describe geographic entities are closely related to the application domain.

In the cadastral example, the physical persons have the following thematic attributes: name, sex, age, profession, telephone, etc.; while the parcels: land use, vegetation type, etc.

In practice, the identifier is usually treated as a thematic attribute. In here, a clear distinction from the thematic attributes is chosen, due to the property of uniqueness.

Additionally, the thematic dimension accommodates all *multimedia* data that may accompany geographic entities in an application. This data consists of sound, images, video, virtual and augmented reality content; and describes the geographic entities. For example, some multimedia data that might accompany the building in Figure 1.2 includes photos of the exterior before and after a reconstruction or a video tour of the interior.

## 1.4 Dynamic Geographic Entities

Each geographic entity either physical (e.g., parcel, river, road, person) or conceptual (e.g., company, municipality, poverty) has an identifier and other spatial (geometric) and thematic (non-spatial) attributes. Other than the identifier, attribute values might vary over *time*. For example, a parcel may have its boundary or owner altered. A person (owner) may have the address changed, while the age changes continuously. The traffic load in an avenue is subject of continuous change throughout a day. All entities with attributes that change values over time are called *dynamic entities*.

In fact all entities are dynamic in nature, as they are part of a continuously changing world, while time is a dimension that is tightly connected to space. This connection is shown through an example in Figure 1.5. A sea vessel changes location as sailing in the Aegean Sea. The dashed line represents the vessel's trajectory in a 2D (space) and a 3D (space+time) representation. However, in practice some geographic entities are usually treated as *static entities*. This is related to a time interval during which the entity attributes remain unchanged or any change is not of interest to the domain.

It is a common practice in handling spatio-temporal phenomena to examine the changes of spatial and thematic attributes of entities over time. This consideration highlights the *behavior* of an entity over time, namely *entity change*. Change takes various forms, all falling within two general categories: (a) *life*, and (b) *motion*.

Life refers to any change that affects the existence status of entities, whereas motion refers to changes in the spatial and/or thematic attributes of living entities.

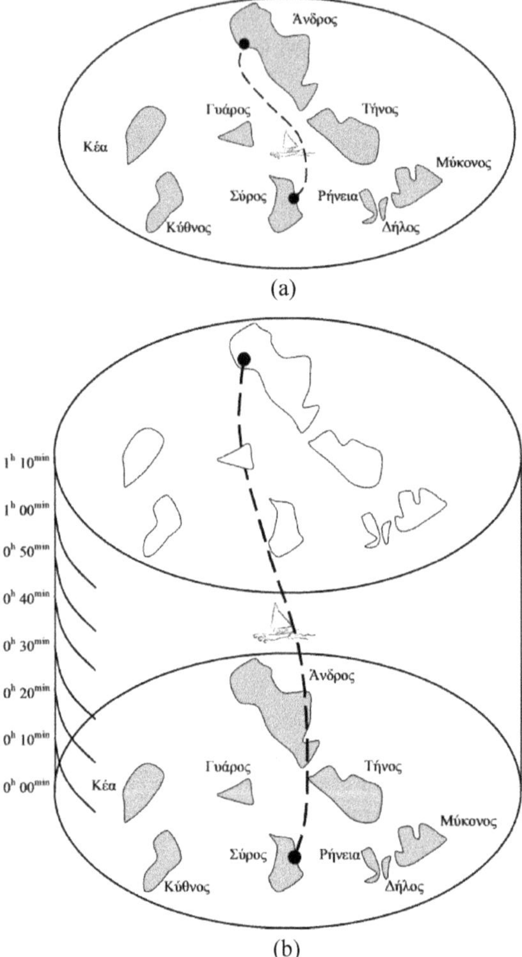


**Fig. 1.5** Space: North Cyclades (a); and space-time: sailing across the North Cyclades in the Aegean Sea (b) (names of the islands in Greek).

### *1.4.1 Life and Motion of Geographic Entities*

The *life* and *motion* of individual entities, physical or conceptual, dominate our existence and perception of the environment that surrounds us. An entity may appear or disappear (e.g., a residential zone or a forest), may merge with other entities or split into two or more new entities (e.g., collection of parcels, or towns). These changes concern the *life* of an entity.

On the other hand, an entity may move with or without simultaneous changes in its form (e.g., air pollution, a boat). Additionally, the thematic attributes of an entity may change (e.g., parcel owner) without any simultaneous change in its spatial attributes. All these changes concern the *motion* of an entity. Figure 1.6 represents some common types of life and motion changes.

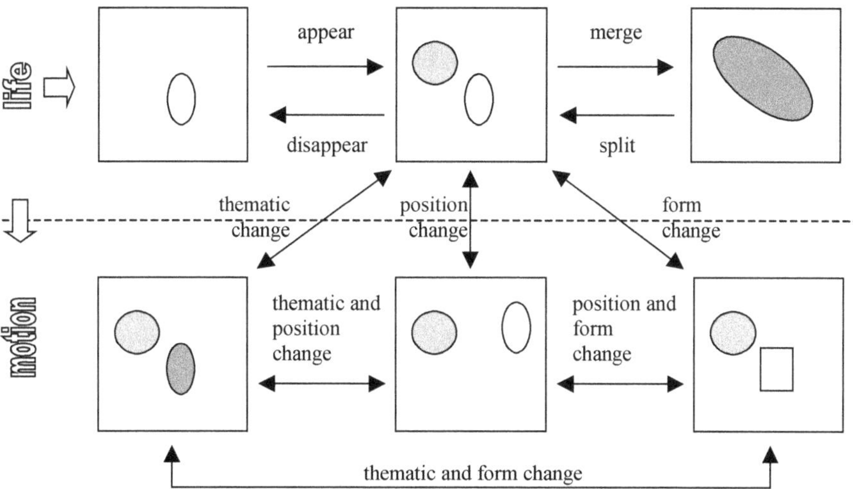


**Fig. 1.6** Life and motion of a geographic entity (based on Frank's model of changes).

## 1.4.2 Temporal Relations between Geographic Entities

The discussion in the previous Section focuses on a geographic entity and the changes over time in regards to its life and motion. When more than one dynamic entity is examined, their *temporal relations* can be considered. These relationships are similar to the spatial relations (e.g., topological) discussed previously.

Temporal relations are generally numerous and may be complex. They are based on the fact that time is continuous and linearly ordered. Hence, for any couple of instant (of zero duration) events, the one will be coincidental with or earlier than the other. For events with non-zero duration, a temporal overlap is also possible to occur.

When temporal relations are of interest, what is most commonly examined is either their ordering in time (e.g., which person or which building is the oldest) or their life overlap if any (e.g., was the building under construction when Company A bought the adjacent parcel?). Obviously, there are applications that may request more complex temporal relations. Chapter 9 examines temporal relations in more detail.

## 1.5 Data Quality

Geographic data are usually collected by applying error prone methodologies. Hence, the attributes of geographic entities (e.g., parcel area, street width, person age, etc.) are assigned values that decline more or less from the real ones.

For a better understanding and exploitation of geographic data, a measure of its *quality* is needed. This measure includes both positional and attribute accuracy of entities, the completeness and logical consistency among data elements. Chapter 9 examines the data quality dimension closely.

## 1.6 Geographic Information

*Data* by itself has a limited value. In order to make it useful, data must be transformed into information (Figure 4.1). *Information* is the outcome of the organization, representation, analysis, and interpretation of data; so that a problem can be solved or a situation (phenomenon) can be understood.

Figure 1.7 shows an example of land classification based on the vegetation type. Data shown in the table are the actual measurements for a sample of nine points in space. Obviously, this data is difficult to understand and interpret. However, a thematic map can be derived from this data, which depicts successfully the vegetation distribution in space to the user.

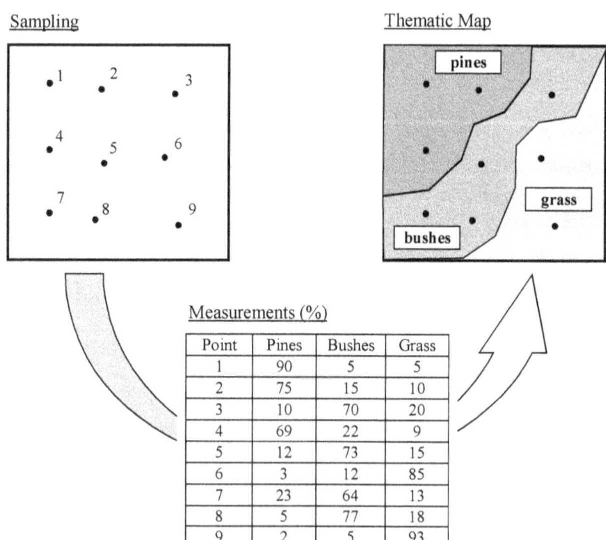


| Point | Pines | Bushes | Grass |
|-------|-------|--------|-------|
| 1 | 90 | 5 | 5 |
| 2 | 75 | 15 | 10 |
| 3 | 10 | 70 | 20 |
| 4 | 69 | 22 | 9 |
| 5 | 12 | 73 | 15 |
| 6 | 3 | 12 | 85 |
| 7 | 23 | 64 | 13 |
| 8 | 5 | 77 | 18 |
| 9 | 2 | 5 | 93 |

**Fig. 1.7** Example of classification based on the vegetation type. Based on a sample of point measurements, a thematic map is generated.

## 1.7 Levels of Measurements

The assignment of values to the attributes of geographic entities based on a rule is called *measurement*. There are four basic levels or scales of measurements: (a) *nominal*, (b) *ordinal*, (c) *cardinal*, and (d) *rational*; which are described next.

### *1.7.1 Nominal Level*

The *nominal* level of measurements assigns identifiers or other qualitative values to the attributes of geographic entities. These values are not subject of any ordering. Typical examples are place names, person identifiers (e.g., social insurance numbers – SIN), names assigned to people or public buildings, etc.

The nominal level of measurement is implemented through arithmetic values, text (alphanumeric characters), or colours. Notice that even in case of pure arithmetic values, the performance of any arithmetic operations makes no sense. For example, the addition of two SINs makes no sense.

### *1.7.2 Ordinal Level*

The *ordinal* level of measurements introduces the notion of *ordering*. The entities are compared to each other and can be ordered based on the values assigned to an attribute. Usually this ordering is done through the correspondence to predefined categories (classes). An example might be the ordering of soils based on moisture into the following categories: "1=dry, 2=light wet, 3=wet, 4=very wet". Another example might be the ordering of road segments into: "1=national road, 2=highway, 3=provincial road, 4=rural road", based on some specifications.

The performance of arithmetic operations or the extraction of ratios from the categories above makes no sense, provided that category (value) 2 is not twice as big as category (value) 1. There is no indication about the difference between two successive categories (values) in the ordering scheme. For example, there is no indication whether the difference in moisture between the categories "dry" and "light wet" is equal to the one between the categories "wet" and "very wet".

### *1.7.3 Cardinal Level*

In the *cardinal* level of measurements (also referred to as *interval* level) the differences between the measurements are clear and measurable. This is achieved through the assignment of values from a (linear) scale with an *arbitrary origin* (null value) and an *arbitrary unit* of measurement (step).

An example of the cardinal level of measurements is the scale of temperature in degrees Celsius. Assume four regions A, B, C, and D with temperatures 10, 20, 30, and 45 degrees Celsius, respectively. It can be stated that the difference in temperature between the regions A and B is equal to the difference between the regions B and C (Celsius scale); or that the difference in temperature between the regions A and B is less than the difference between the regions C and D.

However, it can never be assumed that region B is twice as warm as region A, provided that the 20 degrees Celsius do not correspond to weather conditions that are double as warm as the 10 degrees Celsius. This is because the temperature origin of 0 degrees Celsius has been chosen arbitrarily. The same condition applies to all scales with an arbitrary origin that lacks of a physical meaning.

As a consequence, deriving ratios at the cardinal level makes no sense. If this makes sense, the rational level of measurements is in use, which is described in the following Section.

The schemes used to measure the time also belong to the cardinal levels of measurements, provided that the origin (time=0) was arbitrarily chosen. Time is measured based on: Gregorian calendar (A.D.), or the Islamic calendar (1A.H.=622AD), etc. Hence, any sentence of type "year 2000 is double of year 1000" is meaningless, as year 0 is a convention in all these calendars without any physical meaning.

## *1.7.4 Rational Level*

The *rational* level is the most complete level of measurements. It is based on scales with an *absolute origin* (null value) which has a physical meaning, and an *arbitrary unit* of measurement.

An example of the rational level of measurements is the scale of weight. The zero weight (origin) is not chosen arbitrarily (like the zero temperature in degrees Celsius), and it has a physical meaning, given that a weight of 0Kgr is equal to a nil force. Obviously, the extraction of ratios makes sense. If one value is multiple of another one, the former represents the corresponding multiple quantity. For example, one person of 100Kgr weights double of (twice as much as) another person of 50Kgr. Other examples of attributes which are assigned values based on the rational level of measurements are the population of a town, the age of a person, etc.

## *1.7.5 An Example: Assignment of Values to the Athletes of a Race*

An example for the levels of measurements often found in literature refers to the assignment of values to the athletes of a Marathon race. Based on the *nominal* level of measurements, each athlete is assigned a number e.g., 238 or 589. This number serves as identification only. Additionally, each athlete is characterized as "man" or "woman" based on gender. In other words, the nominal measurements do not correspond to a scale; instead they are handled as sets (Figure 1.8).

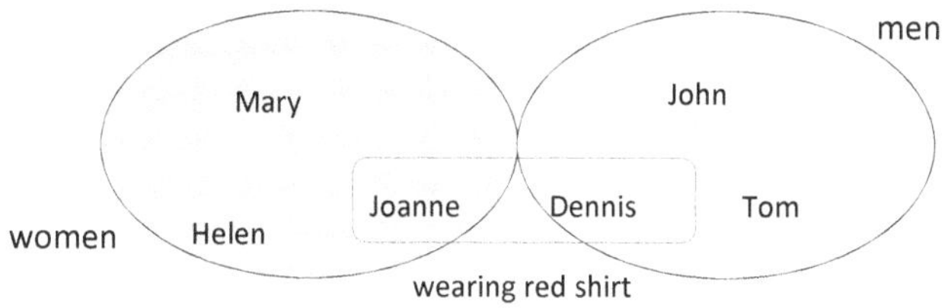


**Fig. 1.8** The nominal level of measurements. The athletes are grouped into sets.

Applying the *ordinal* level of measurements, the athletes are ordered based on their finish sequence, as it is shown in Table 1.1. In this Table there is no indication about the distance between two successive measurements, e.g., how close was the finish between Joanne and Mary.

**Table 1.1** Ordering of athletes based on their finish.

| Winners (Order) | Women | Men |
| --- | --- | --- |
| 1$^{st}$ | Joanne | Dennis |
| 2$^{nd}$ | Mary | Tom |
| 3$^{rd}$ | Helen | John |

The *cardinal* level of measurements assigns the athletes the time of their finish (Figure 1.9). This makes the interpretation of differences between the values feasible. Specifically, if the difference in finish time between two athletes is short, then they had finished close to each other (e.g., Tom versus Joanne) and vice-versa.

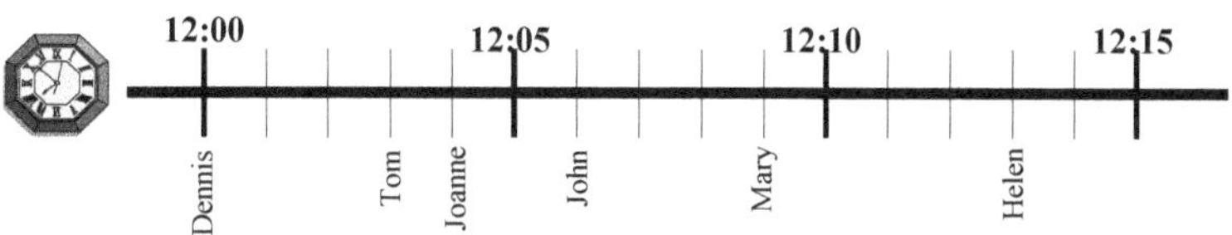


**Fig. 1.9** The cardinal level of measurement. Time of finish for each athlete.

The *rational* level of measurements is derived by subtracting two cardinal measurements; e.g., the time of start for each athlete is subtracted from the corresponding time of finish. This results in the duration of the race for each athlete (Figure 1.10). The speed of each athlete can also be derived and therefore any rational comparison makes sense.

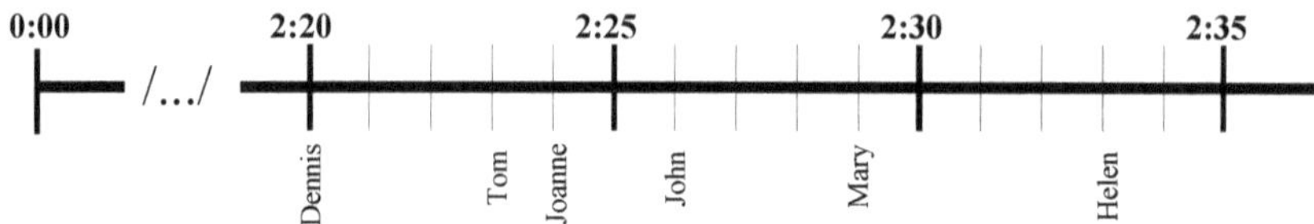


**Fig. 1.10** The rational level of measurements. Race duration for each athlete.

### *1.7.6 Other Levels of Measurements*

The four levels of measurements (Table 1.2) described in the previous Sections are not exhaustive. In geosciences, the management of data with peculiarities is commonly needed. For instance, data may be *directed* or *cyclic*, e.g., the azimuth or the geographic longitude. In the azimuth, the value (angle) following $359°$ (degrees) is $0°$. The average of two directions $359°$ and $1°$ is $180°$. This means that two directions that nearly coincide with the North, give as an average the direction towards the South! This problem is similar to the one of the computer virus Y2K. This problem rose because the year after (19)99 was year (20)00 and not (19)00. All computer systems that were recording the year with two digits faced that problem on the 2000-01-01.

Another problem arises from the expression of angles in degrees-minutes-seconds, e.g., $80°35'22''$. If appropriate programming is missing, the addition of one minute in the angle $32°59'$ will result $32°60'$ instead of $33°00'$. This problem is avoided with the expression of angles in decimal degrees, e.g., $32°59' = 32.9833...°$, prior to the execution of any arithmetic operation.

**Table 1.2** A description of the four levels of measurements.

| Level | Properties | Valid operations |
| --- | --- | --- |
| Nominal | Identification, no subject of any ordering | equation (=) |
| Ordinal | Ordering without any notion of spaces between the classes. | ordering (>, <) |
| Cardinal | Contnuous measurements, with specific distances from an arbitrary origin. | addition, substraction, scale change |
| Rational | Contnuous measurements, with specific distances from an absolute origin. | multiplication, division (rates) |

## 1.8 Units of Measurements

All values assigned to the attributes of a geographic entity are either *qualitative* (nominal and ordinal levels) or *quantitative* (cardinal and rational levels) in nature.

For the quantitative measurments, both the origin (null value) and the unit of measurement are defined. The value (derived value) that is assigned to an attribute of an entity is the result of the comparison of a quantitative measurement (magnitude) against a homogeneous *unit of measurement*, as follows:

$$value = \frac{magnitude}{unit\ of\ measurement}$$

The units of measurements for quantities are defined by various international or national standards. One of the most commonly used standard is the International Units System (S.I.). Table 1.3 shows some basic units of measurements in geosciences.

**Table 1.3** Quantities and the corresponding units of measurements.

| Quantity | Units of Measurements |
| --- | --- |
| Length | 1 meter (m), 1 mile, 1 yard |
| Angle | 1 degree (deg), 1 grade (grad), 1 radian (1 rad) |
| Area | 1 square meter ($m^2$), 1 hectare (=10000 $m^2$) |
| Volume | 1 cubic meter ($m^3$) |
| Temperature | 1 degree Celsius (°C) |
| Pressure | 1 atmosphere (Atm), 1 mm Hg |
| Time | 1 second (sec), 1 minute (min), 1 hour (h) |

## 1.9 Geographic Data Sources

Geographic data sources are classified into two categories: (a) *primary sources*, and (b) *secondary sources*. The execution of a project seeks the use of data that is collected and processed by others in the past, such as maps, statistic data, etc. This data falls in the category of secondary sources. Should this data be unavailable, unreliable, expensive, or confidential, new data must be collected directly from the ground, by applying surveying and/or census methodologies. This new data falls in the category of primary sources.

The use of existing data usually costs less. In many cases, the updating and correction of secondary source data from primary source data is imposed. For instance, assume the scenario of a geographer who undertakes a project in Nova Scotia's Cape Breton Island in 2014. The geographer makes use of a map (secondary source) of scale 1:25.000 compiled by the national cartographic agency, last revised in 1999. The geographer might need to collect some primary data in order to update the content of the map (e.g., survey and add a new road constructed in 2005) and/or enrich the content with features that do not appear in a typical topographic map, but are of interest to the project (e.g., the traffic load in the road network).

The most important primary and secondary resources for geographic data are the following:

- *Geodetic measurements:* Geodetic measurements (topography and geodesy) are primary sources of geographic data. They record the spatial (geometric) dimension of geographic entities, through direct and indirect measurements.
- *Census and sampling:* These methodologies also fall into the primary sources. They record demographic data by applying survey methods.
- *Air-photographs and satellite images:* The aerial photographs and satellite images constitute a rich primary source of geographic data. By applying photogrammetric, remote-sensing and photo-interpretation methodologies the extraction of spatial and thematic data for the study area becomes feasible.
- *Existing maps:* All existing maps constitute a worthy secondary source of geographic data. Data depicted on a traditional map refers to both the spatial and thematic dimension of geographic entities.
- *Other data:* This data may be quantitative and/or qualitative in nature and refers to social, economic or environmental phenomena. It is usually the outcome of past projects available to the user.

Geographic data is available in analog or digital form. During the last few decades digital data become more and more available. Nowadays, the largest organizations collecting and disseminating geographic data worldwide adopt digital technologies. Depending on the country, the basic geographic data providers are: the National Cartographic Organizations, Military (Department of Defense), Cadastral and Mapping Agents, Statistical Services, Technical and Economic Chambers, other private agents, etc.

Lately, the Earth browsers (Google Maps, Bing Maps, etc.) are rapidly being transformed into the main resource of geographic data and background maps for geovisualizations and web maps. This data is usually very reliable and free to use. The Volunteered Geographic Information (VGI) is also a recent initiative and constitutes a very important resource of geographic data. VGI is the harnessing of tools to create, assemble, and disseminate geographic data provided voluntarily by individuals. Some examples VGI sources are Wikimapia, OpenStreetMap, and Google Map Maker.


## References and Further Reading


Aronoff, S., 1989. *Geographic Information Systems: A Management Perspective*. WDL Publications.

Burrough, P.A., and McDonnell, R.A., 1998. *Principles of Geographic Information Systems*. Oxford University Press.

Frank A., Raper, J., and Cheylan, J.P. (Eds), 2001. *Life and Motion of Socio-Economic Units*. GISDATA Series, No. 8. Taylor and Francis.

Longley, P.A., Goodchild, M.F., Maguire, D.J., and Rhind, D.W., 2001. *Geographic Information Systems and Science*. Wiley.

Maguire, D.J., Goodchild, M.F., and Rhind, D.W. (Eds), 1991. *Geographic Information Systems: Principles and Applications*. Longman Scientific and Technical.

Malczewski, J., 1999. *GIS and Multicriteria Decision Analysis*. John Wiley & Sons.

Stefanakis, E., and Sellis, T., 1997, Towards the design of a DBMS repository for the application domain of GIS: requirements of users and applications. In *Proceedings of the 18th International Cartographic Conference*, Stockholm, Sweden, pp. 2030-2037.

Walford, N., 1995. *Geographical Data Analysis*. Wiley.

# 2 Measuring the Earth

## 2.1 Introduction

People have been concerned about the *shape and size of the Earth* since antiquity
(Figure 2.1). Egyptians performed the first systematic measurements on the earth's
surface to be able to redefine their farms after the frequent flooding of the Nile
River. Phoenicians and Greeks proposed the spherical earth. Eratosthenes (276-
192 BC) was the first who estimated the radius of the earth with a high degree of
precision.

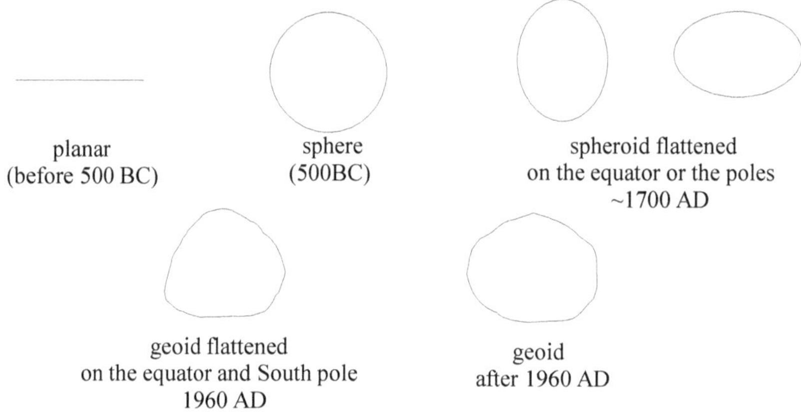


**Fig. 2.1** The evolution of the perception for the Earth's shape.

Every year, on the summer solstice, at local noon in the Ancient Egyptian city
of Swenet, the Sun rays were reflecting directly off the water surface of a deep
well. At the same time, Eratosthenes, measured the angle of the Sun rays ($\varphi$) on a
vertical tower in Alexandria, located 800km away (Figure 2.2).

Assuming that the Sun rays in Swenet and Alexandria were parallel (provided
the long distance between the Sun and the Earth), Eratosthenes equated this angle
($\varphi$) with the spherical angle of the arc (S) connecting the well with the tower. The
computation of the earth's radius was achieved through the formula $R = S/\varphi$. The
result was exceptionally precise, given the errors in the measurement of the well-
tower distance (S) and the angle ($\varphi$), through the length of the tower's shadow.

After the 17th century AD, European astronomers, mathematicians and physi-
cists examined closely the earth's shape and size. The actual size and shape of the
earth was realized only after the technological evolutions in microwaves and elec-
tronics, the manufacturing of high precision instruments, the development of computers, and the evolution in satellite technologies, photogrammetry, etc., which
took place in the second half of the 20th century.

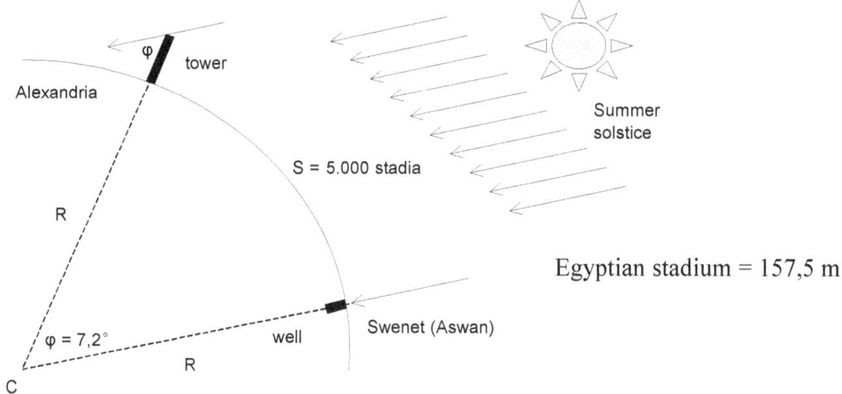


**Fig. 2.2** Computation of the Earth's radius (R = S/φ) by Eratosthenes.


## 2.2 Earth's Shape and Size

The *Earth's surface* (land surface) is characterized by an irregular shape with large geomorphological formations (Figure 2.3). The tops of the mountains reach up to 9,000 meters above the sea and the depths of the oceans reach 11,000 meters. Obviously, the height difference between these points is very small compared with the dimensions of the earth (approx. radius: 6,400 kilometers).

Due to the rotation, the land has taken a spheroid shape flattened at the poles, and approached by an *ellipsoid* (Figure 2.4). In reality, the shape of the earth differs from the ellipsoid, because of the (a) heterogeneous material, and (b) uneven distribution of the masses.

The shape of the earth is approached better by an imaginary surface, the surface of the *geoid*. The geoid is the equipotential surface of earth's gravity field, which is best adapted to the *mean sea level (MSL)*. In other words, the geoid is the surface formed, by the sea, calm in the middle level, also being extended under the continents (Figure 2.5).

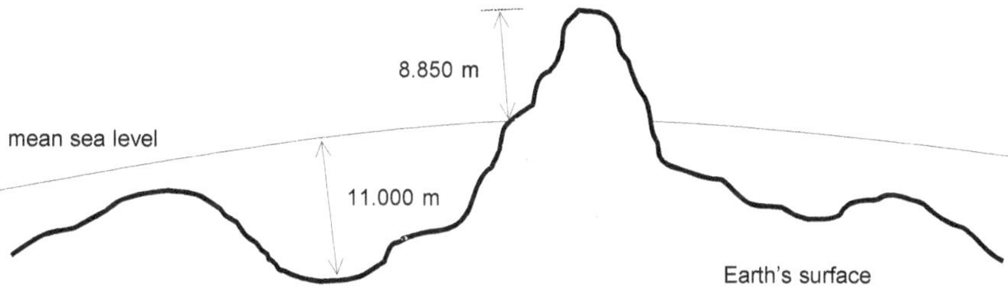


**Fig. 2.3** The formations of the Earth's (land) surface.

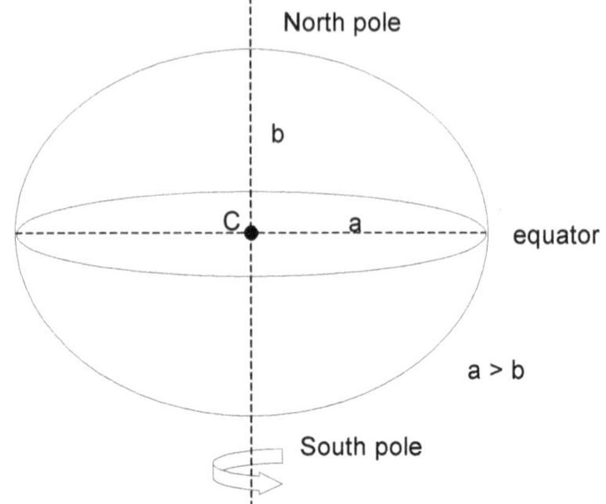


**Fig. 2.4** The spheroid (ellipsoid) as a mathematical model of the Earth.

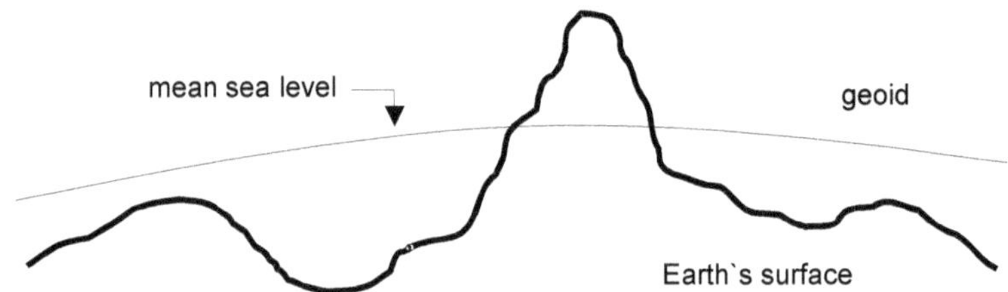


**Fig. 2.5** The surface of the geoid (the mean see level as an approximation of the geoid).

Specifically, the geoid is defined as the equipotential surface traction and rotation of the earth, which approximates the mean sea level (MSL) with an accuracy of ±1m, corrected for the effects of changes in the water density, waves, tides and currents.

The surface of the geoid by definition is perpendicular to the direction of gravity; i.e., the *vertical*. Therefore, there is a one-to-one correspondence between each point P of the earth's surface with a single point $P_G$ of the geoid (Figure 2.6).

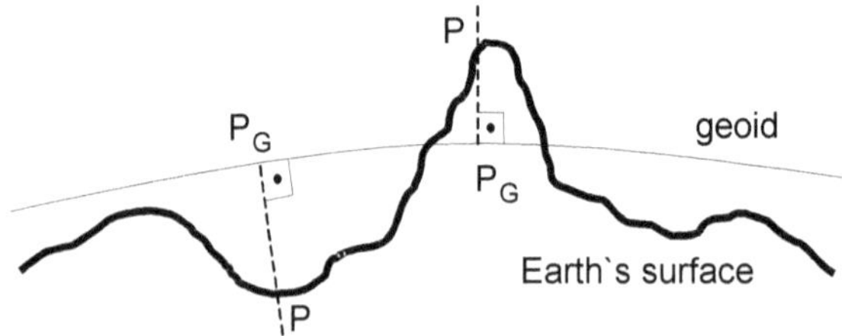


**Fig. 2.6** The Earth's surface and the geoid.

The geoid is an irregular wavy surface, which cannot be described mathematically. Its surface can be determined point-by-point, with: (a) surveying and astronomical observations, (b) measurements of gravity, and (c) analysis of the orbits of artificial satellites.

In practice, the geoid is approximated by an ellipsoid (Figure 2.7) with a maximum deviation of less than 100m throughout the earth (Figure 2.8). The determination of the geometric elements of the ellipsoid, which best approximates the geoid, is a key problem in geodesy.

Table 2.1 (see Figure 2.4) shows the parameters of some popular ellipsoids. The measure of flattening f, e.g., 1:300 means that in a model of earth's ellipsoid with equatorial radius a=30cm, the pole lies away from the equatorial plane b=29,9cm.

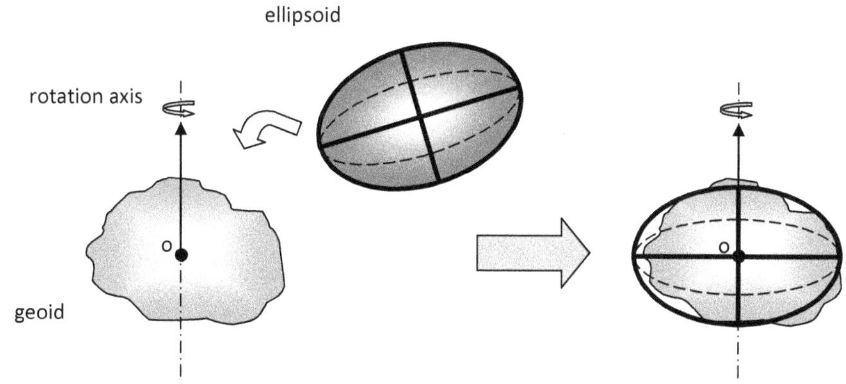


**Fig. 2.7** Mounting an ellipsoid on the geoid.

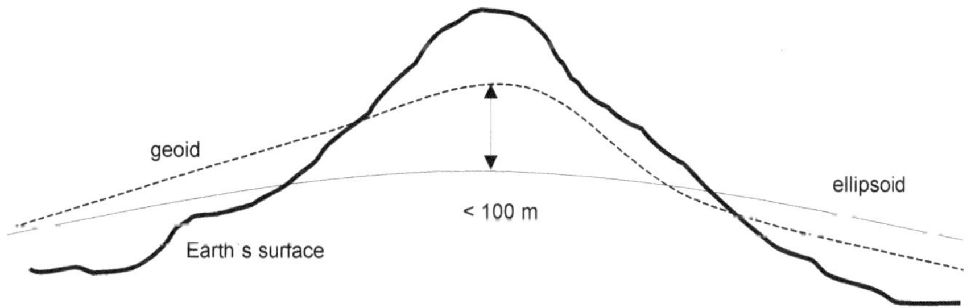


**Fig. 2.8** Approximation of the geoid by an ellipsoid.

**Table 2.1** The most commonly used ellipsoids (Figure 2.4).

| *Ellipsoids:*<br>*Parameters* | Bessel<br>1840 | Clarke<br>1866 | Hayford<br>1909 | Krassowsky<br>1942 | GRS 67<br>1967 | GRS 80<br>1980 |
|---|---|---|---|---|---|---|
| a | 6.377.397,15 | 6.378.206,4 | 6.378.388,00 | 6.378.245 | 6.378.160,00 | 6.378.137,0 |
| b | 6.356.078,96 | 6.356.583,8 | 6.356.911,94 | 6.356.863 | 6.356.774,51 | 6.356.752,3 |
| f = (a-b)/a | 1:299,2 | 1:295,0 | 1:297,0 | 1:298,3 | 1:298,2 | 1:298,257 |

Figure 2.9 shows the arrangement of the geoid, ellipsoid and earth's surface (land) in an area. Notice that the order may be reversed in other parts of the earth

(see Figure 2.8). For each point P on the land the following measures apply: (a) the elevation of the geoid, (b) the elevation of the point, (c) the geometric altitude, and (d) the deviation of the vertical.

*Geodetic height* (N) at point P is the separation between the geoid and the reference ellipsoid at this point. *Altitude* (H) of point P is the distance from the point to the geoid (i.e., the mean sea level). This distance is measured along the vertical line, which is perpendicular to the geoid. *Geometric altitude* (h) of point P is called the perpendicular distance from the point to the reference ellipsoid. *Deviation of the vertical* ($\theta$) at point P is the angle between the perpendicular vector to the ellipsoid and the vertical (i.e., the perpendicular to the geoid) at the point. As shown in Figure 2.9, because of this deviation, the geometric height (h) is not equal to the sum (H+N).

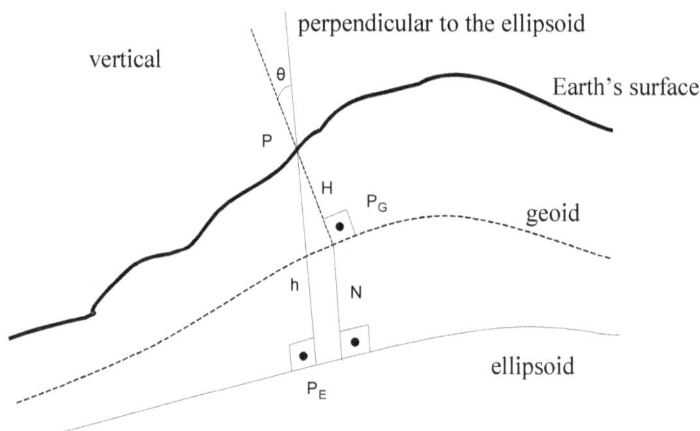


**Fig. 2.9** Earth's surface, geoid and ellipsoid.

Figure 2.10 shows a map of the geoid. The contour lines represent the separation of the geoid from a reference ellipsoid (geodetic heights N) in absolute values (meters). Figure 2.11 also shows the geodetic heights through hypsometric tints. Light grays represent the positive heights (i.e., places where the geoid lies above the ellipsoid), while the dark grays represent the negative heights (i.e., places where the geoid lies below the ellipsoid). The biggest positive heights are found in Iceland (A) and Philippines (B), while the largest negative heights in India (C) for a global ellipsoid (see Figure 2.7 and Section 2.8).

In practice the surface of the geoid is approximated by simple mathematical surfaces to facilitate the measurements and calculations. The surface used to approximate the geoid depends on the extent of the area of interest and the accuracy requirements.

In particular, for *areas within a radius of 10km* the land can be assumed as *flat* (a tangential plane at the centroid of the area) without introducing substantial errors in measurements. Figure 2.12 summarizes this configuration. A plane is tangent to the point O of the earth (here represented by a sphere). Assuming a flat land, a point A on the spherical surface is mapped to its orthogonal projection point A'. Thus, the spherical arc OA (s) corresponds to the straight segment OA' (d). The difference s-d (which is indicative of the error due to the flat approximation of the spherical surface), is less than half of a centimeter for distances up to 10km away from the point O (see table in Figure 2.12).

For *areas of radius up to 100km*, the earth can be considered as *spherical* (Figure 2.13). The radius R of the sphere is calculated for different latitudes (φ) by applying the Euler's formula:

$$R = R(\varphi) = \sqrt{\frac{\left(a^2\cos(\varphi)\right)^2 + \left(b^2\sin(\varphi)\right)^2}{\left(a\cos(\varphi)\right)^2 + \left(b\sin(\varphi)\right)^2}}$$

where a, b are the axes of the ellipsoid (see Figure 2.4 and Table 2.1) and φ the latitude.

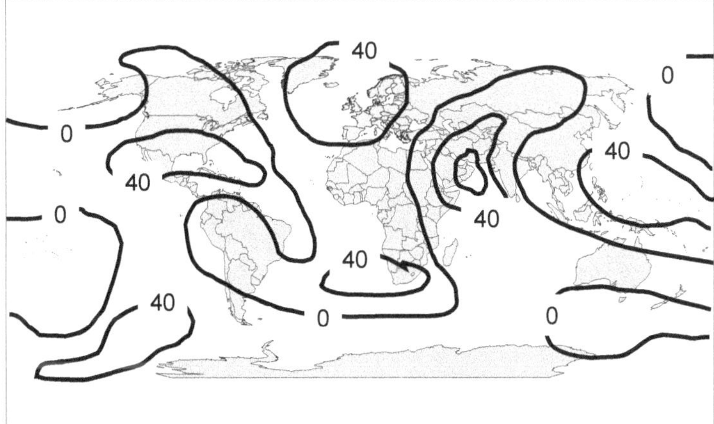


**Fig. 2.10** Map of the world with contour lines representing the geodetic heights in absolute values.

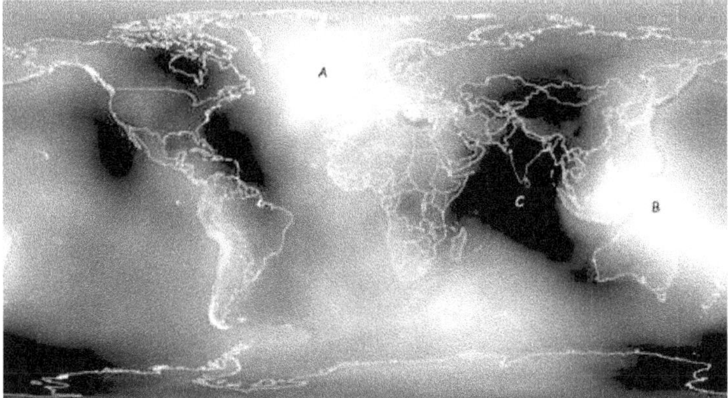

**Fig. 2.11** The elevation of the geoid. With light gray the positive geodetic heights (the geoid lies above the ellipsoid), while with the dark grey the negative geodetic heights (Wikipedia).

Figure 2.14 shows the radius of the spherical approximation for the GRS80 ellipsoid at different latitudes ($\varphi$). Usually, for practical applications in surveying, an average radius of R=6.371km is used.

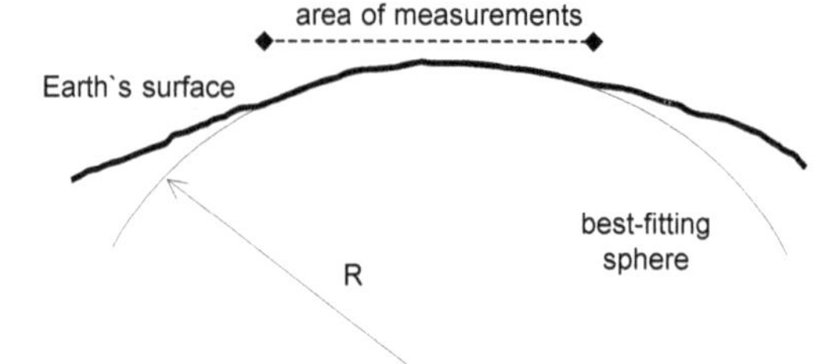


| s (m) | φ (deg) | d | s-d (m) |
|---|---|---|---|
| 0 | 0,00000 | 0,000 | 0,000 |
| 1000 | 0,00899 | 1000,000 | 0,000 |
| 5000 | 0,04497 | 4999,999 | 0,001 |
| 7500 | 0,06745 | 7499,998 | 0,002 |
| 10000 | 0,08993 | 9999,996 | 0,004 |
| 12500 | 0,11242 | 12499,992 | 0,008 |
| 15000 | 0,13490 | 14999,986 | 0,014 |
| 20000 | 0,17986 | 19999,967 | 0,033 |
| 25000 | 0,22483 | 24999,936 | 0,064 |
| 30000 | 0,26980 | 29999,889 | 0,111 |
| 50000 | 0,44966 | 49999,487 | 0,513 |
| 100000 | 0,89932 | 99995,894 | 4,106 |

**Fig. 2.12** Calculation of the errors due to the Earth`s curvature.

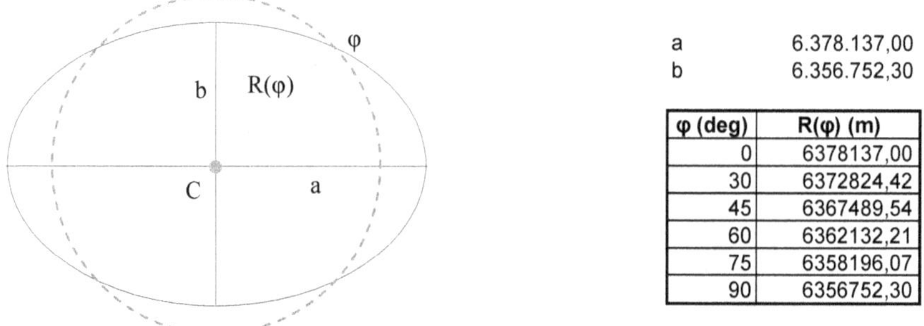


**Fig. 2.13** Approximation of the Earth with a sphere.



|   | |
|---|---|
| a | 6.378.137,00 |
| b | 6.356.752,30 |

| φ (deg) | R(φ) (m) |
|---|---|
| 0 | 6378137,00 |
| 30 | 6372824,42 |
| 45 | 6367489,54 |
| 60 | 6362132,21 |
| 75 | 6358196,07 |
| 90 | 6356752,30 |

**Fig. 2.14** Computation of the best fitting sphere of the GRS80 ellipsoid at various latitudes.

Finally, for *areas with an extent greater than 100km* (in radius) the *ellipsoid* approximation is chosen (Figure 2.4). As it will be shown later in this Chapter (Section 2.8), the choice of the ellipsoid parameters depends on the area of interest; as the best match of the ellipsoid to the geoid is sought. Each country, state or province chooses the reference ellipsoid that best approximates the local shape of the geoid, thus defining its geodetic reference system or *geodetic datum*. The center of the ellipsoid does not necessarily coincide with the earth's center of mass, while the small axis is parallel to the conventional axis of rotation of the earth with accuracy better than 1 ppm.


## 2.3 Coordinate Systems

*Coordinate system* is a set of rules for determining the position of points in space, using an ordered set of numbers, the *coordinates*. The coordinates needed to describe the location of a point range from one to three (or four when taking into account the time) for mapping purposes. This number determines the *dimension* of the coordinate system (Figure 2.15).

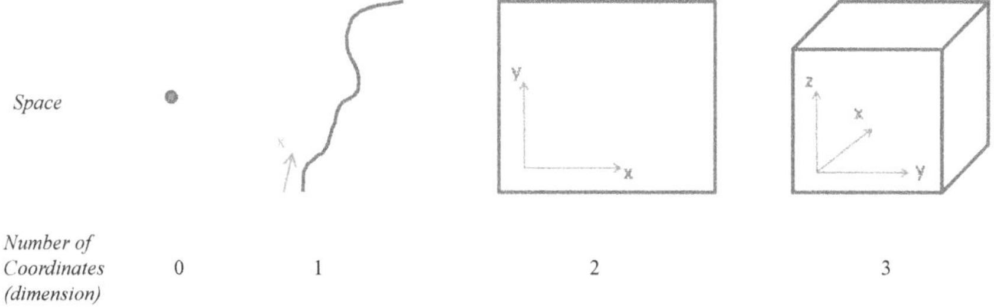


**Fig. 2.15** Spaces of dimension ranging from 0 to 3.

Each coordinate is assigned a label, such as *X*, *longitude* or *easting*; and has a reference line, called *axis*. The axes of a coordinate system can intersect at right or oblique angles. Each axis has a *unit* of measurement (such as meters or degrees) and an *origin* (zero point) for the measurement of coordinates.

Overall, the establishment of a coordinate system includes the definition of (Figure 2.16): (a) the *axes*, (b) the *unit of measurement* per axis (usually it is the same for all axes), and (c) the *origin* for each axis (usually placed at the intersection of the axes).

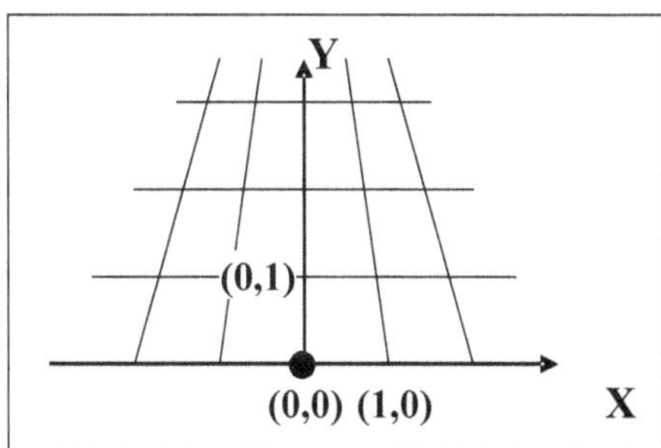


**Fig. 2.16** An example coordinate system in a two-dimensional space.

All points with the same value for one coordinate define a *grid* line. The axes are lines of this grid. The lines of the grid in the Cartesian coordinate system are straight lines, parallel to each other for a specific dimension, and perpendicular to those of other dimensions. To avoid negative coordinates in certain positions of the coordinate system a *false origin* can be established (Figure 2.17).

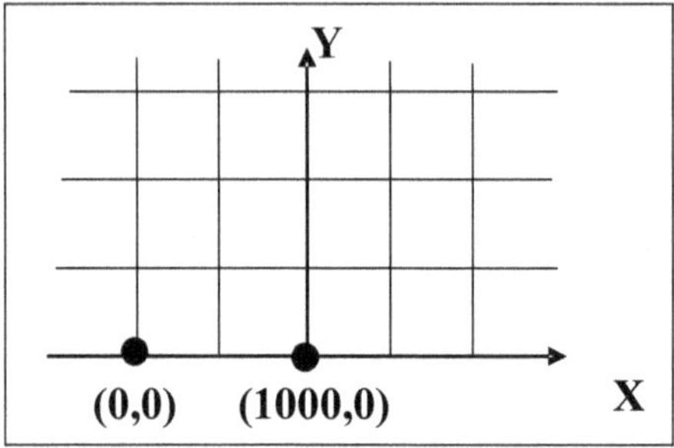


**Fig. 2.17** Transition of the origin and the false origin.

The three-dimensional coordinate systems are characterized by a *polarity* that describes the arrangement of the axes. Consider the three axes X, Y, and Z. In a *left-handed* system, the transition from the X axis to the Y axis is the rotation of the left hand with the thumb aligned to the Z axis (Figure 2.18a). In *right-handed* system, this is similarly achieved with the right hand (Figure 2.18b). A two-dimensional system can also be characterized by polarity, if the third axis is considered.

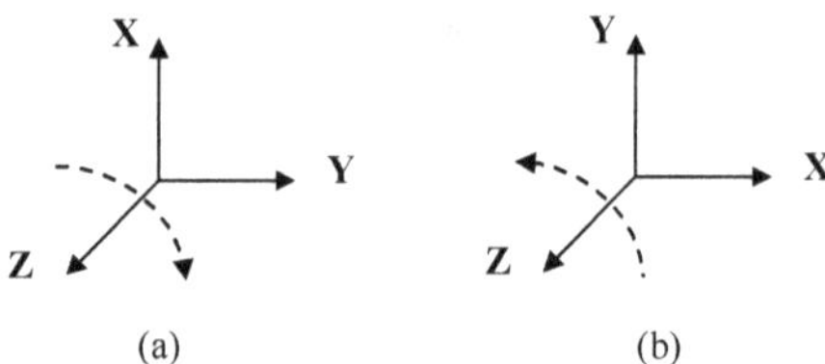


**Fig. 2.18** The left-handed (a) and the right-handed (b) three-dimensional Cartesian coordinate system.

A coordinate system is associated with a reference surface, which is usually a plane, sphere, or ellipsoid. Systems that apply the plane are called *planar* systems, while those applying the sphere or ellipsoid are called *spherical* or *ellipsoidal* systems, respectively. A coordinate system is defined on the reference surface to describe all locations on that surface. On the other hand, the reference surface is usually the origin of the third dimension (Figure 2.19).

In geosciences three coordinate systems are commonly used: (a) the *Cartesian* or rectangular, (b) the *polar*, and (c) the *geographic* coordinate system. The following Sections briefly describe these systems.

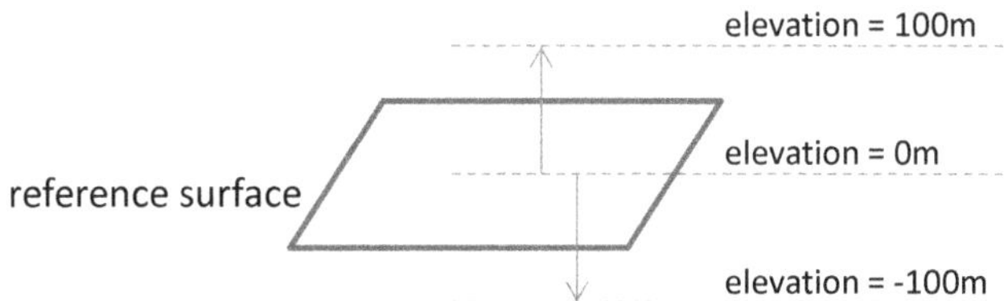


**Fig. 2.19** Measuring the third dimension of space as a vertical distance from the reference surface.

## 2.3.1 Cartesian Coordinate System

A *Cartesian coordinate system* is defined by axes that are orthogonal to each other. Therefore, the grid system consists of rectangles. It is the most common coordinate system in geosciences. This system has a planar surface as a reference surface.

In the two dimensional space two axes (labeled as X and Y) are defined, while in the three dimensional space three axes (labeled as X, Y, and Z or H) are defined. Examples of the Cartesian coordinate system in the two- and three-dimensional spaces are shown in Figure 2.20.

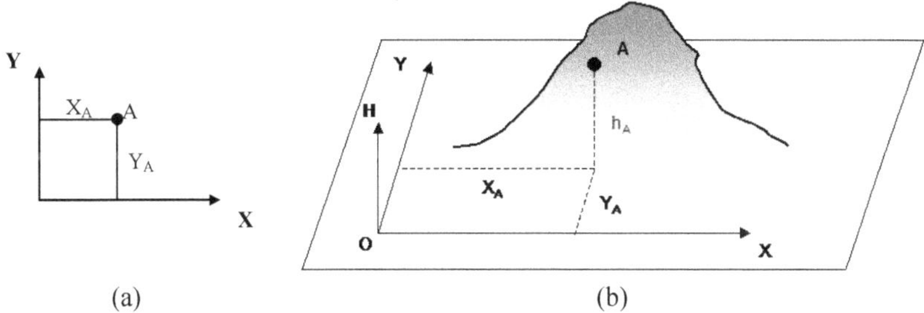


**Fig. 2.20** The Cartesian system in the space of two (a) and three (b) dimensions.

## 2.3.2 Polar Coordinate System

In the *polar coordinate system* of a two-dimensional space, one coordinate is the distance, while the other is the azimuth (heading angle). These coordinates are labeled as d and θ, respectively. The reference surface is the plane. The polar coordinate system is often used in topography (surveying).

An example of this system is shown in Figure 2.21. The conversion of the polar coordinates to Cartesian and vice versa, is a problem of trigonometry. Specifically, if $(d_A, \theta_A)$ the polar and $(X_A, Y_A)$ the Cartesian coordinates of point A, the formulas that relate them are as follows:

$$X_A = d_A \cdot \sin\theta_A \qquad\qquad d_A = (X_A{}^2 + Y_A{}^2)^{1/2}$$
$$Y_A = d_A \cdot \cos\theta_A \qquad\qquad \theta_A = \arctan(X_A/Y_A)$$

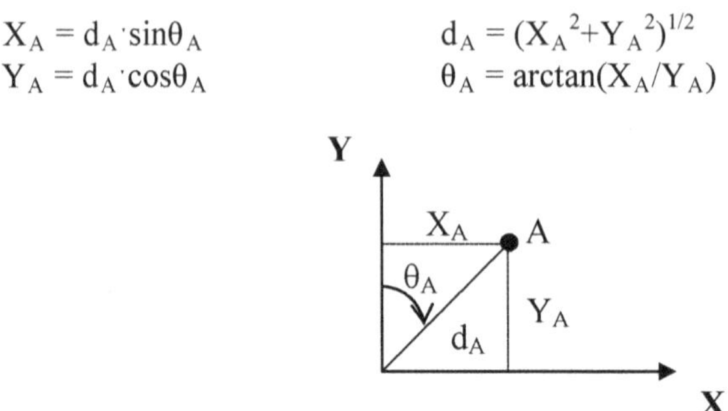

**Fig. 2.21** The Cartesian and polar systems in the two-dimensional space.

## *2.3.3 Geographic Coordinate System*

The *geographic coordinate system* is applied on spherical or ellipsoidal reference surfaces. This system is used as a reference for the locations of the earth's surface, when the latter is modeled by a sphere or an ellipsoid (spheroid). This space is usually referred to as *2.5-dimensional space*.

The position on the earth's surface is determined by two angular measurements, the *latitude* ($\varphi$) and *longitude* ($\lambda$), which together compose the geographic coordinate system ($\varphi,\lambda$). An example geographic reference system is given in Figure 2.22. There are significant differences in the latitude, as it is defined on the sphere and the ellipsoid. On the other hand, the definition of the longitude is identical for both reference surfaces.

### 2.3.3.1 Latitude on the Sphere

The *latitude* $\varphi$ is an angle measured at the center O of the sphere and is defined by the equatorial plane and the radius-line of a point P on the surface of the sphere (e.g., the angle EqOP in Figure 2.23, where OEq on the equatorial plane).

The equator is the origin for the latitudes of the globe and therefore its latitude value is $0°$. North and south of the equator the latitude increases to the value of $90°N$ at the North Pole, and $90°S$ at the South Pole of the earth. In computations involving latitude coordinates, these of the northern hemisphere are positive (ranging from $0°$ to $+90°$), while those of the southern hemisphere are negative (ranging from $0°$ to $-90°$).

### 2.3.3.2 Latitude on the Ellipsoid

On the ellipsoid, two angular measurements can be used to describe the *latitude* of a point P (Figure 2.24): (a) the *geocentric latitude* $\psi$, which is the angle EqOP, measured at the center of the ellipsoid, between the plane of the equator and the line segment OP; and (b) the *geodetic latitude* φ, which is the angle EqMP, where M is the intersection of the perpendicular (vertical) to the tangent (at the meridian sector; Figure 2.38) on P with the equatorial plane. In geodesy and cartography the geodetic latitude is used to describe the *latitude* φ of a point P on the ellipsoid. More details on the geometry of the ellipsoid are given in Section 2.5.

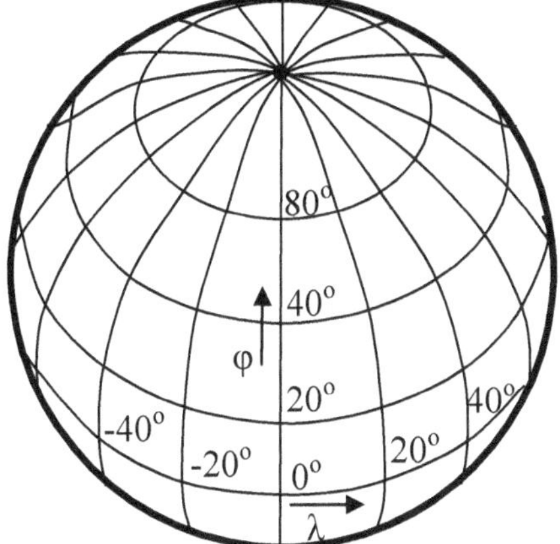


**Fig. 2.22** The geographic reference system.

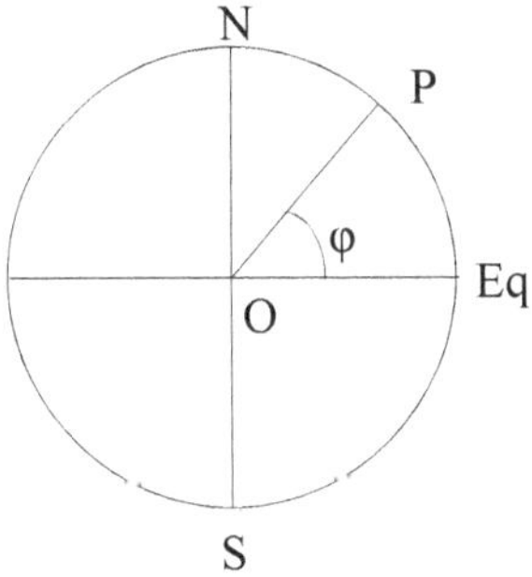


**Fig. 2.23** The geographic latitude φ on the sphere.

### 2.3.3.3 Longitude

Figure 2.25 shows two planar surfaces, which cross the center of the earth and are perpendicular to the equator. These two planes intersect along the axis NOS (the conventional rotation axis of the earth). Additionally, each of them intersects the sphere (or the ellipsoid respectively) along two great circles (or ellipses respectively). One plane (NPS) passes through the point P. The other (NGS) passes through the reference point G, which is the *origin of longitudes*.

The *longitude* λ is defined as the dihedral angle between these two planes, measured on the plane of the equator. This is the angle COD in Figure 2.25. The latitude is a directed angle and labeled as east (E) or west (W) depending on the relative position of the points C and D. In calculations, all eastings from C are positive (ranging from $0°$ to $+180°$), while all westings are negative (ranging from $0°$ to $-180°$).

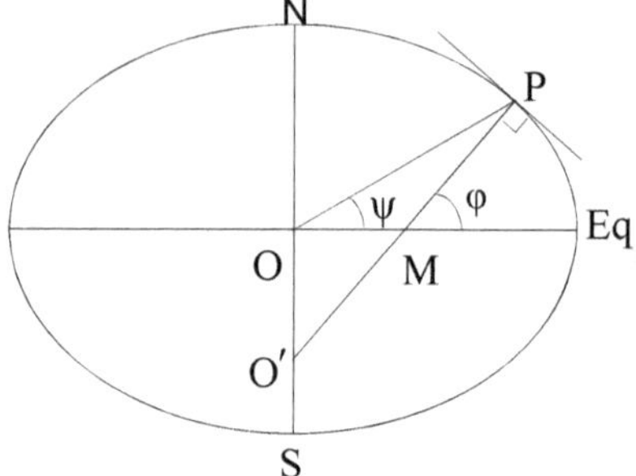

**Fig. 2.24** The geocentric (ψ) and geodetic-geographic latitude (φ) on the ellipsoid.

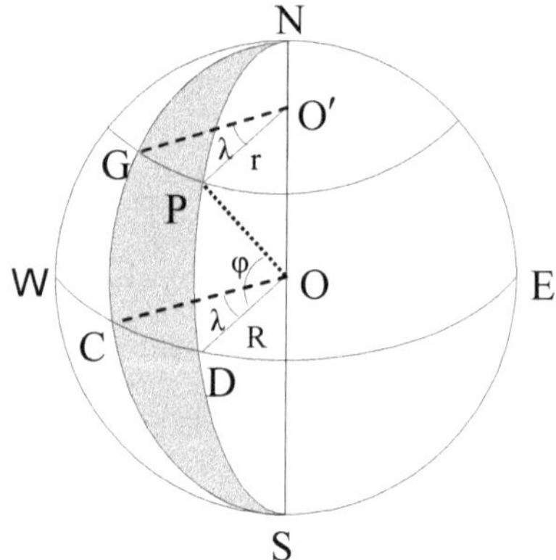

**Fig. 2.25** The geographic longitude (λ).

### 2.3.3.4 Parallels and Meridians

The set of points having the same latitude (φ) define a circle on the sphere or the ellipsoid (Figure 2.26a). The plane containing the circle is parallel to the equator and is called parallel latitude plane or *parallel*. The parallel with latitude (φ) on the sphere is a circle with radius (Figure 2.25): r=Rcosφ, where R is the radius of the sphere.

The set of points having the same longitude (λ) lie on the same plane (Figure 2.26b) and form a semicircular arc on the sphere or a semi-ellipsis on the ellipsoid.

This arc is called *meridian*. All meridians pass through the *Poles*. The meridians and parallels intersect each other at right angles and compose the *geographic grid* (Figure 2.26c).

The origin for measuring the longitude is the position of the Royal Observatory of Greenwich, near London. The meridian passing through that position is called the prime meridian or *Greenwich meridian* and its longitude is $0^{\circ}$.

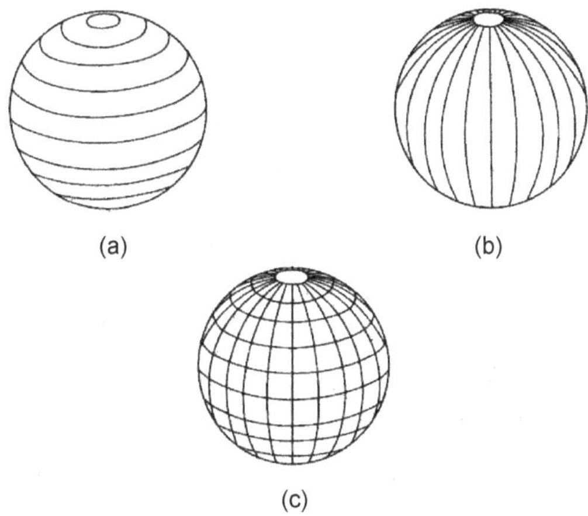


**Fig. 2.26** The grid of parallels (a) and meridians (b). The geographic grid (c).

## 2.4 Spherical Geometry

The sphere is often used as an approximation of the geoid, due to its simple geometry. In the following subsections, the basic lines on the surface of the sphere as well as some elements of *spherical trigonometry* are presented.

### 2.4.1 Basic Lines on the Spherical Surface

Let us assume that the earth is spherical, the center of the earth's mass coincides with the center of the sphere (K), and the rotation axis of the earth coincides with the North-Pole – South-Pole diameter of the sphere (NS axis) (Figure 2.27).

#### 2.4.1.1 Great and Small Circles

The locus of the intersection of the sphere with a plane is a circle (Figure 2.28). The size of the circle varies. If the plane crosses the center of the sphere, the circle gets a maximum radius (Figure 2.28b), which equals to the radius of the sphere (R). This circle is called *great circle.*

If the plane intersecting the sphere does not pass through its center point, the locus of the intersection is a circle with a radius smaller than R (Figure 2.28a) and is called *small circle.*

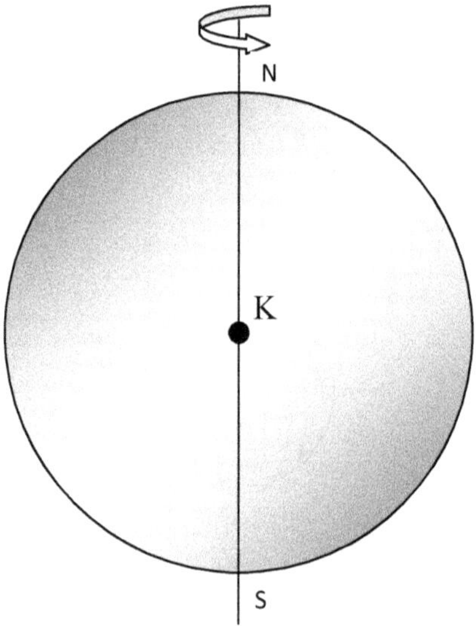


**Fig. 2.27** The assumption of the spherical Earth.

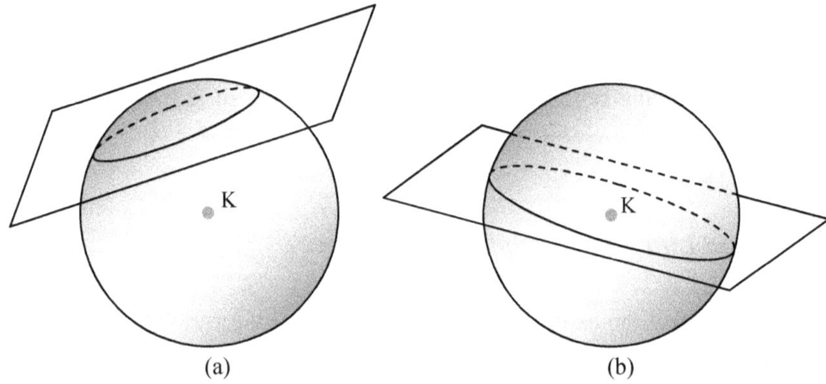


**Fig. 2.28** The locus of the intersection of a plane with a sphere is a circle: (a) small circle: if the plane does not cross the sphere center, and (b) large circle: if the plane crosses the sphere center.

Obviously, the line of the *equator* is the locus of the intersection of the sphere by a plane crossing its center (K) and being perpendicular to the rotation axis of the earth (NS). The (other) *parallels* of the geographic grid result from the inter-section of the sphere with planes that are parallel to the equator's plane (and not crossing K). On the other hand, the *meridians* result from the locus of the intersection of the sphere with planes containing the rotation axis of the earth.

Therefore, from the grid of meridians and parallels (Figure 2.29), all the meridians and the equator are great circles; while all parallels (except the equator) are small circles.

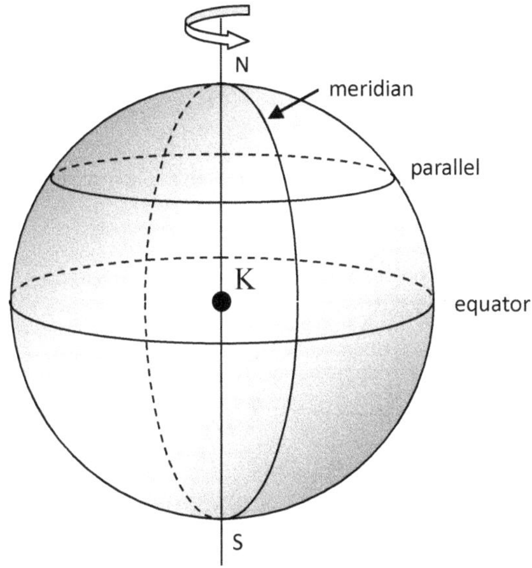


**Fig. 2.29** The geographic grid (meridians and parallels).

### 2.4.1.2 Shortest Routes on the Sphere

Consider two points A and B on the surface of a sphere of radius R (Figure 2.30). Obviously, there are an infinite number of lines on the surface of the sphere connecting the two points. The one with the minimum length is the great circle passing through the points A and B. This property is particularly important for determining the shortest paths on the surface of a sphere.

Hence, to identify the shortest route from point A to point B, it suffices to find the locus of the intersection of the sphere with a plane defined by points A, B and K, where K is the center of the sphere (Figure 2.30b – three non-collinear points define a plane in space).

Based on the above property, assume an airplane traveling between points A, B, C, D and E in Figure 2.31. The shortest flight dictates the following routes:

- From A to B: the great circle (dashed line) AB and not their common parallel.
- From A to E: the common meridian.
- From C to D: the line of the equator.
- From E to C: the great circle (dashed line) EC.

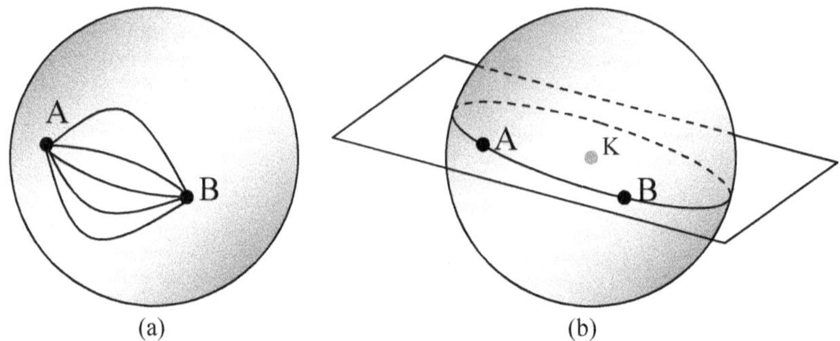

**Fig. 2.30** Determining the shortest course between A and B on a spherical surface.

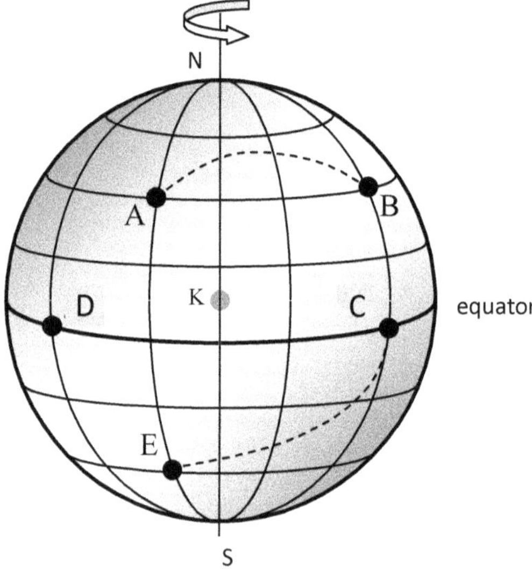

**Fig. 2.31** Shortest routes on the sphere.

### 2.4.1.3 Loxodrome

Each line on the surface of the sphere with a constant heading angle (i.e., azimuth) is called *loxodrome*. Figure 2.32 shows an example loxodrome from point A to point B.

Generally, the loxodrome is not a great circle; with the exception of the equator and the meridians. On the other hand, all lines comprising the geographic grid (meridians and parallels) are loxodrome lines.

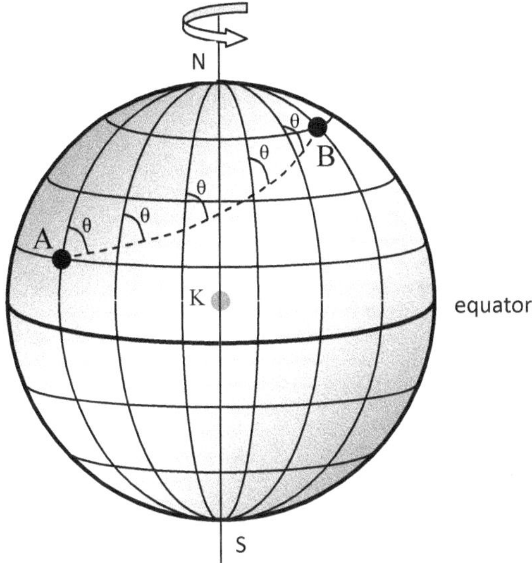



**Fig. 2.32** Example loxodrome.


## 2.4.2 Elements of Spherical Trigonometry

*Spherical trigonometry*, in correspondence with planar (Euclidean) trigonometry, is used to calculate distances and angles on the surface of the sphere. *Spherical triangle* is every three-sided surface ABC on the sphere surface, whose sides are arcs of great circles (Figure 2.33a,b).

The arcs are described in terms of angular distances at the center of the sphere (Figure 2.33c). Let $K_c$ the centric angle of the arc c, and R the radius of the sphere. Then, the length of the arc c is given by equation:

$$c = K_c \cdot R \quad \text{(the angle } K_c \text{ expressed in radius)}$$

As in planar trigonometry, a spherical triangle can be solved when three of its elements (e.g., two sides and the angle between them or three sides, etc.) are known. Consider the spherical triangle ABC (Figure 2.34) with sides a, b, c and angles A, B, C, respectively. The sides a, b, c correspond to three centric angles ($K_a$, $K_b$, $K_c$, see Figure 2.33), while the angles A, B, C correspond to dihedral angles of the adjacent great circles. For example, the angle C is equal to the dihedral angle defined by the planes ACK and BCK (Figure 2.33).

In spherical triangles, the sum of the internal angles (A+B+C; Figure 2.34) varies from $180°$ to $540°$.

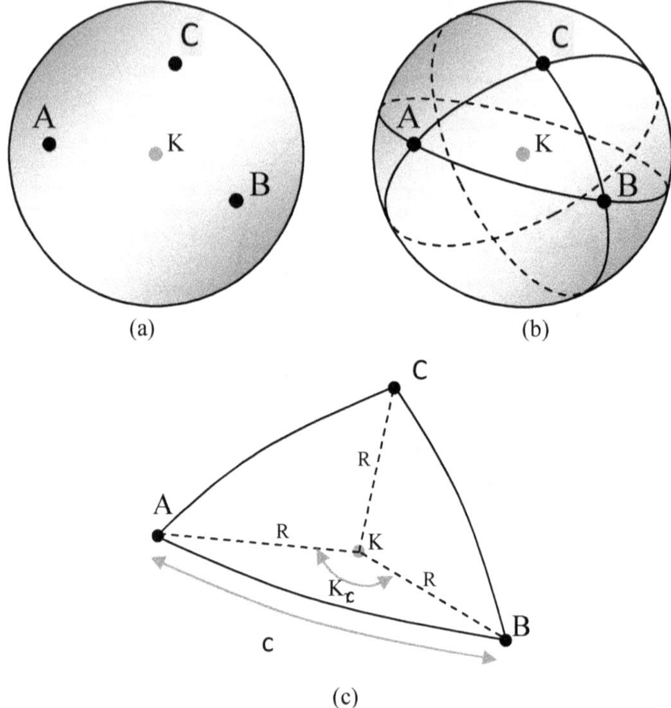

Fig. 2.33 Definition and elements of a spherical triangle.

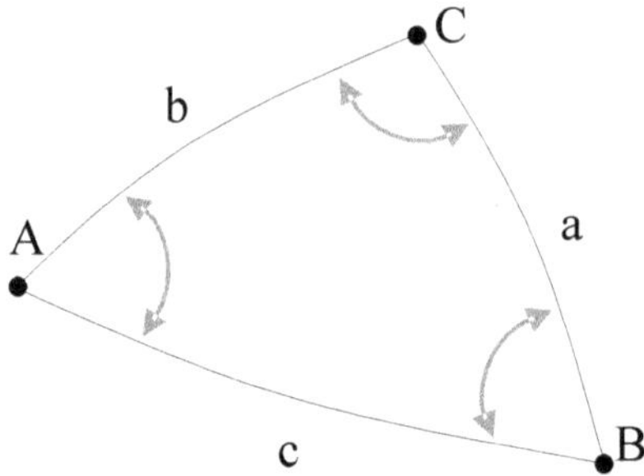

Fig. 2.34 A spherical triangle.

### 2.4.2.1 Law of Cosines

Let ABC a spherical triangle (Figure 2.34) with the two sides b and c, as well as the angle A between them known. The third side can be computed by the *law of cosines* as follows:

$$\cos a = \cos b \cdot \cos c + \sin b \cdot \sin c \cdot \cos A$$

On the other hand, if the three sides (a,b,c) of the triangle ABC are known, the angles can be computed by the following formula (here for the angle A):

cosA = (cosa – cosb·cosc) / (sinb·sinc)

## 2.4.2.2 Length of the Shortest Path

If the geographic coordinates of two points $A(\varphi_A,\lambda_A)$ and $B(\varphi_B,\lambda_B)$ are known, the law of cosines can be used to determine the length of the arc of the great circle connecting them. In other words, the length of the shortest path from A to B on the sphere can be computed.

Specifically, the two points A, B along with the Pole N form a spherical triangle (Figure 2.35). The problem is reduced to solving the spherical triangle ABC in Figure 2.36 (where C the Pole N in Figure 2.35).

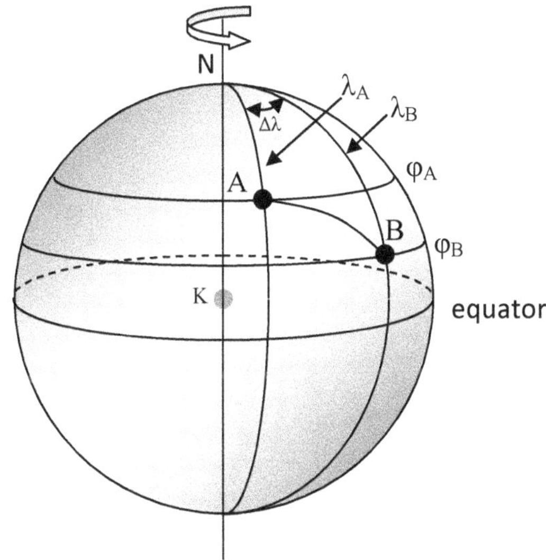


**Fig. 2.35** Computation of the shortest path between A and B on a sphere.

The sides a and b are equal to the differences $90°-\varphi_B$ and $90°-\varphi_A$, respectively. The angle C is equal to the difference $\lambda_B-\lambda_A$. Side c which is wanted, coincides with the arc AB of the great circle connecting A and B. From the law of cosines:

$$\begin{aligned}
\cos c &= \cos a \cdot \cos b + \sin a \cdot \sin b \cdot \cos C = \\
&= \cos(90°-\varphi_B)\cdot\cos(90°-\varphi_A) + \sin(90°-\varphi_B)\cdot\sin(90°-\varphi_A)\cdot\cos(\lambda_B-\lambda_A) = \\
&= \sin\varphi_B\cdot\sin\varphi_A + \cos\varphi_B\cdot\cos\varphi_A\cdot\cos(\lambda_B-\lambda_A)
\end{aligned}$$

Hence:

$$c = a\cos(\ \sin\varphi_A\cdot\sin\varphi_B + \cos\varphi_A\cdot\cos\varphi_B\cdot\cos(\lambda_B-\lambda_A)\ )\ \ [rad]$$

Therefore, the length of the great circle from A to B is equal to:

AB = c·R

Finally, the azimuth of the path AB at A can also be derived from the law of cosines as follows:

$$\cos A = (\cos(90°\text{-}\varphi_B) - \cos(90°\text{-}\varphi_A)\bullet\cos c) / (\sin(90°\text{-}\varphi_A)\bullet\sin c) =$$
$$= (\sin\varphi_B - \sin\varphi_A\bullet\cos c) / (\cos\varphi_A\bullet\sin c)$$

As the great circle is not generally a loxodrome, the heading angle (azimuth) of movement from A to B changes along the great circle.

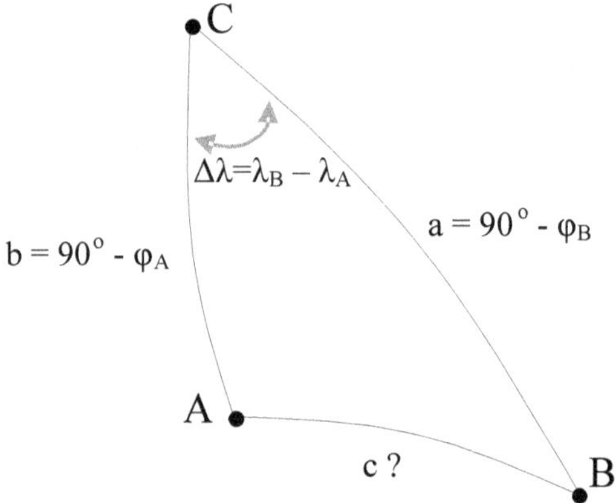


**Fig. 2.36** Solving the spherical triangle ABN in Figure 2.35 (C=N).

## 2.5 Geometry of the Ellipsoid

This Section provides some basic elements with regards to the geometry of the *ellipsoid*. The ellipsoid is a solid obtained by rotating an ellipse (a, b) around its minor axis b (Figures 2.4 and 2.37). The basic elements of this solid are:

- The *axes*:

  a (major axis) and b (minor axis)

- The *flattening* (f):

$$f = \frac{a-b}{a}$$

- The *first eccentricity* (e):

$$e = \sqrt{\frac{a^2 - b^2}{a^2}}$$

- The *second eccentricity* (e′):

$$e' = \sqrt{\frac{a^2 - b^2}{b^2}}$$

The *equation* of the ellipsoid is given by the formula:

$$\frac{X^2}{a^2} + \frac{Y^2}{a^2} + \frac{Z^2}{b^2} = 1$$

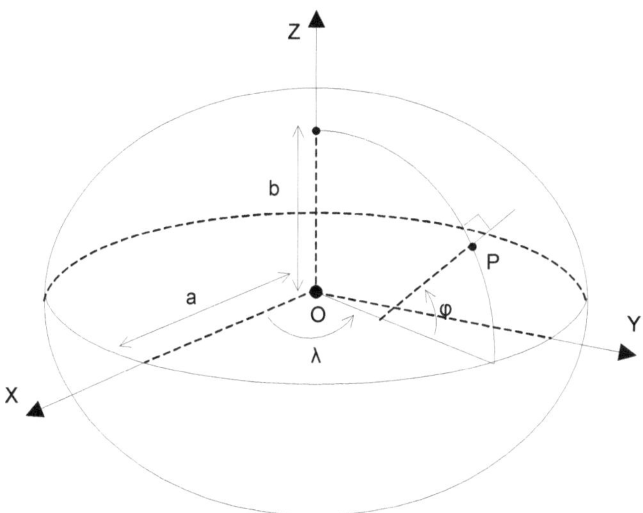

**Fig. 2.37** The geometry of the ellipsoid.

The flattening and the eccentricities are critical for the shape of the ellipsoid. If the axes a, b are equal in length (a=b), then the ellipsoid degenerates into a sphere of radius R=a=b. The difference in length between the axes a and b in the ellipsoid models for the earth (see Table 2.1) is about 22km.

At a point P on its surface the ellipsoid has three curvatures (Figure 2.38): (a) curvature of the *meridian intersection* $K_m$, (b) curvature of the *parallel circle* $K_p$, and (c) curvature of the *main vertical intersection* to the *meridian intersection* $K_N$. These curvatures are computed by the following equations:

$$K_m = -\frac{a(1-e^2)}{\sqrt[3]{1-e^2\sin^2\varphi}}$$

$$K_p = \sqrt{\frac{a^2\cos^2\varphi}{1-e^2\sin^2\varphi}}$$

$$K_N = \frac{a}{\sqrt{1-e^2\sin^2\varphi}}$$

Obviously, for the sphere (e=0):

$$K_m = K_N = a = R \quad \text{and} \quad K_p = a{\cdot}\cos\varphi$$

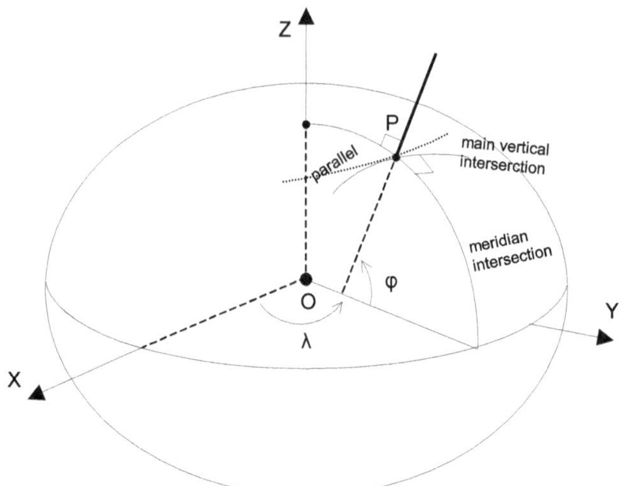


**Fig. 2.38** The three curvatures at point P on an ellipsoid.

Two points on the ellipsoid can be connected through an infinite number of ellipsoid lines. Two of them are of interest in earth sciences: (a) the *vertical section*
to the ellipsoid, and (b) the *geodesic line.*

Regarding the *vertical section*, let two points $P_1$, $P_2$ on the earth's surface and
$P_1{}'$, $P_2{}'$ their traces on the ellipsoid (Figure 2.39, also see Figure 2.9). If $\varphi_1 \neq \varphi_2$
and $\lambda_1 \neq \lambda_2$, then the lines $P_1 P_1{}'$ and $P_2 P_2{}'$ are *non-coplanar*, i.e., they do not lie on
a single plane.

Hence, the line $P_1 P_1{}'$ along with the point $P_2{}'$ define a plane (*first vertical section*), which intersects the ellipsoid along a planar ellipse segment (section 1-2).
Similarly, the line $P_2 P_2{}'$ along with point $P_1{}'$ define a second plane (*second vertical section*), which intersects the ellipsoid along a planar ellipse segment (section
2-1). It turns out, with reasonable accuracy, that the lengths of the two vertical sections are the same, as are the angles $\theta_1$ and $\theta_2$ formed between the two sections at
their ends, i.e., $\theta_1 = \theta_2$.

*Geodesic line* is a unique line connecting two points $P_1$, $P_2$ of the ellipsoidal
surface with the minimum length (Figure 2.40). The geodesic line is surrounded
by the two vertical sections, while for short distances (<100 km), it practically coincides with them.

The calculation of the length of the vertical sections and the geodesic line is obtained by solving the corresponding linear and surficial integrals, which is beyond
the goals of this document. The reader may refer for details to specialized textbooks on mathematics and/or geodesy.

Finally, the geodesic line through two points $P_1$ and $P_2$ can be drawn if from
the middle point of the chord joining the points a perpendicular to the ellipsoid is
brought. The intersection point belongs to the geodesic line. Repeating the process
recursively a condensation of geodesic line can be obtained.

Geodesic lines are open lines (Figure 2.41) lying within a zone $(-\varphi_{max}, \varphi_{max})$, which depends on the value of the azimuth at the equator $A_{eq}$.

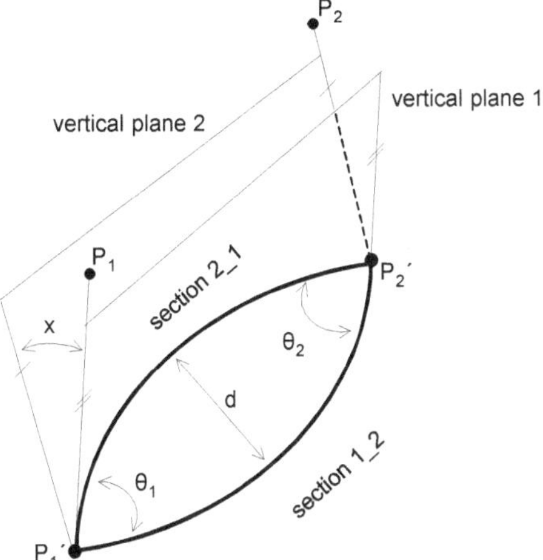


**Fig. 2.39** The vertical sections connecting two points on the ellipsoid.

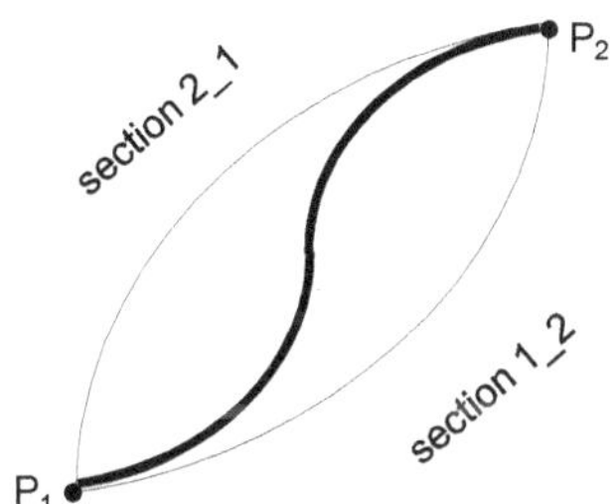


**Fig. 2.40** The two vertical sections and the geodesic line between two points of the ellipsoid.

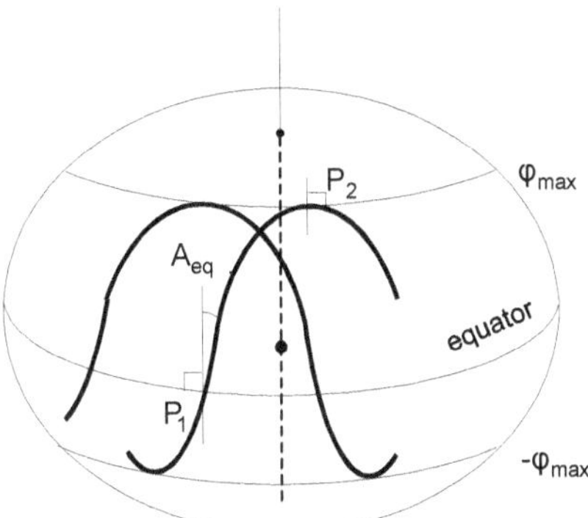


**Fig. 2.41** Open geodesic lines.

## 2.6 Conversion between Geographic and Cartesian Coordinates

In applications of geodesy and mapping, the geographic (ellipsoid) coordinates are mostly used to describe a position. On the other hand, there are applications, especially those related to space and satellite navigation, which adopt the Cartesian coordinates, usually in a geocentric three-dimensional reference system (Figure 2.37).

For the conversion of geodetic (geographic) coordinates ($\varphi,\lambda,h$), where h the geometric altitude (see Figure 2.9), to the Cartesian coordinates (X,Y,Z), the following formulas are applied:

$$X = (K_N + h)\cos\varphi \cdot \cos\lambda$$

$$Y = (K_N + h)\cos\varphi \cdot sin\lambda$$

$$Z = [K_N(1-e^2) + h]\sin\varphi$$

For the inverse conversion the corresponding formulas are as follows:

$$\tan\lambda = \frac{Y}{X}$$

$$\tan\varphi = \frac{Z + e^2 K_N \sin\varphi}{\sqrt{X^2 + Y^2}}$$

$$h = \begin{cases} X\sec\lambda \cdot \sec\varphi - K_N \\ Y\mathrm{cosec}\lambda \cdot \sec\varphi - K_N \end{cases}$$

There are two equivalent ways for the calculation of h. With regard to $\varphi$, the calculation is achieved by iteration starting with an initial value (the second part of the equation) until the desired accuracy is reached.

The above formulas are very useful when datasets from various resources adopting different geodetic datums need to be combined.

## 2.7 Reference System for the Elevation

"Olympus Mountain has a peak of altitude 2,950 meters." The altitude of the peak refers to its vertical distance from the sea. However, the sea surface is not constant but varies periodically and follows a phenomenon called *tide*. The tide is the result of a combined action of the weather (air pressure, winds, etc.) and the gravitational forces exerted by the Moon and the Sun to the Earth's water masses. Therefore, the challenge is to determine the reference surface at zero altitude (elevation origin) (Figure 2.19).

As mentioned previously, the reference surface for the *elevation* is the geoid, which practically coincides with the mean sea level (MSL) in a place. Therefore, the elevation (altitude or height) of a point is defined as the vertical distance of the point from the MSL.

The *mean sea level* (MSL) results from long-term observations of the sea level in one place using a special instrument, called *tide gauge*. This way, it is possible to specify the elevation origin for a place. To determine the elevations of the earth's surface specific methods, called *leveling*, are applied.

Notice that, in practice other surfaces may be defined as reference of elevations. For example, in nautical maps, the (negative) depths of the sea floor have as origin the *lowest astronomical tide* (LAT), which is a surface beneath which the sea level lies rarely. LAT is the lowest forecast of sea level for any astronomical formation and average climatic conditions. In the constructions, an arbitrary origin for the elevation is usually settled in the site. Then all points of the structure are assigned height values based on this false origin, which is a hypothetical, zero surface, passing through this origin and is parallel to the geoid (or MSL).

## 2.8 Geodetic Reference Systems

To identify each point location on (in or around) the earth's surface it is necessary to establish a coordinate system. Since we are dealing with a three-dimensional space, a three-dimensional coordinate system is needed, e.g., the *3D Cartesian system (X,Y,Z)*. In this system, the origin is placed at the center of the earth's mass, the Z axis coincides with the (conventional) axis of rotation of the earth, while the X-axis lies on the equatorial plane and crosses the zero meridian of Greenwich. The Y-axis also lies on the equatorial plane and crosses the meridian of 90° East (Figure 2.42 and Figure 2.37).

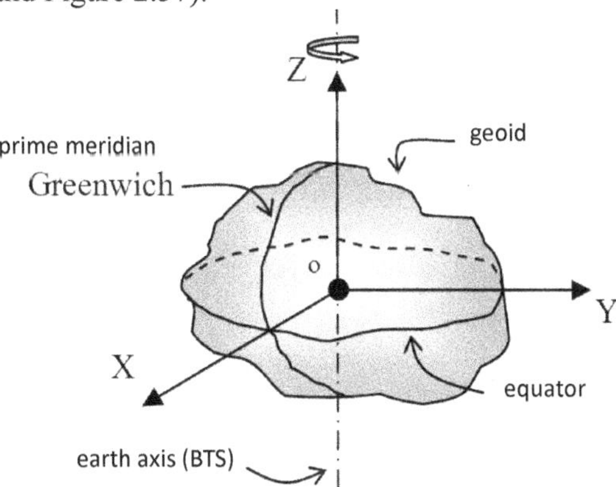


**Fig. 2.42** The geocentric Cartesian coordinate system.

The three-dimensional Cartesian coordinate system, described above, is called *geocentric Cartesian coordinate reference system*. This system, although it can describe the location of each point of the earth, it is not always a convenient system for applications in classical geodesy and cartography.

One reason is that these applications are limited on the earth's surface and in a small area around it, while the Cartesian coordinate system describes the entire space inside the earth and the outer space. So, as a coordinate system is much broader than the needs of classical geodesy and cartography. In addition, the elevations in geodetic and cartographic applications are measured in relation to the mean sea level (MSL), while the geocentric system cannot record them explicitly.

For the purposes of geodetic and cartographic applications, more appropriate coordinate systems have been designed and implemented. These systems are called *geodetic coordinate reference systems* or *geodetic datums*.

For the measurements of elevations (heights) a *vertical datum* has been defined, with the geoid (practically the MSL) as reference of heights and depths; and all measurements made along the vertical.

For the horizontal (surface) measurements, a mathematical solid (sphere or ellipsoid) is theoretically mounted on the earth. On this solid it is possible to describe the position of each point through horizontal (surface) coordinates. The mathematical solid, mounted appropriately (see next) on the Earth and having attached the coordinate system defines the *horizontal datum*.

Obviously, an ellipsoid by itself is not a geodetic datum, as it can be mounted anywhere and in any orientation. To establish a horizontal datum, the ellipsoid needs to be mounted in a geocentric Cartesian coordinate system. An example is the mounting of the ellipsoid of Clarke (see Table 2.1), so that its center coincides with the origin of the geocentric Cartesian system and its small axis with the Z axis (Figure 2.43). Notice that the ellipsoid, as defined in a horizontal datum, is often the surface of the reference for the heights, thus simultaneously sets the vertical datum.

Geodetic datums are either *global* or *local*. *Global datums* attempt to minimize the differences between the ellipsoid and the geoid throughout the earth's surface (Figures 2.10, 2.11 and 2.44a). *Local datums*, on the other hand, minimize these differences in a particular region (e.g., a province, state, or continent) (Figure 2.44b).

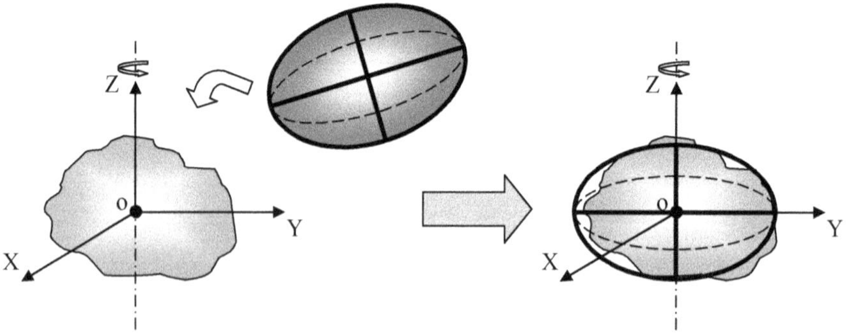


**Fig. 2.43** Mounting an ellipsoid on the geocentric Cartesian coordinate system.

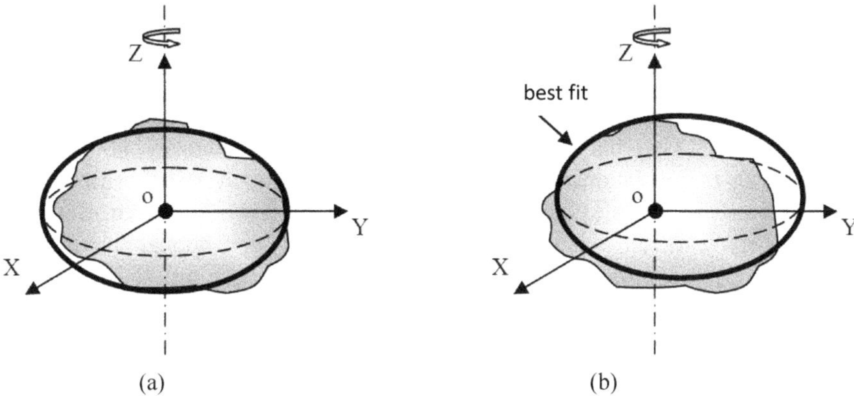



**Fig. 2.44** Geodetic datums: global (a) and local (b).

In *global datums* (e.g., GRS80; Table 2.1), the center of the ellipsoid is fixed at the origin of the geocentric system (center of mass of the earth) and the small axis coincides with the Z axis (axis of rotation of the earth). With regards to the geographic coordinates (Figure 2.45), the origin for the longitudes ($\lambda$) is the level of XZ (which coincides with the meridian of Greenwich), while the origin for the latitude ($\varphi$) is the level of XY (which coincides with the equatorial plane).

The location of a point A on the earth's surface or the surrounding area is defined by three coordinates ($\lambda_A, \varphi_A, h_A$), laid down by drawing the perpendicular from the point to the ellipsoid (Figure 2.45). Obviously, the values of these parameters directly depend on the position, orientation, and dimensions of the ellipsoid (the datum). Hence, the position of a point is described by the three coordinates plus the geodetic datum parameters (position, orientation, and size of ellipsoid).

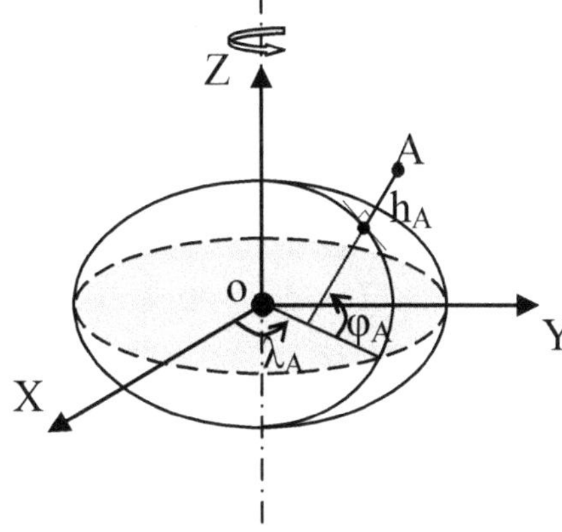



**Fig. 2.45** Geographic coordinates in global datums.

In *local datums*, the ellipsoid can be displaced from a geocentric system to achieve optimal matching to the area of interest. Generally, the local datum is customary to only shift the ellipsoid, so as to maintain the short axis always parallel to the conventional axis of rotation of the earth, with an accuracy of better than 1ppm.

Notice that neither the center of the earth's mass nor the rotation axis are stationary in time. They both move and change location continuously. For practical purposes, so that a stationary reference is established, the average position of the rotation axis (North Pole) during the period 1900-1905 was measured. Based on this, the conventional axis of rotation of the earth (Conventional International Origin - CIO) was introduced as the origin for all geodetic datums.

## References and Further Reading

Anderson, J.M, Mikhail, E.M., 1998, *Surveying – Theory and Practice*. McGraw-Hill.

Chen, Y.Q., and Lee, Y.C., 2001. *Geographical Data Acquisition*. Springer.

Mikhail, E.M., Ackerman, F., 1976, *Observations and Least Squares*. University Press of America.

Torge, W., 1980. *Geodesy – An Introduction*. Walter de Gruyter.

# 3 Map Projections

## 3.1 Introduction

After the Earth is measured and all geographic data is collected (Chapters 1 and 2) and possibly analyzed (see next Chapters), there is a need to visualize it. The most common medium for visualizing geographic data is the map, which in most cases is a planar surface, e.g., the traditional paper map or the digital map shown on the monitor of a computer system. Hence, there is a need to project the curved surface of the earth on a flat medium. This is accomplished by applying a *map projection.*

A map projection refers to the depiction of the earth's surface on a planar surface. Specifically, it refers to a transformation of the geographic coordinates $(\varphi,\lambda)$ of an area to map coordinates. This transformation is expressed mathematically as follows (Figure 3.1): $(X,Y) \leftarrow f(\varphi,\lambda)$; where $(X,Y)$ the Cartesian coordinates of the planar surface (map).

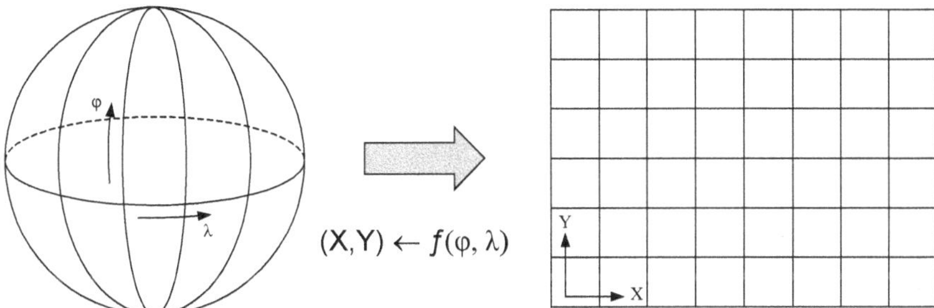


**Fig. 3.1** Map projection transformation.

## 3.2 Projection Surfaces

Map projection transformations usually involve complex mathematical calculations. This is mainly because of the difference in form between the nearly spherical shape of the earth (ellipsoid) and the flat surface of a map. This difference is responsible for introducing *distortions,* which should be minimized so that the map is accurate and usable.

Assume the earth as a small glass globe with all geographic features drawn on its surface with black ink. Consider a light source mounted at the center thereof. After wrapping the globe with a transparent rubber sheet and switching on the light, all geographic features are projected on the sheet. Using a pen, all features of interest can be traced on the sheet. The sheet is a replica of the globe. However, the sheet as a "map" is intractable due to its spherical shape.

For practical reasons maps are planar. They can be folded or printed in book pages. Obviously, it is possible to cut the rubber sheet and make it planar by pulling or compressing it. As such, a plane map representation of the globe can be produced. However, these changes lead to deformations. The scale does not remain constant throughout the map and the shape of the features (cartographic entities) is deformed. This occurs because it is not possible to deploy a sphere (or ellipsoid) onto the plane without distortions.

Imagine a watermelon and the need to make its surface flat (Figure 3.2). Initially, the watermelon is spherical and far different from the plane as regards to its shape. Let's cut it in half. This results to two hemispherical surfaces, which again do not match with the plane. Let's cut it further until a small portion is generated. The peel of this portion is still curved. Should someone push it down (against a table) to match with the plane, the peel breaks. This break introduces distortions; provided that any two points which were adjacent and lying either side of the section, they now fall far apart on the plane (table).

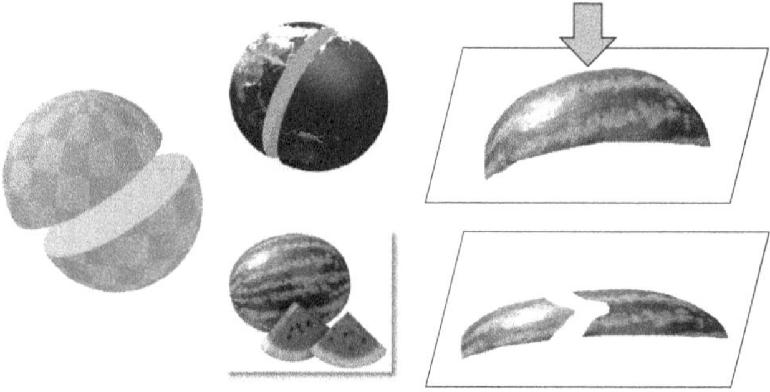

**Fig. 3.2** The watermelon analogy.

There are some curved surfaces which can be deployed onto a plane without distortions (Figure 3.3). Two of them are the *cylinder* and the *cone*. These two surfaces (along with the plane) have widely been used in cartography as an intermediate surface, also called *projection surface*, to project the spherical surface (Earth) onto a plane (map).

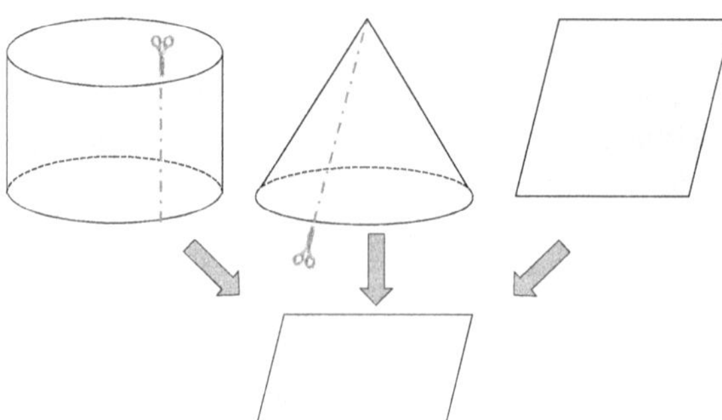

**Fig. 3.3** Deploying the cylinder, cone, and plane onto a planar surface.

Figure 3.4 shows how the glass globe could be wrapped by these projection surfaces. The cylinder or cone could be tangent onto the sphere along a line, which is either a great circle (e.g., the equator) or a small circle (e.g., any parallel but the equator), respectively. The plane could be tangent at any point of the sphere.

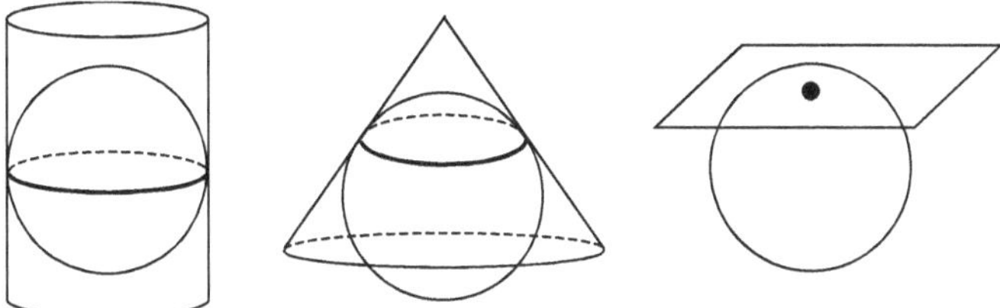

**Fig. 3.4** Wrapping the globe with the projection surfaces.

Figure 3.5 shows how a cylindrical projection surface (e.g., a rice paper) could wrap the glass globe discussed previously. By unfolding the cylinder a flat map is generated. Obviously, as the cylinder (similarly with the cone or plane) is a rough approximation of the globe, the features as projected and drawn on it (rice paper) will be geometrically distorted. In other words, the deformation cannot be avoided.

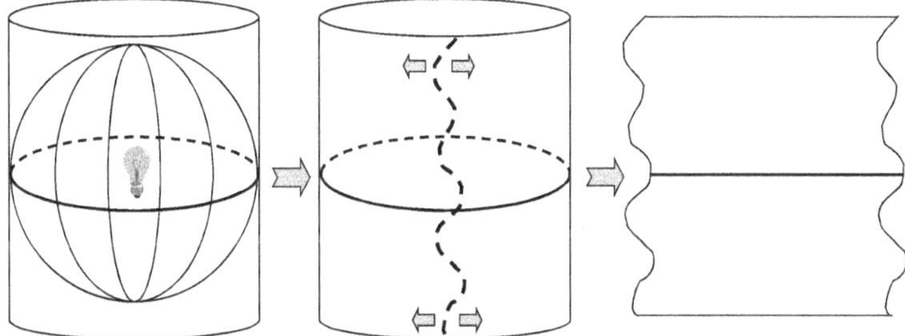

**Fig. 3.5** A rice paper in cylindrical shape is used to wrap the glass globe and create the flat map.

In reality, the maps are not created using light globes, but analytical methods. Depending on the projection surface adopted, map projections are classified into (a) *planar* or *azimuthal*, (b) *cylindrical*, (c) *conical*, and (d) *other* projections. There are various techniques available to calculate or reduce the distortions. Some of them are described in the following Sections.

## 3.3 Map Distortions

A common practice to study the deformations of a map projection is to consider that the earth's surface is covered by many small circles (Figure 3.6) and examine their shape and size on the projection surface. In general, due to the distortions, these circles will be represented as ellipses with variable sizes on the map surface.

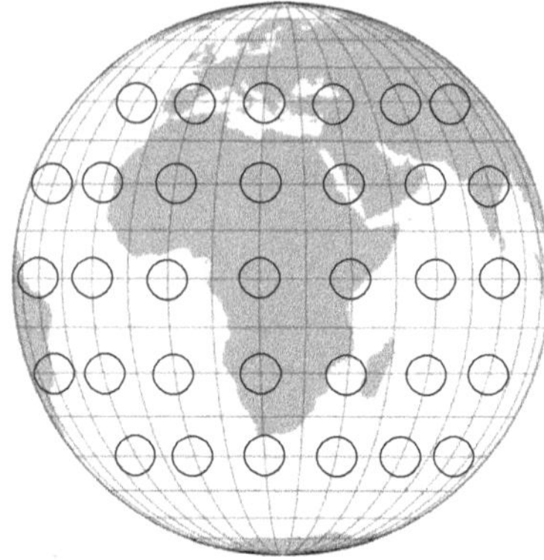

**Fig. 3.6** The Earth globe covered by a dense network of small cirlcles.

Figure 3.7 shows how these circles are represented in an azimuthal projection (stereographic projection: plane tangentially placed at the North Pole) and in a cylindrical projection (Mercator projection: cylinder tangentially placed along the equator line). Obviously, the magnitude of deformation increases with the distance from the tangent point or line.

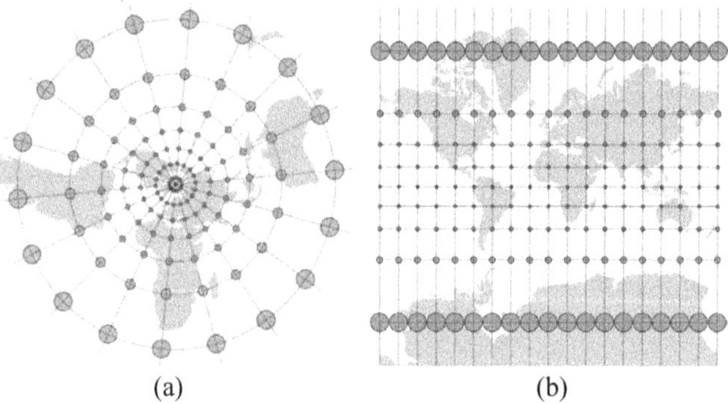



**Fig. 3.7** The form of the circles (Figure 3.6) in the stereographic (a) and Mercator (b) projection.

Figure 3.8 shows the distribution of distortions in the basic configuration of the three categories of map projections (based on the projection surface). Dark areas represent low distortions, while light areas represent high distortions.

Alternatively, the three categories of projections (azimuthal, cylindrical and conical) can be applied by having the projection surface (plane, cylinder and cone, respectively) intersecting the globe (Figure 3.9), instead of being tangent to it (Figure 3.4). The advantage of this *secant configuration* is that it increases the matching of the projection surfaces with the sphere. The azimuthal projection intersects the sphere along a circle (e.g., a parallel), while the other two (cylindrical and conical) intersect the sphere along two circles (e.g., two parallels).

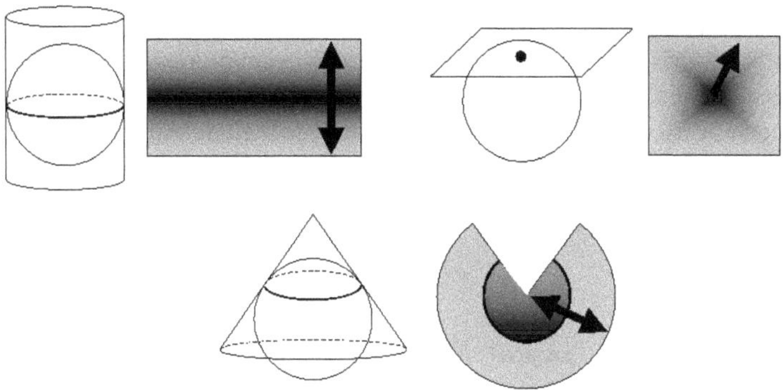

**Fig. 3.8** The distribution of errors in the tangent cylindrical, azimouthal, and conical projections.

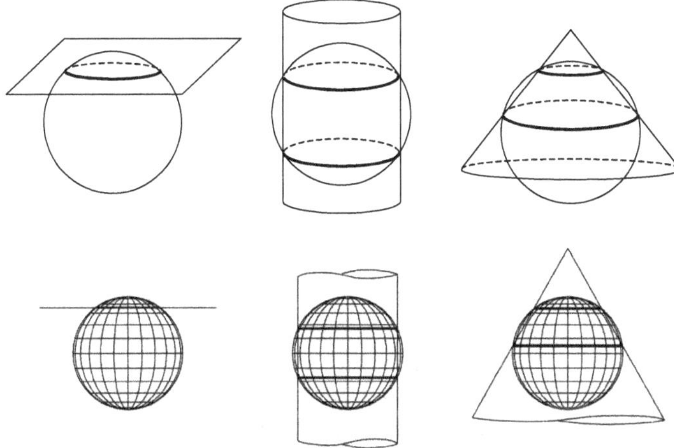

**Fig. 3.9** A secant configuration of the azimuthal, cylindrical, and conical projections.

There is no map (projection) free of distortions. The goal is to minimize (but not eliminate) the deformation within a narrow band of the earth's surface projected onto a map. Considering the circles and ellipses as projected on the projection surface (Figures 3.6 and 3.7), three types of distortions may occur:

The first refers to the *area* of the projected regions as the circles on the map have a different size than those on the earth's surface. The second refers to the *forms* (angles and directions) as the circles are projected with ellipses of various orientations on the map. Finally, the third refers to the *lengths*, as the circles are represented in different width and/or height on the map.

Generally, it is not possible to project the spherical (ellipsoid) surface on the plane and preserve all *areas*, *forms* and *lengths* (Figure 3.10). However, some projections may preserve the (small) areas unchanged. These are called *equivalent* projections. Others can preserve the lengths along some directions and are called *equidistant* projections. Note that it is impossible to create a projection with a uniform scale throughout the whole map. Finally, some projections preserve the forms (e.g., shapes of continents and countries) in small areas and are called *conformal* projections. It is impossible to create a projection that preserves the angles in all its extent. However, it is possible to preserve the angular measurements from one or two points on the map to all other points.

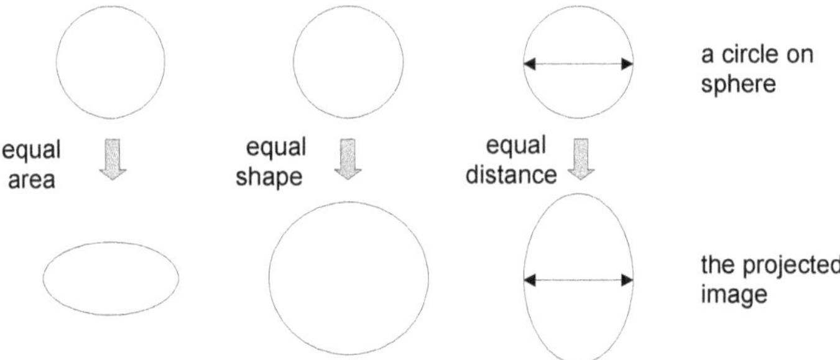


**Fig. 3.10** Retain the areas, forms (angles), and lengths in a map projection.


## 3.4 Categories of Map Projections

The classification of map projections can be based on different criteria, such as (a) the shape of the projection surface, (b) the mounting of the projection surface on the globe, (c) the properties of the projection, (d) the implementation of the projection, etc.

The shape of the projection surface can vary. As mentioned above, three surfaces are commonly adopted in cartography (Figure 3.3): the plane, the cone and the cylinder, to form the: *azimuthal, conical,* and *cylindrical* projections, respectively.

After the selection of the projection surface, it should be decided how this surface will be mounted on the globe. There are three alternatives: (a) the projection surface be tangent to the globe at a point (for azimuthal) or along a circle (in conical and cylindrical); (b) the projection surface intersect the globe along one or two circles; and (c) the projection surface consist of multiple sub-surfaces (e.g., non-coplanar planes) which share multiple lines and points with the globe. These alternatives lead to the *tangent, secant,* and *multisurface* projections, respectively.

Depending on the orientation of the projection surface in relation to the globe the projections are further classified into (Figure 3.11): (a) *orthogonal*, if the normal vector of the plane (in azimuthal) or the axis of the projection surface (in conical and cylindrical) is co-linear to the rotation axis of the Earth; (b) *transverse*, if they are perpendicular to each other, and (c) *oblique*, if they intersect in another angle.

As described in the previous Section, map projections can be classified into *equivalent, equidistant,* and *conformal* depending on the measures they preserve, i.e., areas, distances, and forms (directions), respectively.

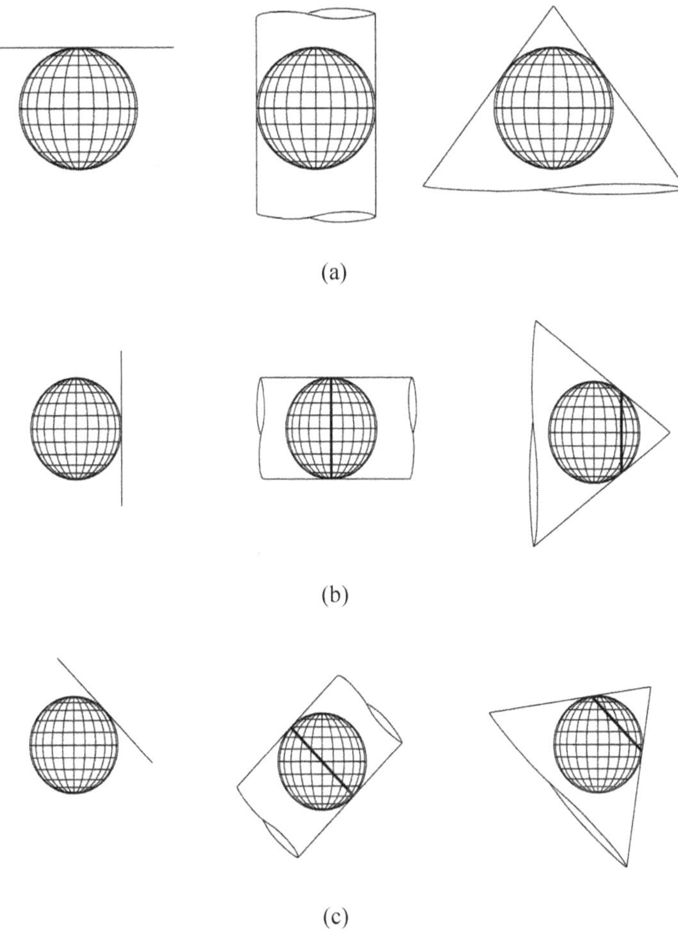

(a)

(b)

(c)

**Fig. 3.11** Orthogonal (a), transverse (b), and oblique (c) projections.

Finally, map projections can be classified based on the implementation method. The method applied to implement a projection can be: (a) *geometric*, (b) *semi-geometric*, or (c) *conventional*.

The geometric (or projective) method simulates the projective rays. The projection of the earth is done by applying the rules of the projective geometry. In semi-geometric method, only a family of lines is projective, while others are induced analytically. Examples of such projections are the pseudo-cylindrical and pseudo-conical projections. In pseudo-cylindrical projections the geographic grid consists of straight lines representing the parallels and curved lined representing the meridians. The conventional method is not based on the projection of light. This method first defines the geographic grid (e.g., a grid of squares, circles, or ellipses) and then specifies the mathematical projection functions.

In the past, many projections have been proposed. Some example projections along with their properties are:

- *Mercator* projection: conformal, cylindrical, tangent;
- *Transverse Mercator* projection: conformal, cylindrical, transverse and tangent (along a meridian);
- *Albers* projection: equidistant, conical, secant (two parallels);
- *Lambert II* projection: conformal, conical, secant (two parallels);
- *Lambert III* projection: equivalent, cylindrical, tangent;
- *Stereographic* projection: conformal, azimuthal, tangent; and
- *Eckert II* projection: equivalent, semi-cylindrical.


## 3.5 Selection of the Map Projection

The choice of the most appropriate map projection depends on: (a) the location, (b) the shape of the area to be mapped, and (c) the application (use) for which the map is intended.

Based on knowledge of the distribution of errors (Figure 3.12), it is possible to choose the most appropriate projection for an area on earth. Figure 3.13 shows some examples. If the region is elongated and follows roughly a great circle of the earth, the cylindrical projection is a good choice. If the region follows a meridian, such as Chile, then a transverse cylindrical projection is appropriate. On the other hand, if the region is elongated along a parallel, except the Equator, like Canada, then a secant conic projection is a good choice. For nearly circular areas, such as France, an azimuthal or conical projection in a tangent or secant configuration can be applied.

The application for which the map is intended to also affects the decision on the map projection to use. For example, sailors need conformal projections (e.g., Mercator projection), because these preserve the directions (azimuths). A cartographer who wants to create a visualization to highlight the relative sizes of the countries in Europe needs an equivalent projection. Topographic maps have been compiled in conformal projections exclusively; while thematic maps usually apply equivalent projections.


## 3.6 Common Projections

In this Section the basic properties and mathematics of some representative map projections are discussed briefly.

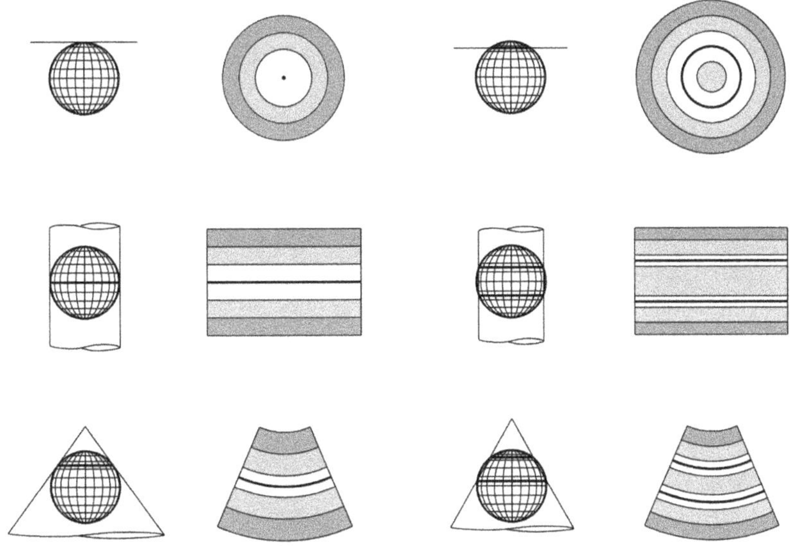

**Fig. 3.12** The distribution of errors in some common configurations (lighter areas correspond to smaller distortions).

**Fig. 3.13** Example countries (Wikipedia).

## 3.6.1 Azimuthal Projections

In an orthogonal *azimuthal projection* the plane surface is tangent at one Pole, the parallels are represented as concentric circles, and the meridians as radii of these circles (Figure 3.14). The meridians and parallels intersect each other at right angles. The directions from the center (Pole) are preserved, while the spacing between the parallels varies depending on the configuration (e.g., gnomonic or stereographic projection).

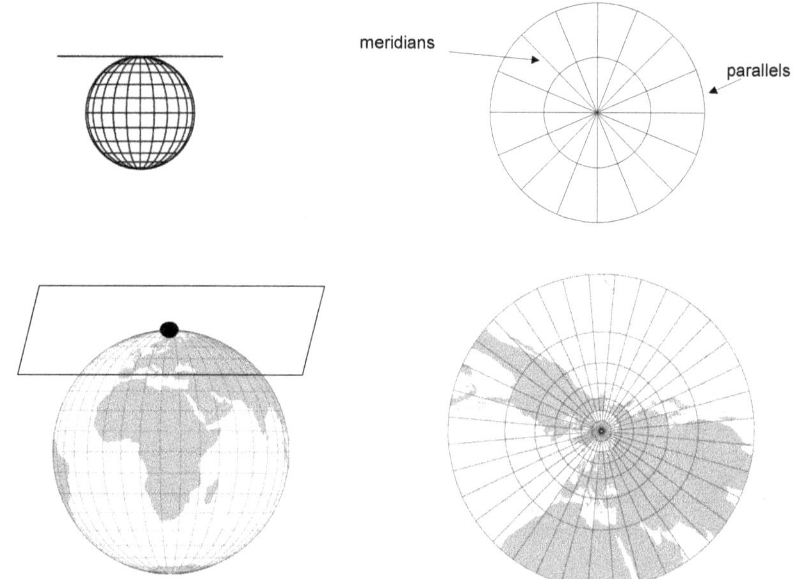



**Fig. 3.14** The geographic grid in orthogonal and tangent azimuthal projection.


### 3.6.1.1 Gnomonic Projection

*Gnomonic* is a tangent azimuthal geometric projection. The light source is located at the center of the sphere (Figure 3.15). Therefore, when the plane is tangent at one Pole the equator is never shown as it is projected to infinity. Obviously, the distortions become very large when moving away from the Pole. Because of the large deformations, the gnomonic projection is rarely used for areas with latitude smaller than 30° in an orthogonal configuration.

The projection is neither conformal nor equivalent (as the deformation along the meridians is different from that of parallels) nor equidistant, while the scale is variable along all directions (Figure 3.15).

The main advantage of the gnomonic projection is the depiction of great circles on the sphere as straight line segments on the map (Figure 3.16). This property has many applications in geosciences, including navigation, and occurs for any orientation of the projection plane (orthogonal, transverse or oblique; Figure 3.11).

Figure 3.17 presents the geometry of the orthogonal gnomonic projection. For simplicity, the solid of a sphere with radius R is used as an approximation of the earth. Similar (but more complex) formulas apply to the ellipsoid approximation.

The radius of the parallel at latitude φ on the projection is given by the formula:

$$r = R \tan(90 - \varphi)$$

Hence, the radius of the parallel of $\varphi=45°$ on the projection is equal to the radius of the sphere (R).

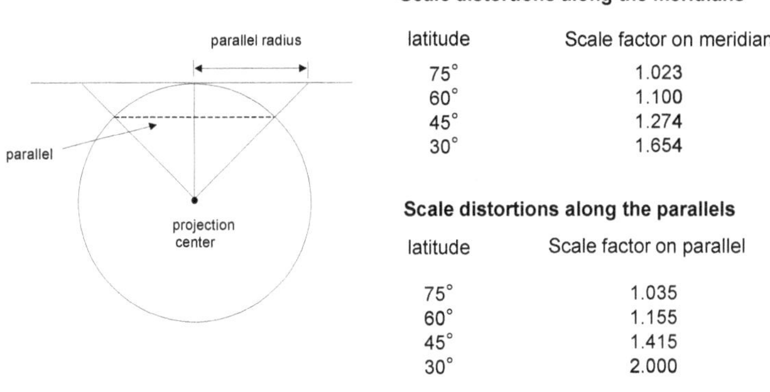


**Fig. 3.15** The configuration of the gnomonic projection and the scale distortions along the parallels and the meridians.

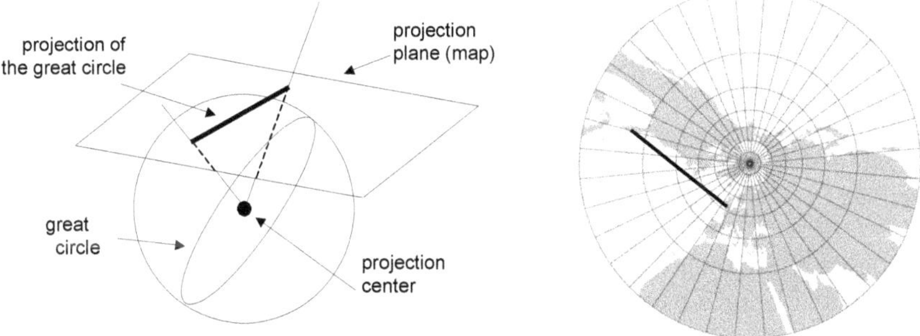


**Fig. 3.16** The projection of any great circle on a gnomonic projection is always a straight line.

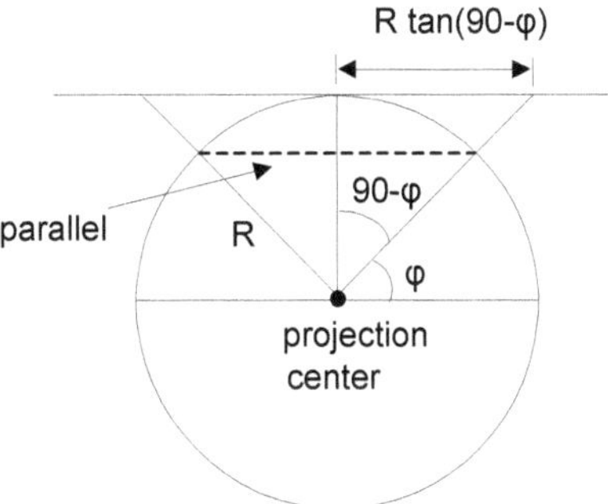


**Fig. 3.17** The geometry of the gnomonic projection.

### 3.6.1.2 Stereographic Projection

The *stereographic* projection (Hipparchus 150 BC) has a similar configuration to the gnomonic projection. The projection center lies at the anti-diametrical point of the point of tangency (Figure 3.18). So, if the plane touches the sphere at the North Pole, the projection center will be located at the South Pole.

As the projection center moves further away from the projection plane (relatively to the gnomonic projection) the deformations along both the parallels and meridians are reduced. Unlike gnomonic projection, the equator can be shown in a stereographic projection.

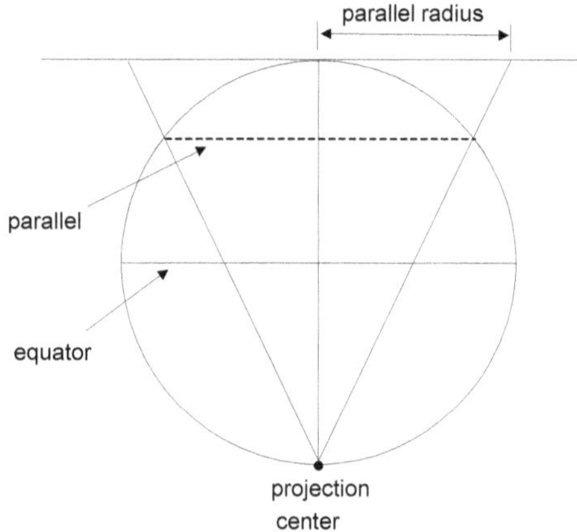


**Fig. 3.18** The configuration of the stereographic projection.

Figure 3.19 presents the geometry of the stereographic projection. The radius of a parallel of latitude φ on the projection is equal to:

$$r = 2R \tan[(90-\varphi)/2]$$

Hence, the radius of the parallel of φ=45° on the projection is roughly equal to: 0,8R (Figure 3.19), where R is the radius of the sphere.

## *3.6.2 Conical Projections*

In *conical projections* the projection surface is a cone which is tangent along a small circle of the sphere (Figure 3.20). The projected grid has a similar form with a grid in azimuthal projections. The parallels are concentric circular arcs (not circles) in an orthogonal conical projection; while the angles between the meridians are proportionately smaller than the actual longitude differences (due to the convergence of meridians).

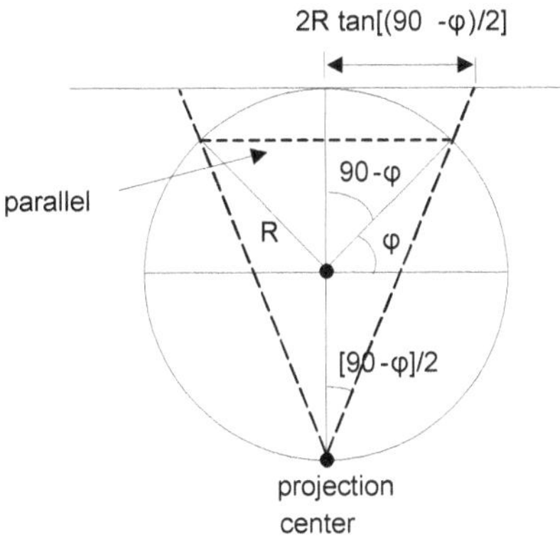


**Fig. 3.19** The geometry of the stereographic projection.

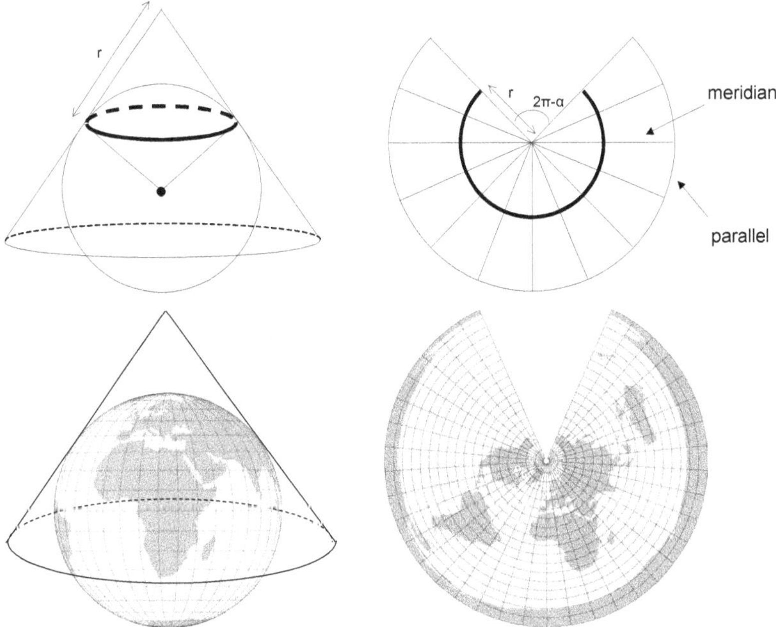


**Fig. 3.20** The configuration of a conical projection.

As shown in Figure 3.20, an arc opening of a <360° represents a circle of 360°. The size of the angle a depends on the parallel along which the cone is tangents to. The value of the angle a increases with the diminishing size of the parallel (in higher latitudes). The Pole in a conical projection can be viewed as a point or as a line (e.g., in an equidistant conical projection; Figure 3.22).

Figure 3.21 presents the geometry of the conical projection. The radius of the tangent parallel on the projection is given by the formula:

$$r = R \,/\, \tan\varphi$$

while the angle a is equal to (Figure 3.21):

$$a = 2\pi \, \sin\varphi$$

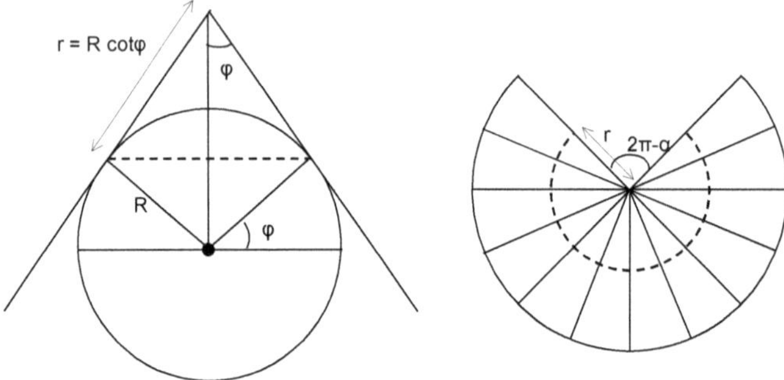


**Fig. 3.21** The geometry of the orthogonal conical projection.

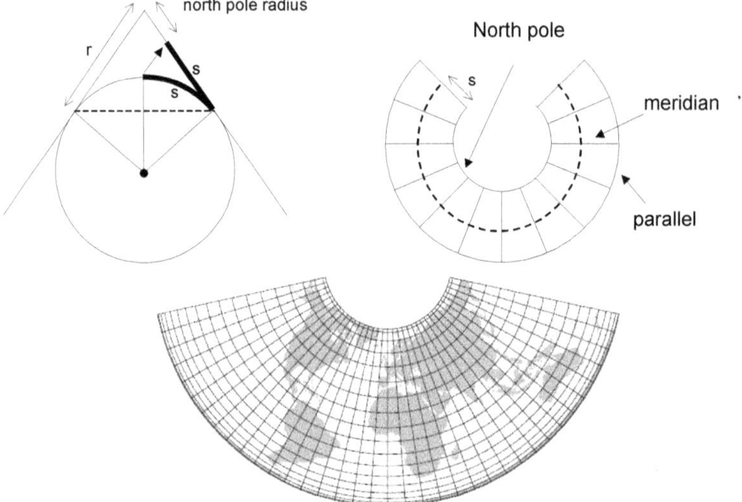


**Fig. 3.22** The equidistant conical projection. Parallels are represented by equally spaced concentric circles on the projection, while the North Pole is represented by a circular arc in order to preserve the distances.

## *3.6.3 Cylindrical Projections*

In an *orthogonal* cylindrical projection the meridians and parallels define two sets of parallel lines intersecting each other at right angles (Figure 3.23). All parallels have a length equal to that of the equator, while the spacing between the parallels varies depending on the configuration. On the other hand, the spacing between the meridians is constant. The Poles are represented by straight lines. Obviously, the grids in transverse and oblique cylindrical projections are not orthogonal.

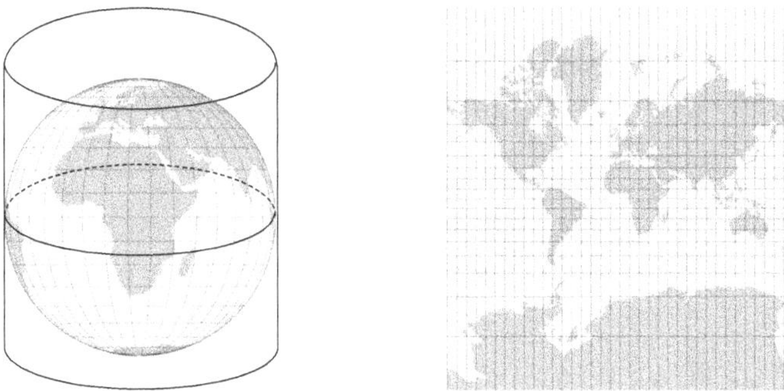

**Fig. 3.23** The configuration of a cylindrical projection.

### 3.6.3.1 Mercator Projection

The *Mercator* projection (Mercator 1569) is a conformal cylindrical projection. The scale factor of meridians when approaching the Poles increases and becomes equal to that of the corresponding parallel (to preserve the shapes and directions). The deformation of the areas increases dramatically toward the Poles (Figure 3.24).

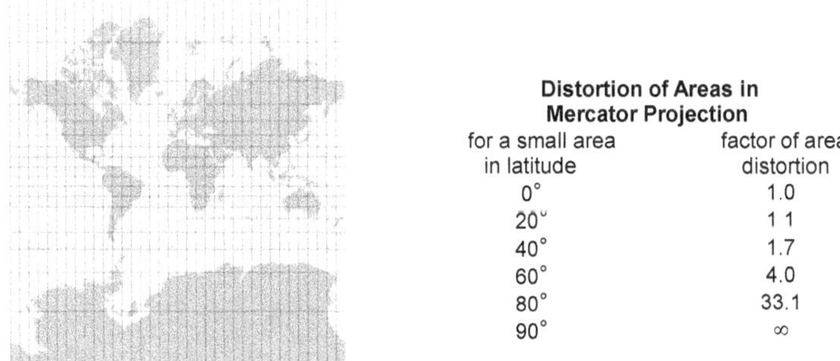

**Distortion of Areas in Mercator Projection**

| for a small area in latitude | factor of area distortion |
|---|---|
| 0° | 1.0 |
| 20° | 1 1 |
| 40° | 1.7 |
| 60° | 4.0 |
| 80° | 33.1 |
| 90° | ∞ |

**Fig. 3.24** Distortion of areas in Mercator projection.

Although conformal, the Mercator projection does not maintain the shapes of large areas. For example, a triangle formed by two meridians, the equator and the Pole is represented as a rectangle on the projection. The Mercator projection is a semi-geometrical projection, as the meridians can be projected geometrically, but not the parallels (as they are shifted; see below).

The lines of constant azimuth, i.e., *loxodromes*, are represented by straight lines (Figure 3.26c), i.e., lines that intersect the meridians are at constant angles, which are equal to the actual azimuth values. This property of the Mercator projection made it very popular in shipping and navigation for over four centuries.

The Mercator projection is also widely used in applications, where there is a need for an accurate projection of directions (azimuths); for example, in maps depicting the vectors of winds, ocean currents, etc. Due to the large distortions of areas (Figure 3.25), Mercator projection is not an option in thematic cartography (population or land distribution), although it has been used in the past for propaganda.

**Fig. 3.25** Very large distortions in Mercator projection. In reality, Africa is 14 times and China 4 times larger than Greenland.

Figure 3.26 presents the geometry of the Mercator projection. The length of a parallel on the sphere is: $2\pi R\cos\varphi$, while on the projection this length is constant and equal to: $2\pi R$ (equal to the circumference of the equator). Hence, the *scale distortion factor* is equal to:

$$\frac{2\pi \cdot R}{2\pi \cdot R\cos\varphi} = \sec\varphi$$

The distance between two parallels lying close to each other on the sphere is:

$$dY = R\, d\varphi$$

This distance is magnified by the coefficient $\sec\varphi$ to obtain a conformal projection (Figure 3.26b), as follows:

$$dY = R\sec\varphi\, d\varphi$$

By integrating this formula, the distance Y of each parallel (of latitude $\varphi$) from the equator on the projection is equal to:

$$Y = \int_0^\varphi R\cdot\sec\varphi\cdot d\varphi = R\ln\tan\left(45° + \frac{\varphi}{2}\right)$$

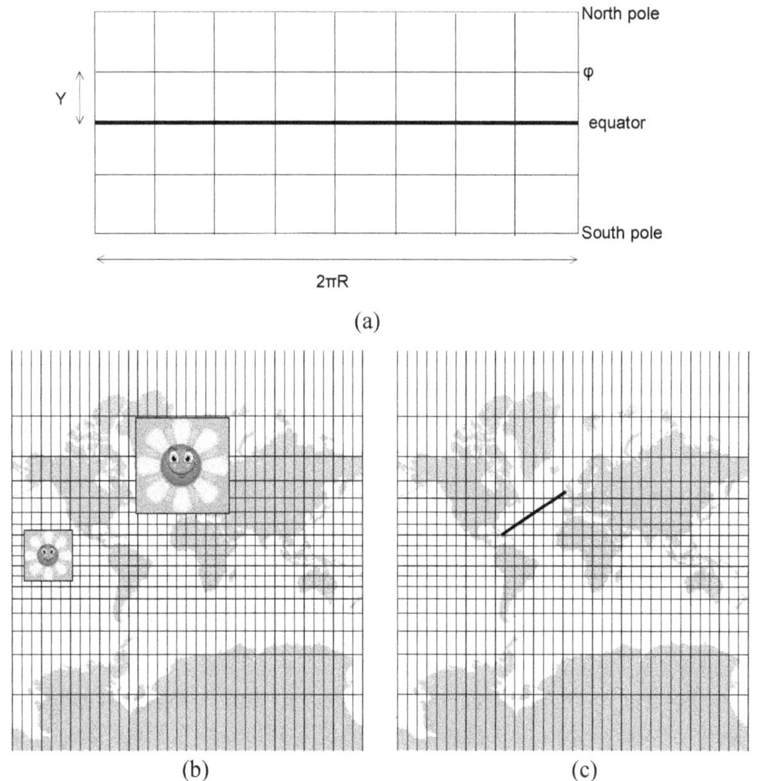



**Fig. 3.26** Mercator projection: the geographic grid (a); conformal projection of the Earth (b); and a loxodrome line (c).

## 3.7 Navigation with Gnomonic and Mercator Maps

Both Mercator and Gnomonic projections have widely been used in combination to support the navigation with compass over the last four centuries (Figure 3.27). The shortest course between two points on the sphere can be found through a straight line on the gnomonic projection.

This line corresponds to a great circle arc, with a variable azimuth, which cannot be implemented by a *compass*. On the other hand, this line can be approximated by a set of loxodromes on the Mercator projection by transferring some representative points (landmarks; e.g., P0,P1,P2,P3) from the Gnomonic to the Mercator projection. These points in pair form a set of loxodromes which approximate the great circle. These loxodromes can be implemented in navigation with a compass.

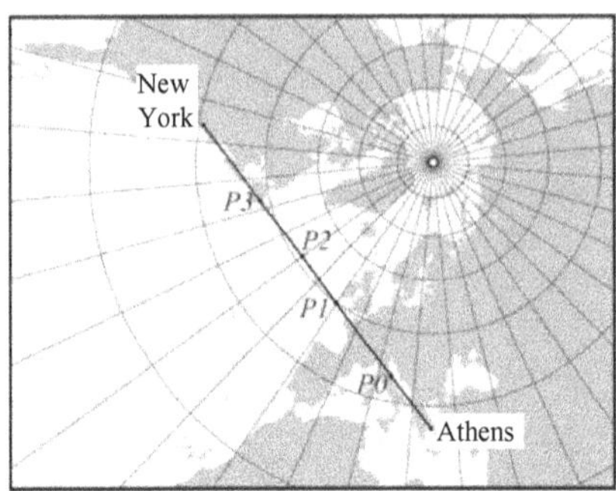
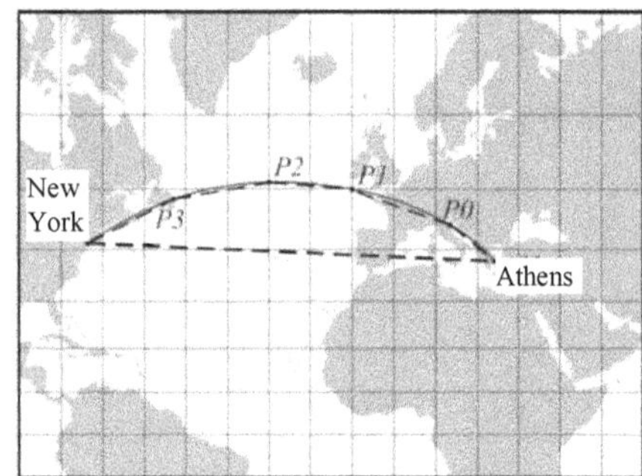


**Fig. 3.27** The Gnomonic and Mercator Projections. The shortest course from Athens to New York approximated by four loxodromes (Athens–P0, P0–P1, P1–P2, P2–P3, P3–NewYork).

## 3.8 Universal Transverse Mercator

*Universal Transverse Mercator (UTM)* is a common coordinate and projection system. It applies a multiple projection which is based on the transverse cylindrical configuration (Figure 3.11b). This offers a minimization of distortions along and around a meridian.

UTM system applies a secant transverse Mercator projection (Figure 3.28a) originally on the ellipsoid of Hayford (a=6,378,388m, b=6,356,912m, f=1/297.0), replaced later by GRS80 (a=6,378,137m, b=6,356,752m, f=1/298.257) (see Table 2.1). A transverse Mercator map is conformal, i.e., it preserves the shapes of small areas of the ellipsoid on the plane.

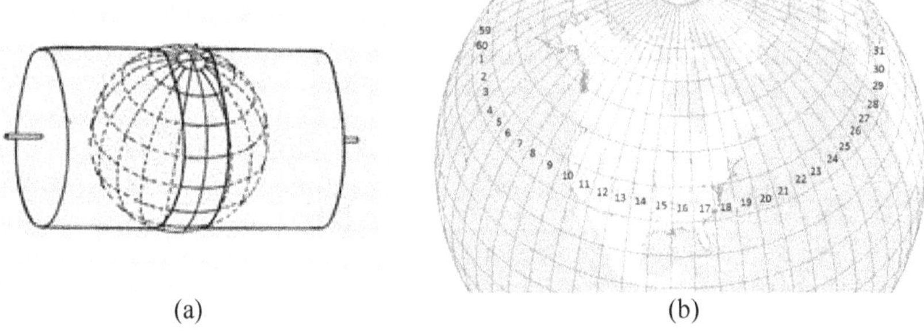

(a)        (b)

**Fig. 3.28** Zones in Universal Transverse Mercator System (Wikipedia).

In the UTM-6° system, the entire surface of the earth is divided into 60 zones (local systems) of 6° width (in longitude) each. Zone 1 covers longitude 180° to 174°W; zone numbering increases eastward to zone 60 that covers longitude 174° to 180°E. Figure 3.28b show the zones in the Pacific, N. America, and the Atlantic. A UTM zone has a range of latitudes from -80° to +84°. Polar areas are not mapped (they are projected in the UTP system; which is not included in the book).

The scaling factor along the central meridian is 0.9996, while the origin is assigned a false easting of 500.000m to avoid negative values. The deformation increases with the square of the distance from the central meridian up to 500ppm (Figure 3.29).

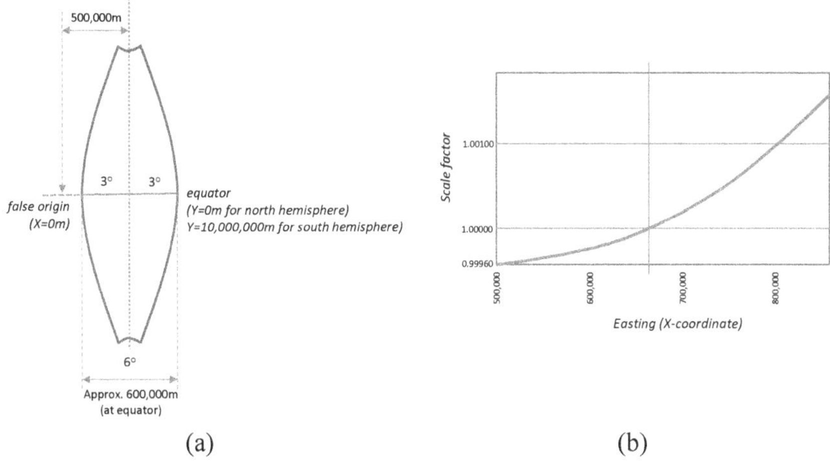



**Fig. 3.29** Coverage and distribution of errors in a UTM zone.

# References and Further Reading


Lee, Y.C., 1992. *Geographic Information Systems*. Lecture Notes. Department of Geodesy and Geomatics Engineering. University of New Brunswick.

Maling, D.H., 1989. *Measurements from Maps: Principles and Methods of Cartometry*. Pergamon Press.

Porter, W., and McDonnell, Jr., 1979. *Introduction to Map Projections*. Marcel Dekker.

Robinson et al., 1995. *Elements of Cartography*, 6th Edition. Wiley.

# 4 Geographic Information Systems

## 4.1 Introduction

*Data* is raw material and unorganized facts that need to be processed. When data is processed, organized, structured, or presented in a given context so that to make it useful, it is called *information*. In other words, the processing, analysis, and reporting of data in a meaningful way is called information (Figure 4.1).

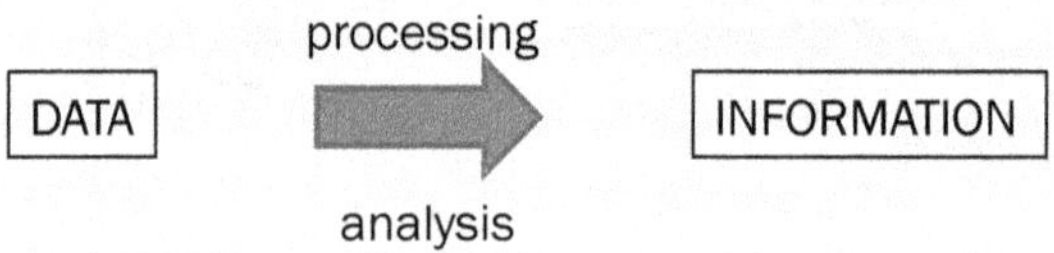


**Fig. 4.1** From raw data to information.

For example, the following data: { (2,2), (4,3), (6,4), (8,5) }, summarizes a "straight line" (Figure 4.2). The "straight line" (information) is more comprehensive than the collection of the eight numbers or the four two-dimensional points (data).

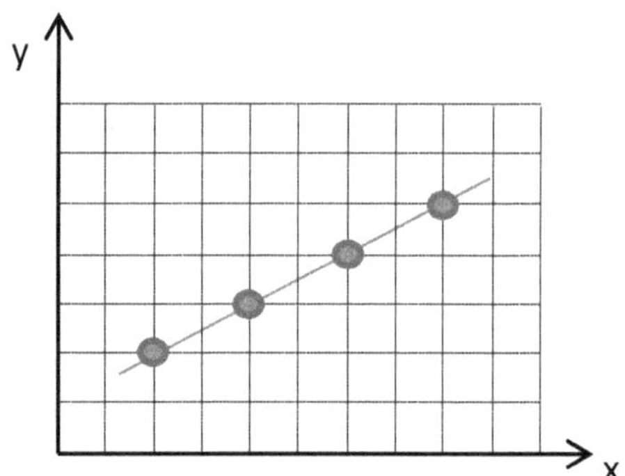


**Fig. 4.2** A collection of four points forming a straight line.

Each collection of tools, which are intended to derive information from data, composes an *Information System (IS)*. In the past, information systems did not make use of computers. An example of such a system is the collection of folders (or cards), which can still be found in libraries, hospitals, and schools. This system, like any other information system, consists of (Figure 4.3): (a) *hardware*, (b) *software* (operations), (c) *data*, and (d) *users* (humans).

In the analog information system of a hospital, the hardware consisted of the drawers and the cards. The operations (software) were dealing with the updates of the cards content and the removal, or insertion of new cards. The data consisted of all patients records. Finally, the users were all the people involved, i.e., the doctors, nurses, and patients.

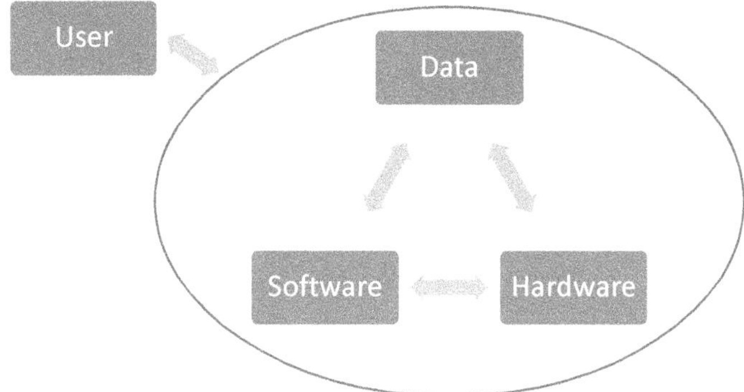


**Fig. 4.3** The basic components of an Information System.

A modern information system is implemented in a computer system and intends to meet three requirements:

- derive useful information and add value to data;
- assist the users in taking decisions upon complex problems; and
- share data and information with other systems (interoperability).

The Sections that follow briefly present the components and types of Information Systems, then, Geographic Information Systems (GIS) are introduced.

## 4.2 Information Systems Architecture

In a modern Information System there are four main components: (a) *user interface*, (b) *database*, (c) *data manager*, and (d) *data analyzer*. The user interface acts as a coupling between the system and the user (or another program). It receives user commands and reports the results back to the user. The database accommodates the system data. The data manager is responsible for the management of the data residing in the database. Finally, the data analyzer performs all analytical operations needed to meet its scope: extract useful information from the data.

All four parts may be organized in a system based on different architectures. A simplified architecture is shown in Figure 4.4. The user interacts with the system via the user interface. A request is passed to either the data manager or the data analyzer, depending on whether it pertains to data management or involves data analysis, respectively. The data analyzer has access to the database through the data manager (in order to retrieve all required data for the analysis). The results of the management and/or analysis of the data are reported to the user (or the external program) via the user interface.

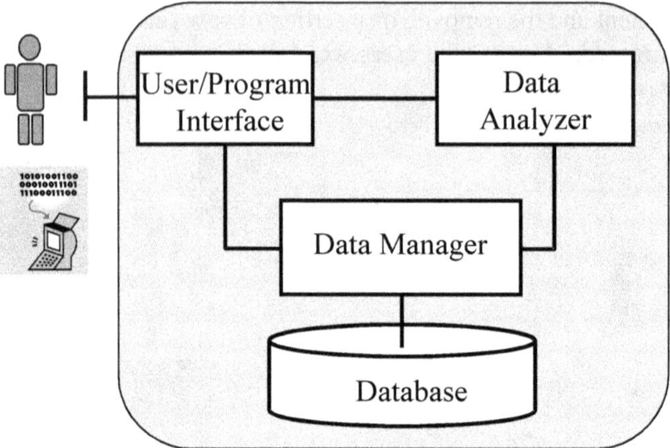


**Fig. 4.4** A simplified architecture of an Information System.


## 4.2.1 User Interface

The *user (or program) interface* is the communication channel between the user
and the system. The backbone component of this channel is the communication
*language* between the two.

On the one hand, the interaction of the information system with other *software
programs* is a matter of standardization. It involves data sharing (in a common
format) and command activation (execution of services). On the other hand, the
interaction of the information system with *humans* involves the definition of a system language. Using this language the users are able to make the system perform
management and analysis operations on the data. An example of such a command
expressed in the system language might be:

```
NEAREST      (       SELECT GEOMETRY
                     FROM HOUSES
                     WHERE CITY = "FREDERICTON"            )
TO RIVER     (       SELECT GEOMETRY
                     FROM RIVERS
                     WHERE NAME = "SAINT JOHN"             )
```

This command searches for the nearest house to the Saint John River in Fredericton. The command consists of three parts. The first part is enclosed in the first
parenthesis and retrieves the location of all houses in the city of Fredericton. This
is a management operation. The second part is enclosed in the second parenthesis
and retrieves the riverbank of the Saint John River. This is a management operation as well. Both operations will be carried out in the data manager. The third part
is a function (NEAREST) which acts over the data retrieved from the previous two

parts. The outcome is the house that lies closest to the river. Nearest is an analytical operation and is carried out by the data analyzer.

## 4.2.2 Database

The *database* is a collection of files that accommodate the application data, as well as other data (e.g., metadata, indices, analytical model parameters, etc.), which is required from the Information System to operate. The organization of the application data into files is not random, but it tends to (a) minimize data volume, and (b) facilitate its retrieval from the data manager.

The data manager and the database constitute the core of an Information System and they are accountable for the system efficiency, consistency, and effectiveness.

## 4.2.3 Data Manager

The *data manager* accesses directly the database and provides all data management operations: insert, delete, update, and select (retrieve) system data. Data is usually dynamic in nature (it changes over time) and differs from system to system. The operations provided by the data manager are similar in all information systems (applications).

In order for the management system (data manager) to be effective, it must organize the data in the database so that: (a) the storage requirements are minimized (using compression methods), and (b) the execution of data management operations is accelerated (using appropriate structures).

The software tools provided by the data manager form a Database Management System (DBMS). A DBMS is responsible for (a) the definition (construction), (b) the storage (population), and (c) the manipulation of data in an Information System (see Chapter 5).

## 4.2.4 Data Analyzer

The *data analyzer* is the component in an Information System that transforms the data into information. It is a software package (toolbox), which provides analytical operations (functions) that can be activated by the user (physical person or external program) through the interface.

In contrast with the data manager, whose operations can be readily standardized, the operations of data analyzer vary significantly from application to application. In general, the operations (functions) of the data analyzer apply comparisons, mathematical estimations, and statistical analytics.

The wide variety of requirements in analysis operations between applications has led to the development of multiple types of Information Systems. Typically, an Information system incorporates some basic analytical operations and provides the tools for the development and encapsulation of new operations, according to the application needs.

## 4.3 Types of Information Systems

Information systems can be classified based on their scope, architecture, users, etc. Nowadays, a wide variety of information systems are available in the market to meet the requirements of specific application domains. These systems offer a rich collection of management and analysis operations. In addition, they are extensible as the users may develop and incorporate new methods into the core system, while the synergy with other computational systems is also supported.

Some basic categories (types) of Information Systems that are available as commercial or shareware packages are the following:

- *Management Information Systems*: These systems support the business decisions of companies or organizations, by analyzing and enriching business data.
- *Health Information Systems*: These systems support either the diagnosis and treatment of illnesses or the organization and management of medical units.
- *Library Information Systems*: These systems support the searching of documents in one or more (distributed) libraries.
- *Communication Information Systems*: These systems support the communication between people and/or groups of people through computer systems.
- *Control Information Systems*: These systems support the military and enterprise communications.
- *Recreational Information Systems*: These systems support the recreation of individual people or groups of people using multimedia applications (computer games, movies, music, etc.).
- *Spatial or Geographic Information Systems*: These systems support the decision making in problems that involve spatial and geographic data.

## 4.4 Geographic Information Systems

*Geographic Information Systems (GIS)* are computer-based systems designed to support the modeling, management, analysis, display, and dissemination of spatially referenced data at different points in time. These systems have been used widely in administrative and productive activities, involving applications from various domains:

- *Socio-economic* applications (e.g., urban and regional planning, cadastre, archaeology and history, natural resources, market analysis, etc.).
- *Environmental* applications (e.g., forestry, fire and disease spread, etc.).
- *Infrastructure and management* applications (e.g., water and sewage networks, telecommunications and energy, transportation, fleet management, navigation of sea vessels and airplanes, etc.).

The role of GIS in all these applications is to offer the users and decision-makers a set of powerful tools to solve the complicated and usually non-structured spatiotemporal problems. Additionally, these systems must be computationally efficient, in order to be useful in a production environment. In the following subsections, the GIS components are briefly presented.

## *4.4.1 Basic Components of a GIS*

As in any Information System, a GIS user interacts with three major components (Figure 4.3): *hardware, software,* and *data.* Hardware and software have a certain life circle; they are subject to technological advances and tend to be replaced by new and better products. Data is the most expensive GIS component to acquire, because its capture takes a lot of time and effort. In addition, most geographic data is dynamic and requires continuous updates, thus a GIS should be designed and built to deal with. A main function of current GIS is to create a series of comprehensive pictures and situations, states and conditions of data items within a spatial context.

### 4.4.1.1 The Data Component

The nature of geographic data was discussed in Chapter 1. Current GIS packages efficiently handle the spatial and thematic dimension of geographic entities. The temporal changes of the entities are partially handled by current technology.

### 4.4.1.2 The Hardware Component

The hardware component includes: (a) the *processing units* that execute programs; (b) *auxiliary storage units* that maintain data and programs; and (c) *peripheral devices* that are used to enter data (e.g., digitizers, scanners, etc.), generate maps and reports (e.g., terminals, plotters, printers, etc.), and communicate with other work stations and the Internet (e.g., modems, multiplexors, device servers, etc.).

### 4.4.1.3 The Software Component

The software components surround the system hardware with the following three layers (Figure 4.5): (a) *operating system*, (b) *system support programs*, and (c) *application software*.

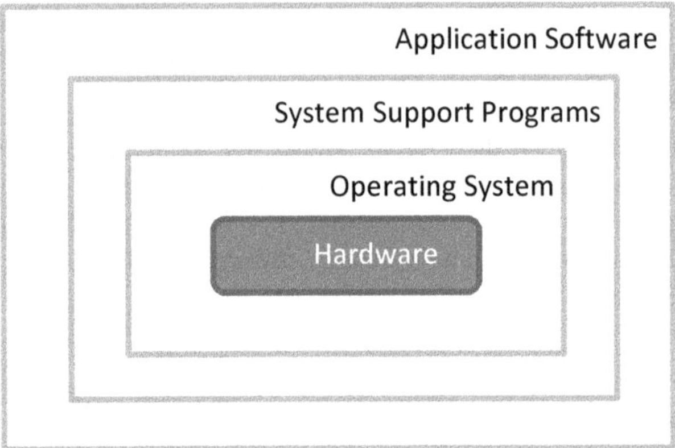


**Fig. 4.5** The layers of the software component in a GIS.

The operating system consists of programs that control the system operations and the communication with all hardware devices connected to the computer. The system support programs perform additional functions such as storage management, program compilation, communication with peripheral devices, etc. The application software is used to perform most of the tasks in a GIS. It consists of multiple integrated programs (modules) designed to implement the modelling, management, analysis, and visualization of geographic data.

The capabilities provided by the application software vary among GIS software vendors. However, they all include a compact set of operations for handling geographic data, which fall into four broad classes as shown in Figure 4.6:

1. *Data maintenance:* This class encompasses a variety of methods for insertion, deletion, and modification of data in a geographic database system. These methods will be elaborated in the following Chapters.
2. *Data selection:* The operations of this class support the selection of geographic entities based on their attribute values. Depending on the dimension(s) involved (Chapter 1; Figure 1.3), operations for data selection fall into the following four categories: (a) spatial selections (they are based on the spatial dimension of geographic entities); (b) thematic selections (they are based on the identifier or the thematic dimension of geographic entities; (c) temporal selections (they are based on the temporal dimension to geographic entities); and (d) mixed selections (they involve more than one dimension of geographic entities, e.g., spatial and/or thematic and/or temporal). Notice that the dimension of data quality should be considered in order to have a measure on the accuracy and quality of the selected set. The selection operations along with the maintenance opera-

tions comprise the set of operations provided by the data manager of an information system (Figure 4.4). These methods will be elaborated in the following Chapters.

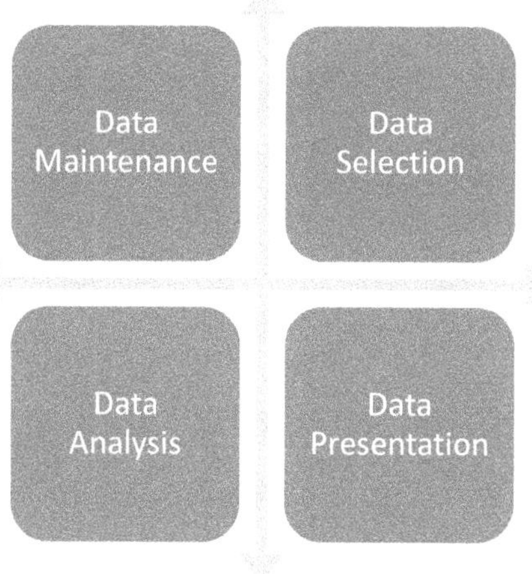


**Fig. 4.6** Basic classes of operations for handling geographic data.

3. *Data analysis:* There is a wide range of basic operations for geographic data analysis, which are heavily dependent on the application domain. Several classifications have been proposed in the past. Data analysis operations process the attributes (spatial, thematic, temporal, and quality) of geographic entities and result in either the generation of new entities or the modification of attribute values assigned to existing entities. In brief, data analysis operations include: (a) classification and generalization operations; (b) measurement operations; (c) overlay operations; (d) neighborhood operations; (e) interpolation operations; and (f) connectivity operations. These operations will be examined in Chapter 11.

4. *Data presentation:* This class encompasses a variety of methods for the presentation of data selection and/or analysis results. The presentation may be of map-like, graph, tabular/text, or multimedia form.

In addition, the application software includes the following tools to support the development of advanced applications: (a) *user command interfaces* (to invoke functions of the operating system, support programs, and run the application software modules); (b) *external programs interface* (to access the GIS components by means of a programming language, such as C++ and Python, and allow the development of complex operations by external programs and applications); and (c) *Web development tools* (a series of tools to disseminate both GIS data and operations over the Web through advanced geospatial web services).

# References and Further Reading

Antenucci, J.C., Brown, K., Croswell, P.L., Kevany, M.J., and Archer, H., 1990. *Geographic Information Systems: A Guide to the Technology*. Van Nostrand Reinhold Press.

Aronoff, S., 1989. *Geographic Information Systems: A Management Perspective*. WDL Publications.

Burrough, P.A., and McDonnell, R.A., 1998. *Principles of Geographic Information Systems*. Oxford University Press.

de By, R.A. (Ed.), 2000. *Principles of Geographic Information Systems – An Introductory Textbook*. ITC Educational Textbook Series.

Garcia Molina, H., Ullman, J.D., and Widom, J., 2000. *Database Systems Implementation*. Prentice Hall.

Lee, Y.C., 1992. *Geographic Information Systems*. Lecture Notes. Department of Geodesy and Geomatics Engineering. University of New Brunswick.

Longley, P.A., Goodchild, M.F., Maguire, D.J., and Rhind, D.W., 2001. *Geographic Information Systems and Science*. Wiley.

Reynolds, G., and Stair, R.M., 2003. *Fundamentals of Information Systems*. Course Technology Publications.

Rigaux, P., Scholl, M., and Voisard, A., 2002. *Spatial Databases with Applications to GIS*. Morgan-Kaufmann.

Worboys, M.F., 1995. *GIS – A Computing Perspective*. Taylor-Francis.

# PART II

# DATABASE SYSTEMS TECHNOLOGY

# 5 Database Systems

## 5.1 Introduction

*Database* is a collection of related data, which represents some aspect of the world, usually referred to as *miniworld*. It is designed, built, and populated with data for a specific purpose (users and applications) to describe an organization or domain. An example database is the collection of records in a personal agenda. The agenda hosts *data* such as names, addresses, and telephone numbers, to describe the relatives, friends, and colleagues. The latter three are *entities* (units or objects; see Chapter 1) of interest to the agenda holder, while data is the values assigned to their *attributes*.

A database can be created, populated, and maintained either manually (e.g., traditional land registry drawers and folders) or automatically (e.g., a computerized cadastral database). The traditional paper map is an analog geographic database. The term geographic (or spatial) refers to the fact that the data on a map is linked to a coordinate system and is multidimensional. The map as a database accommodates entities with spatial reference (e.g., house, road, river, lake), the attributes of these entities (e.g., house location, settlement type: village, town, city, capital city), as well as their spatial relations (e.g., the distance between two cities).

Figure 5.1 shows a simplified paper map. The entities depicted on this map are: cities (point), road, river, railway (line), and lake (polygon). The attributes of these entities are conveyed through the use of appropriate *map symbols* (e.g., the capital city), which are described on the map legend. The recognition of the spatial relations (topological, direction, distance) between the entities is realized by visually comparing their locations (e.g., City 1 is located near the river and south-west of the lake; y-axis indicates the direction of North).

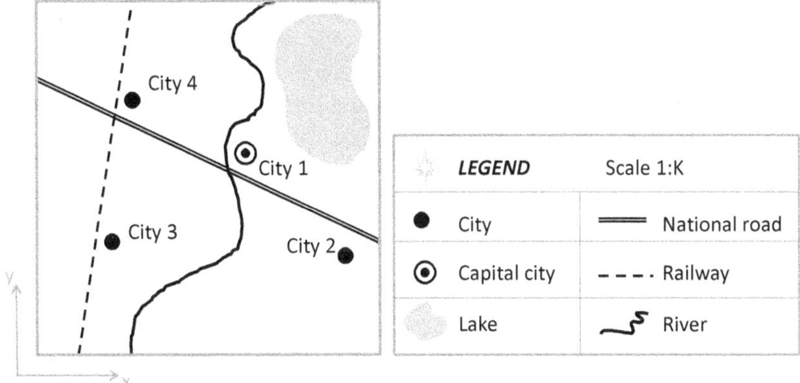


**Fig. 5.1** The maps as a geographic database.

A digital database (e.g., a cadastre or land registry database) is implemented and maintained by a set of software applications, which constitute the database management system. A *Database Management System (DBMS)* is a collection of programs to define, construct, and manipulate a database (see next Section).

The processing of data is usually time-consuming. To achieve efficient processing, a DBMS makes use of appropriate *index* mechanisms. Indices are also present in paper maps. Figure 5.2 presents an index that is often attached to paper maps. This index is called *grid file* (Chapter 13) and in this example it supports the fast search of a city location given its name. The index is a table. The first column in the table orders the city names on the map and the second column reports the cell where each city falls within. Hence, the search is narrowed to a small area on the map and eventually it is accelerated.

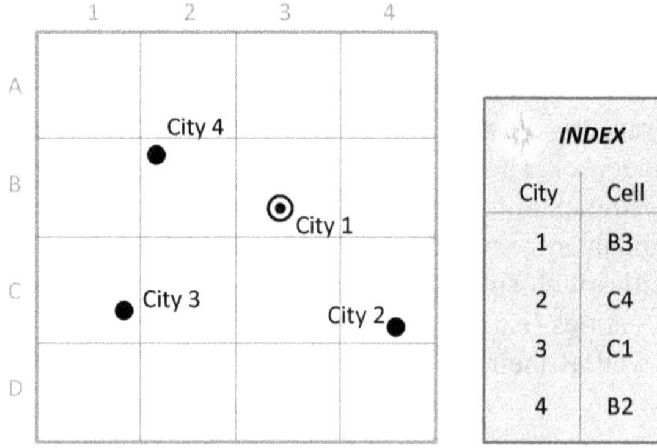


**Fig. 5.2** A typical index structure for paper maps.

A *Database System (DBS)* is a database along with the DBMS for handling its content. Figure 5.3 illustrates the architecture of a DBS. The system consists of four parts: (a) the interface to the users and external programs; (b) the DBMS software, which is further divided into the software for processing the queries coming from the interface and the software to access the data and metadata of the system; (c) the database, where the data is stored; and (d) the database catalogue, which describes and documents the database content with appropriate *metadata* items. The DBMS, before accessing the actual data which is stored in the database, searches the database catalogue to find the physical location of the data items and the index structures on top of this data, if any.

## 5.2 Database Management Systems

A Database Management System (DBMS) is a general purpose software system that facilitates three basic processes in database management (Figure 5.4): (a) *definition*, (b) *construction*, and (c) *manipulation* of a database.

The definition of a database refers to the specification of the data types, structures and constraints of data to reside in the database. The construction of a database includes the population of the database with actual data. Finally, the manipulation of a database involves: the retrieval of data, update of data, and generation of reports from the data. The DBMS is also responsible for facilitating the access and dissemination of the database content to users and external programs. This is usually referred to as *sharing* of the database.

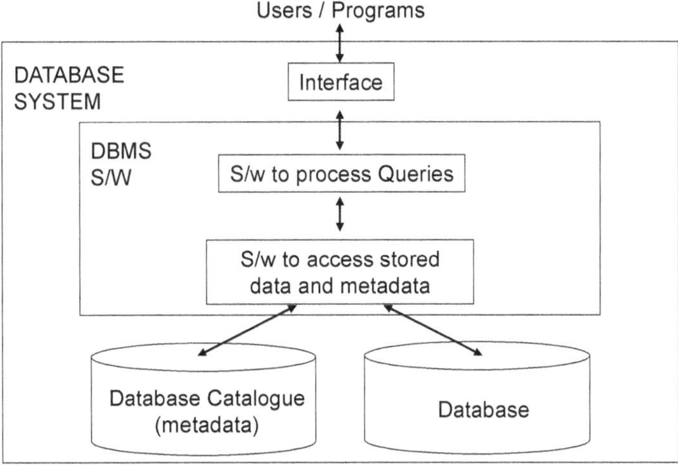


**Fig. 5.3** The architecture of a Database System.

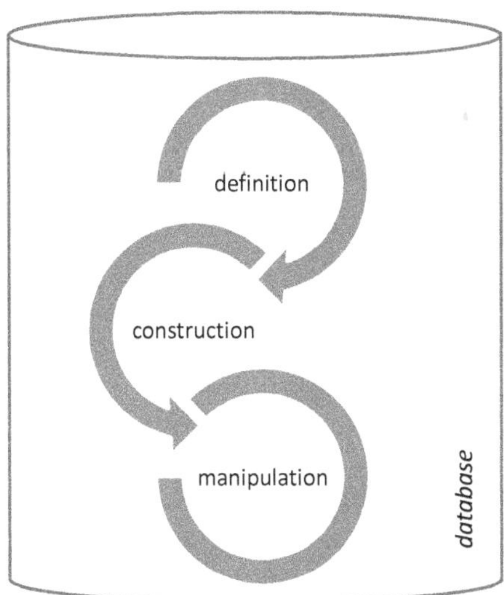


**Fig. 5.4** The basic procedures offered by a DBMS. They all refer to the database.

## *5.2.1 An Example*

Figure 5.5 presents an example situation for a simplified cadastral application. This view, which is usually called *miniworld*, is represented in the seven files in Figure 5.6. Each file accommodates a set of records of the same type. For instance, in the file OWNERS, the data describing the parcel owners (physical persons) is stored. These seven files together constitute the database of the cadastral application.

The *definition* of the database includes: (a) the structure of the records in each file (e.g., the records in the file TITLES have the following items: the parcel identifier PARCEL, the owner identifier OWNER, the percentage of ownership PERCENTAGE, and the date of purchase PDATE); (b) the data type of each item and any constraints that apply on it (e.g., the SURNAME is an alphanumerical string of up to 15 characters, the date is of type mm/dd/yyyy, while the percentage is an integer between 0 and 100; i.e., data constraint); and (c) the shared domains between items (key attributes) of two or more files to assure the combination (join) of their content (e.g., the SIN in the file OWNERS shares the same domain with the OWNER in the file TITLES).

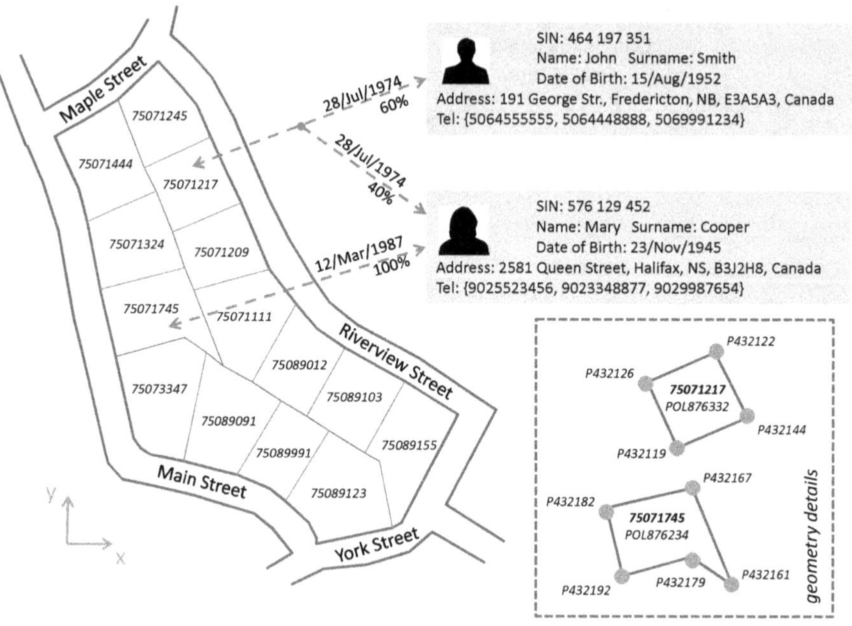


**Fig. 5.5** An example cadastral plan.

The *construction* of the database includes the population of the files with data items. For instance, the data describing the owners is stored in the file OWNERS, while the data describing the parcels is stored in the file PARCELS. Based on the definition, the records in two or more files may be related to each other. For example, the record describing the properties of Mary Cooper in the file OWNERS is related to two records in the file PARCELS through the file TITLES.

The *manipulation* of the database involves search queries and updates of the data items in the files. For instance, the following statements are processed by the DBMS to manipulate the database content: "which parcels are fully or partially owned by Mrs. Mary Cooper?" or "update the address of Mr. John Smith to 115 Prospect street, Fredericton, NB, E3B8J2, Canada". The queries or updates are all expressed in a system language, such as SQL, OQL, etc. (see Chapter 6).

**OWNERS**

| SIN | SURNAME | NAME | DoB | STREET | No | CITY | PROV | ZIP | COUNTRY |
|---|---|---|---|---|---|---|---|---|---|
| 464197351 | SMITH | JOHN | 08/15/1952 | GEORGE | 191 | FREDERICTON | NB | E3A5A3 | CANADA |
| 576129452 | COOPER | MARY | 11/23/1945 | QUEEN | 2581 | HALIFAX | NS | B3J2H8 | CANADA |
| ... | ... | ... | ... | ... | ... | | | ... | ... |

**TELEPHONES**

| OWNER | NUMBER |
|---|---|
| 464197351 | 5064555555 |
| 464197351 | 5064448888 |
| 464197351 | 5069991234 |
| 576129452 | 9025523456 |
| 576129452 | 9023348877 |
| 576129452 | 9029987654 |
| ... | ... |

**TITLES**

| PARCEL | OWNER | PERCENTAGE | PURCHASE_DATE |
|---|---|---|---|
| 75071217 | 464197351 | 60% | 07/28/1974 |
| 75071217 | 576129452 | 40% | 07/28/1974 |
| 75071745 | 576129452 | 100% | 03/12/1987 |
| ... | ... | ... | ... |

**PARCELS**

| ID | USE | BUILT_FACTOR | ADDRESS | POLYGON |
|---|---|---|---|---|
| 75071217 | HOUSING | 1.40 | 542 RIVERVIEW STREET, FREDERICTON | POL876332 |
| 75071745 | PARKING | 1.20 | 323 MAIN STREET, FREDERICTON | POL876234 |
| ... | ... | ... | ... | ... |

**POLYGONS**

| PLID | AREA | PERIMETER |
|---|---|---|
| POL876332 | 1.235 | 142 |
| POL876234 | 1.440 | 169 |
| ... | ... | ... |

**POINTS**

| PTID | X | Y | Z | POLYGON | ORDER |
|---|---|---|---|---|---|
| P432122 | 45678.34 | 8938.89 | 34.20 | POL876332 | 1 |
| P432144 | 45705.56 | 8879.67 | 32.85 | POL876332 | 2 |
| P432119 | 45621.12 | 8845.87 | 31.97 | POL876332 | 3 |
| P432126 | 45592.56 | 8910.91 | 32.88 | POL876332 | 4 |
| P432167 | 45650.33 | 8813.12 | 30.71 | POL876234 | 1 |
| P432161 | 45692.11 | 8726.44 | 28.12 | POL876234 | 2 |
| P432179 | 45653.98 | 8749.92 | 28.65 | POL876234 | 3 |
| P432192 | 45550.19 | 8730.51 | 27.92 | POL876234 | 4 |
| P432182 | 45539.87 | 8802.01 | 29.33 | POL876234 | 5 |
| ... | ... | ... | ... | ... | ... |

**Fig. 5.6** Cadastral data organized in seven files.

## *5.2.2 The Advantages of Database Management Systems*

The use of a Database Management System offers many advantages over the traditional file processing approach. Figure 5.7 presents an example of the traditional file processing. In this example, two agencies (users) the Ministry of Development and the Ministry of Finance need to process cadastral data for their own purposes. Hence, each agency handles its own set of files and has developed and maintained independent applications to process those files. There is no interaction between the two agencies although they both handle similar data items and have common needs.

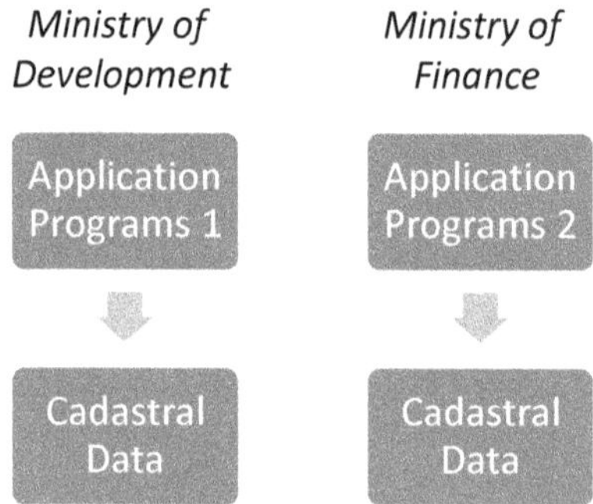


**Fig. 5.7** The traditional file processing approach.

This approach is characterized by a *redundancy* in the data definition and storage as cadastral data items are duplicated in two repositories, one for each agency. Consequently, there is also a redundancy in the efforts to maintain the two datasets up-to-date. Additionally, the data is structured differently in each agency and the corresponding application programs have been customized appropriately.

Figure 5.8 presents a framework which adopts the DBMS technology. In this approach there is a unique repository for cadastral data, which is accessible by multiple users. This approach has the following advantages:

- *Redundant data is eliminated.* The application data is stored in a single repository and is accessible by multiple users (data sharing). This approach offers low storage space requirements provided that the data items are stored once. Additionally, this eases the update process, provided that all updates are carried out into a single repository and are visible to all applications and users.

- *Data is not dependent on the application programs.* The database management system offers a clear definition of the database content. The metadata catalogue describes how the data is structured in the database and what restrictions apply to the data items. This way, the details about data storage and management are hidden to the external programs. The structure of data may be altered or a new index may be introduced to improve the system performance without affecting the external programs. Hence, the development of the application programs is facilitated. Obviously, an application program needs to access the metadata catalogue to locate the required data items in the database and acknowledge their structure and/or index mechanisms.

* *Multiple interfaces to data and operations.* The concurrent access to the database and the execution of operations on the data items by multiple users is supported. At the same time, the database content is secure and remains consistent after each update through the use of appropriate locking mechanisms.

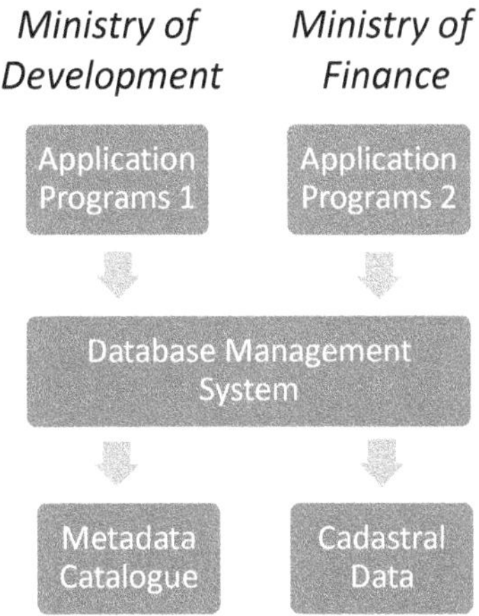


**Fig. 5.8** The DBMS approach for Figure 5.7.

## 5.3 Data Model

*Data Model* is a collection of concepts that can be used to describe the structure of a database. The term *structure* includes: (a) the data types, (b) the relations between the data items, and (c) the constraints that should hold on the data.

Most data models also include a set of basic *operations* for retrievals and updates on the database, as well as a *language* for the definition, construction, and manipulation of the database.

The database models fall into three categories according to the types of concepts they use to describe the structure of a database (Figure 5.9):

1. *Conceptual* or *high-level* data models. They make use of concepts that are very close to the way humans perceive data.
2. *Physical* or *low-level* data models. They apply a series of concepts that describe in fine detail how the data is structured in the database files.
3. *Logical* or *medium-level* models. They are classified between the previous two categories. On one hand, they hide the details on how the data is organized within a computer system, while on the other hand, they can be easily implemented in a computer system using appropriate methodologies.

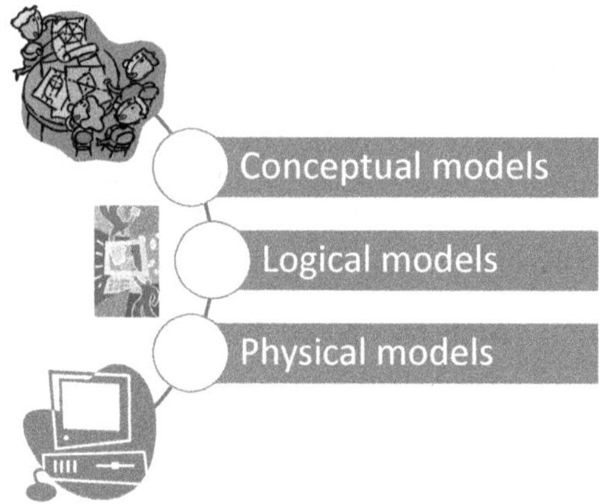


**Fig. 5.9** Types of data models.

## 5.4 Database Schema and Instances

All data models make a clear distinction between the content (the database) and its description. The description of the database is called database *schema*, while the data items that reside in a database at a specific point in time form the database *instance*.

The database schema is specified during the database design process and is not expected to change frequently, provided that each modification on the schema may affect the database design decisions. An example schema of the simplified cadastral database is given in Figure 5.10.

OWNERS

| SIN | SURNAME | NAME | DoB | STREET | No | CITY | PROV | ZIP | COUNTRY |
|-----|---------|------|-----|--------|----|------|------|-----|---------|
|     |         |      |     |        |    |      |      |     |         |

TITLES

| PARCEL | OWNER | PERCENTAGE | PURCHASE_DATE |
|--------|-------|------------|---------------|
|        |       |            |               |

PARCELS

| ID | USE | BUILT_FACTOR | ADDRESS | POLYGON |
|----|-----|--------------|---------|---------|
|    |     |              |         |         |

**Fig. 5.10** The cadastral database schema definition (part of).

The database instance (or snapshot) shows the database state at a particular moment. In a dynamic database, the instance is expected to change often. An example snapshot for the cadastral database schema of Figure 5.10 is shown in Figure 5.11.

OWNERS

| SIN | SURNAME | NAME | DoB | STREET | No | CITY | PROV | ZIP | COUNTRY |
|---|---|---|---|---|---|---|---|---|---|
| 464197351 | SMITH | JOHN | 08/15/1952 | GEORGE | 191 | FREDERICTON | NB | E3A5A3 | CANADA |
| 576129452 | COOPER | MARY | 11/23/1945 | QUEEN | 2581 | HALIFAX | NS | B3J2H8 | CANADA |
| ... | ... | ... | ... | ... | ... | | | ... | ... |

TITLES

| PARCEL | OWNER | PERCENTAGE | PURCHASE_DATE |
|---|---|---|---|
| 75071217 | 464197351 | 60% | 07/28/1974 |
| 75071217 | 576129452 | 40% | 07/28/1974 |
| 75071745 | 576129452 | 100% | 03/12/1987 |
| ... | ... | ... | ... |

PARCELS

| ID | USE | BUILT-FACTOR | ADDRESS | POLYGON |
|---|---|---|---|---|
| 75071217 | HOUSING | 1.40 | 542 RIVERVIEW STREET, FREDERICTON | POL876332 |
| 75071745 | PARKING | 1.20 | 323 MAIN STREET, FREDERICTON | POL876234 |
| ... | ... | ... | ... | ... |

**Fig. 5.11** An instance (snapshot) of the cadastral database (part of).

## 5.5 Database Management System Architecture

In the mid '70s the ANSI-SPARC (American National Standards Institute, Standards Planning And Requirements Committee) suggested the ideal architecture for database management systems (DBMS). The model consists of three *levels*: the internal level, the conceptual level, and the external level; as shown in Figure 5.12.

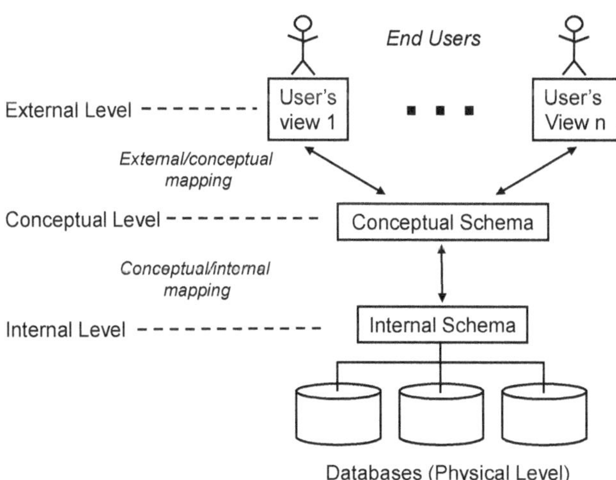


**Fig. 5.12** The ANSI/SPARC model for a DBMS architecture.

The *internal level* is responsible for the physical storage of the data items in the databases. This level describes the data structures at the physical level and the paths to access the actual data items (internal schema). The *internal schema* is defined through the use of a physical data model.

The *conceptual level* describes the database structure for a group of users. This level hides the physical level details, while it focuses on the description of entities, data types, relations, operations, and constraints (conceptual schema). The *conceptual schema* is defined through the use of a conceptual or logical data model.

The *external level* includes a number of external schemas, called *user views*. Each view describes a part of the database, which is of interest to the corresponding group of users. The *external schema* is defined through the use of a conceptual or logical data model.

All three schemas (internal, conceptual and external) are just descriptions of data. The actual data is stored in the databases at the physical level. Each group of users refers to the corresponding external schema only.

The user posts a request to the external schema. The DBMS maps the user request to the conceptual schema (external to conceptual mapping) and then to the internal schema (conceptual to internal mapping), which will process the data items stored in the databases. All extracted and processed data must be converted (invert conversion) to the corresponding user view at the external level.

The objective of the three-layer model is to separate the users' views of the database from the way it is physically represented. This approach has many advantages. Firstly, it allows independent and customized views on the database content as each user is able to access the same data, but have a different customised view on the data. Secondly, it hides the physical storage details from the users and they are allowed to work with the data, without concern for how it is physically stored. Thirdly, the database storage structures may be changed without affecting the users' views and the application programs.

## 5.6 Development of a Database System

Modern Database Systems (DBS) are complicated and their development must be supported by appropriate methodologies and software tools. The development of a DBS consists of the following steps (Figure 5.13):

1. *Definition of the user and application requirements:* In this step the requirements of both the users and the applications are analyzed. This is a critical step for the system to meet their requirements (e.g., operation types, interfaces, indices, etc.).
2. *System design:* This step refers to the design of a system that meets the requirements determined in the previous step. The system design is carried out in three levels, which correspond to the conceptual, logical and physical design. Chapter 6 examines these levels closely.
3. *System implementation:* This step proceeds forward to the system implementation based on the decisions taken in the system design step. The implementation usually adopts a commercial or open source database management system (DBMS), which is capable to meet the design requirements.

4. *Data entry:* After the system is implemented, the population of the database takes place. The data describing the application's miniworld is organized in the system database.
5. *System evaluation:* In this step, the system operation is checked and assessed by the system developer and the potential users. This step may trigger the system design and/or the system implementation steps if the requirements are not met as expected.
6. *System operation:* After the system is approved by the previous step, its life-circle sets off.
7. *System maintenance:* This step is being performed throughout the system life-circle to assure proper system function. The system maintenance is usually the responsibility of the system administrator and involves operations like backups, loggings, granting authorized access to users, etc.

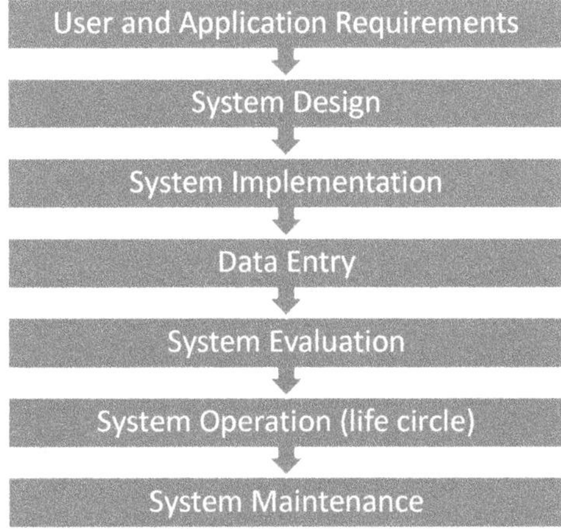


**Fig. 5.13** Development of a Database System.

The Chapters that follow focus on the system design (Step 2), where the most important decisions regarding the system structure and functionality are taken. Specifically, Chapter 6 presents a systematic approach to design a database system (DBS), while Chapter 7 discusses some advanced topics usually applied in the design of geographic database systems.

## References and Further Reading


Date, C.J., 2003. *An Introduction to Database Systems.* Addison-Wesley.
Elmasri, R., and Navathe, S.B., 2000. *Fundamentals of Database Systems.* Addison-Wesley.
Garcia Molina, H., Ullman, J.D., and Widom, J., 2000. *Database Systems Implementation.* Prentice Hall.
Jardine, D. (Ed.) 1977. *The ANS/SPARC DBMS Model.* North Holland.
Ramakrishnan, R, and Gehrke, J., 2002. *Database Management Systems.* McGraw-Hill.

# 6 Database System Design

## 6.1 Introduction

The scope of the database system design process is twofold: (a) to represent appropriately the application data in the system; and (b) to serve the application functionality requirements (i.e., accelerate the processing, minimize the storage requirements). The stages of database system design are (Figure 6.1):

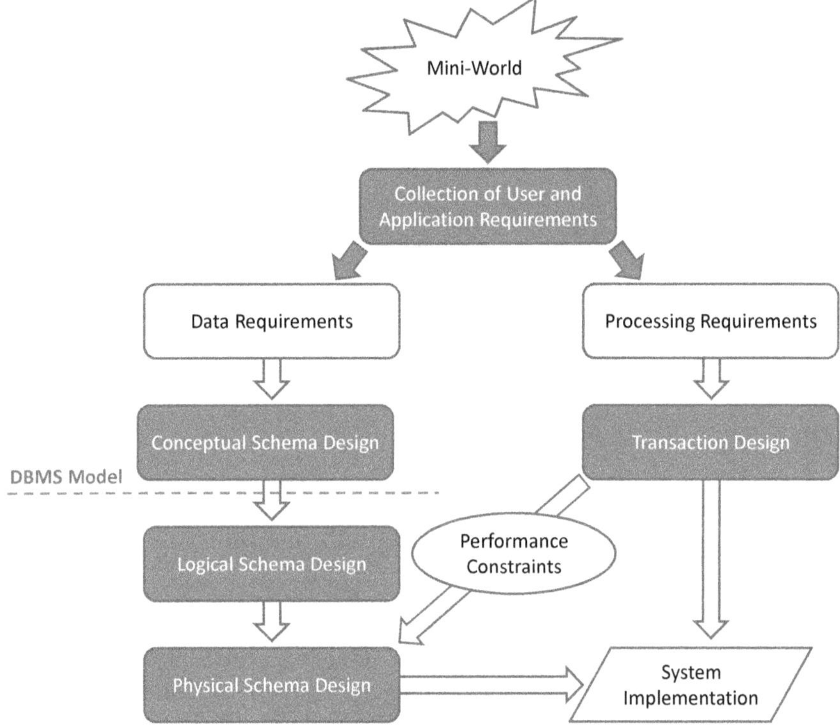


**Fig. 6.1** The stages of the database system design process.

- *Collection and analysis of the miniworld requirements:* In this stage the user and the application requirements are recognized and analyzed. The system requirements with regards to both the data and processes (transactions) are clearly determined. The design steps that follow will take these data and processing requirements into consideration, so that the system to be implemented is able to meet them.
- *Conceptual schema design:* In this stage, the data to be stored in the database are clearly understood. Their semantics and potential constraints are analyzed in detail. This stage is the most important in the design process, provided that

all decisions taken here affect the subsequent design steps, which are more or less a straightforward mapping or implementation of the conceptual schema definition.

- *DBMS model selection:* The database system will be implemented in a Database Management System (DBMS). The mapping of the conceptual schema to the DBMS heavily depends on the DBMS model. There are several options with regards to the DBMS model, such as the relational, object-oriented, or object-relational model. These models are implemented by commercial and open source DBMS software packages (e.g., Oracle, Postgres, MySQL, MS Access). The choice of the DBMS model and software package affects all subsequent stages.

- *Logical schema design:* In this stage, the conceptual schema is mapped to a (logical) schema, which can be implemented in the DBMS model chosen. The logical schema hides the details of the database physical schema.

- *Transaction design:* In this stage, the processing requirements for the system are clearly specified. Based on the application requirements, the expected transactions and processing operations are determined along with their input and output parameters.

- *Physical schema design:* This is the final stage before the implementation of the database system. It is where all the data structures to both accelerate the system performance and minimize the storage requirements are designed. Any decisions are based on the logical schema and the transaction design as determined in the previous steps.

## 6.2 Conceptual Schema Design

The *domain experts* of an application (e.g., cadastre) have a clear understanding of the application requirements. In most cases these requirements are also available in a text document (e.g., the law for a national cadastre). However, the implementation of the application to a database system is carried out by a group of *developers* (analysts and programmers) who are not always familiar with the application requirements and usually face difficulties to fully understand such a document. Hence, there is a need for the domain experts to describe the application requirements clearly to the system developers (Figure 6.2).

The scope of the *conceptual schema* is to provide a clear description of the application requirements, which can be easily understood by both groups, the domain experts and the developers, and is independent on the database system implementation. There is a wide variety of *conceptual models* available in the literature. They all try to provide an enhanced expressiveness with a limited number of constructs, while they all represent the reality in a diagrammatic form.

These diagrams can be easily understood by technologists and end-users; hence all issues related to the system requirements are resolved at an early stage of the system design process and before its actual implementation. One of the most commonly used conceptual model is the *entity-relationship or ER-model*, which is described in the following Sections. More advanced models, such as the enhanced entity relationship model or the UML class diagrams, are also available and will be described in Chapter 7.

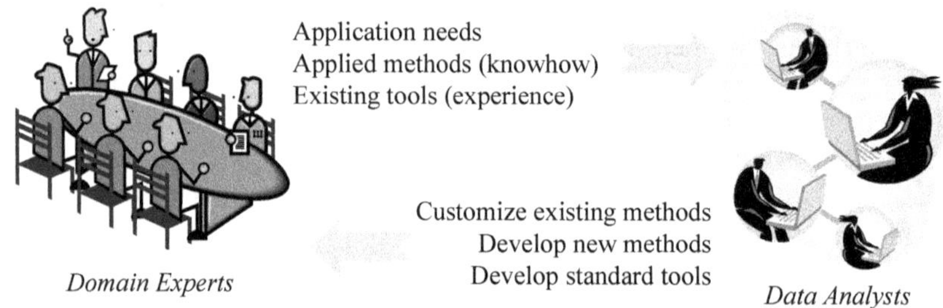


**Fig. 6.2** The interaction between domain experts and data analysts (developers).

## 6.3 The Entity-Relationship Model

The *Entity-Relationship model (ER-model)* is a diagrammatic representation of the miniworld into a set of *entities* and their *relationships*. This is built through a set of constructs, which are represented in a diagram using appropriate notations (symbols). Figure 6.3 shows a common notation used in the ER-diagrams.

As mentioned in Chapter 1, *entity* is a unit with a physical (e.g., a parcel, a person owner) or conceptual (e.g., a company, a school) existence. An entity has properties, which are called *attributes* and describe the entity (e.g., an owner has got a surname, a name, an address). An entity instance comprises the values assigned to its attributes (e.g., the surname of an owner is "Smith").

### *6.3.1 Attributes*

An attribute is either *simple* (e.g., a person's surname) or *composite*. A composite attribute consists of a set of other attributes, e.g., an address consists of the street name, number, postal code, and city name. In addition, an attribute can be *single-valued*, when it is assigned a unique value, e.g., the date of birth; or *multivalued*, when it is assigned more than one values, e.g., a person's telephone numbers.

An attribute can be either *stored* in the database, e.g., a person's date of birth; or *derived* after processing the database content, e.g., the current person's age (today's date minus the date of birth). Finally, the attributes which identify the entity instances are called *key attributes*, e.g., a person's SIN, a parcel's id. An entity instance (e.g., a person) may be identified by a *single* attribute (e.g., the SIN) or a *composition* of more than one attributes (e.g., the name, surname and the date of birth; assuming that the combination of these three values is unique in the database).

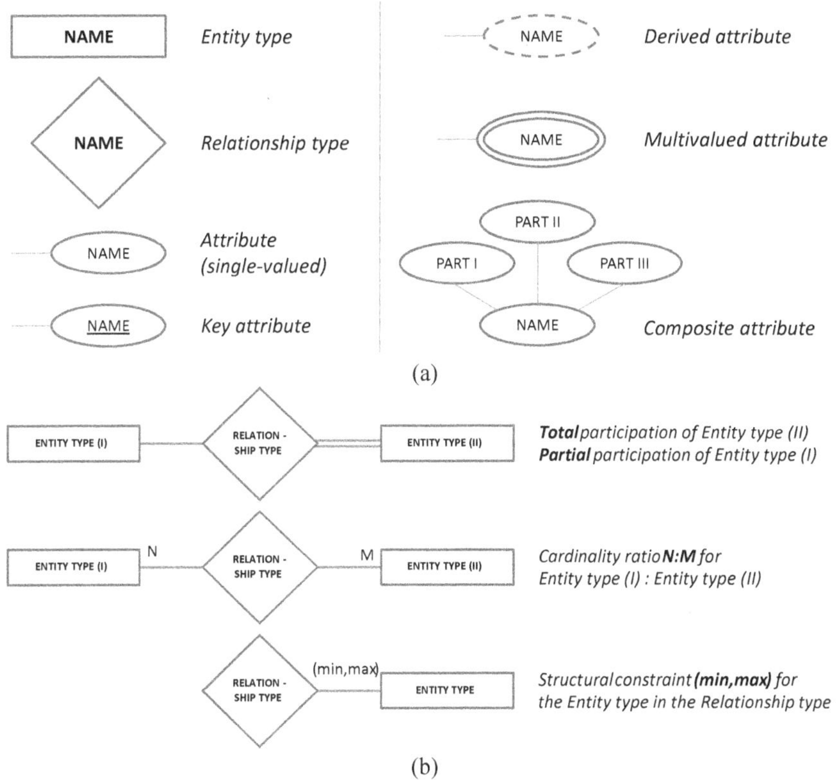



**Fig. 6.3** Basic ER-diagram notation.

## 6.3.2 Entity and Relationship Types

A set of entities sharing common attributes and are of the same type constitute an *entity type* E. For example, the parcels in a cadastral database constitute the entity type: PARCELS. The representation of an entity type in an ER-diagram is through a rectangle, which denotes the name of the entity type (Figure 6.3a). The attributes assigned to an entity type are represented by ellipses also denoting their name. The form of the ellipse outline corresponds to the attribute type (Figure 6.3a).

The entities of one or more entity types may be related through relationships. A *relationship type* R among entity types $E_1$, $E_2$, ..., $E_n$ conceptually relates entity instances from these types. The representation of a relationship type R in an ER-diagram is through a diamond, denoting the name of the relationship type (Figure 6.3a). Both entity types and relationship types can be considered as mathematical sets.

The number of entity types that participate in a relationship type R defines the *degree* of R. Figure 6.4 shows an example of a relationship type R of degree two. This relationship type is called TITLES and relates two entity types: the parcels (PARCELS) with their owners (OWNERS). The instances $p_1$, $p_2$, $p_3$, ... represent the entities in the entity type (set of) PARCELS, while the instances $o_1$, $o_2$, $o_3$, ... represent the entities in the entity type (set of) OWNERS. Finally, the $t_1$, $t_2$, $t_3$, ... represent the relationship instances of the relationship type TITLES. The line segments in Figure 6.4 represent the relationship instances that apply between the entity instances. Specifically, parcel $p_1$ is shared by owners $o_1$ and $o_2$ through the titles $t_1$ and $t_2$. On the other hand, owner $o_2$ has an ownership over parcels $p_1$ and $p_2$ through the titles $t_2$ and $t_3$.

Next, the discussion is limited to relationship types of degree two. Relationship types of degree higher than two will be considered in Chapter 7.

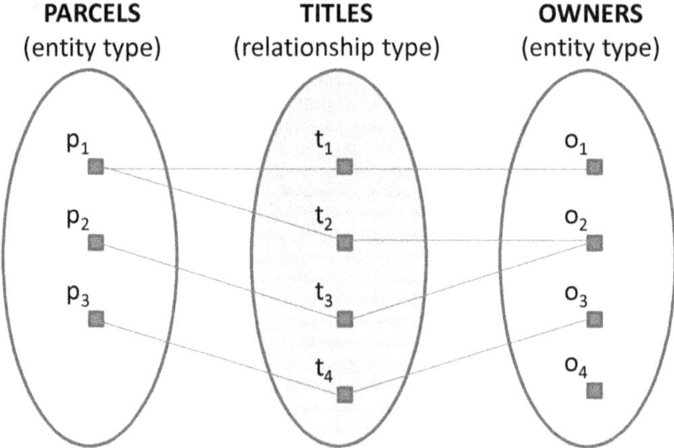


**Fig. 6.4** Example relationship type of degree two (2).

### 6.3.3 Cardinality Ratio and Participation Constraint

A relationship type is characterized by (a) the cardinality ratio, and (b) the participation constraint. The *cardinality ratio* specifies the number of relationship instances that an entity instance can participate in. For instance, in Figure 6.4 the relationship type TITLES between the entity types PARCELS and OWNERS is of cardinality ratio N:M. This means that each parcel is owned by one or more (1...M) owners, while each owner may own one or more (1...N) parcels. As shown in the Figure 6.4, there is at least one instance in both entity types which is connected to more than one instances of the relationship type (i.e., $p_1$ is connected with $o_1$ and $o_2$ through $t_1$ and $t_2$ respectively; similarly $o_2$ is connected with $p_1$ and $p_2$ through $t_2$ and $t_3$ respectively). A relationship type of degree two may have a cardinality ratio of either 1:1, 1:N, or N:M.

The *participation constraint*, on the other hand, specifies whether the existence of an entity instance depends on its relationship to another entity instance through a relationship type R. As shown in Figure 6.4, there is no entity instance in the type PARCELS not being related with at least one instance in the entity type OWNERS through the relationship type TITLES. This can be interpreted as a *domain constraint*: parcels without owners cannot be hosted in the database, or differently parcels with unknown owners are assigned to the State (one instance in the owners' set represents the State). This constraint is called as *total* participation of the entity type PARCELS in the relationship type TITLES. In opposition to this, Figure 6.4 shows that there is an instance ($o_4$) in the entity type OWNERS which is not related to any instance in the relationship type TITLES. Hence, the domain allows the presence of a person (owner) in the cadastral database without having any parcel assigned to the person (the person might have been the owner of a parcel in the past and sold it to another person). In this case, the entity type OWNERS has a *partial* participation in the relationship type TITLES.

The notation of the total and partial participation in the ER-diagram is done through the use of a double or single line segment respectively, connecting the entity type (rectangle) with the relationship type (diamond). On top of the line segment, the corresponding cardinality ratio is denoted (Figure 6.3b). Alternative versions of the ER-diagrams also denote the limits (minimum and maximum values) of the cardinality ratio (*structural constraint*; Figure 6.3b). Figure 6.5 shows some example relationship types of degree two with various cardinality ratios and participation constraints for a community of people (e.g., a nation).

### 6.3.4 Relationship Type Attributes

A relationship type R, like any entity type, can be assigned a set of attributes. For example, the instance $t_1$ in relationship type TITLES (Figure 6.4; see Figure 5.5) can be assigned the following attributes: date of purchase (c.g., 28/Jul/1974), percentage of ownership (e.g., 60%; the rest 40% is assigned to $t_2$). Another example is shown in the last entry (f) in Figure 6.5, where the date of marriage has been assigned to the relationship type MARRIAGE. In the same example, the (0,3) refers to the structural constraint for the participation of each entity type to the relationship type (Figure 6.3). The interpretation of this constraint is that a man or woman may get married up to three times.

## 6.4 Example: The Cadastral Application

Figure 6.6 presents the ER-diagram of a simplified cadastral application. This diagram clearly summarizes the user and application requirements. Next, the domain is described in a natural language (English).

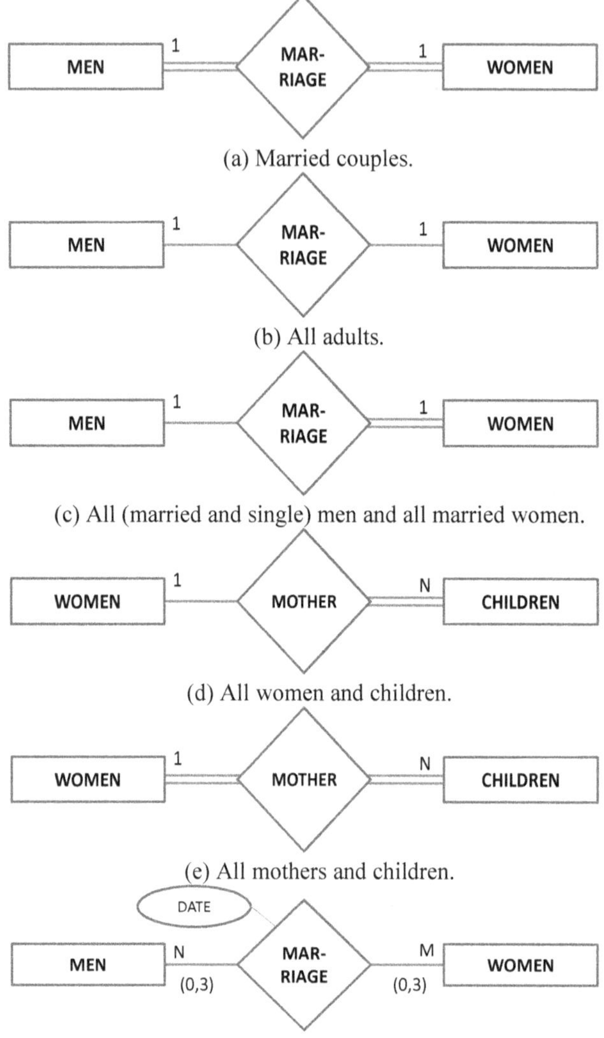


(a) Married couples.

(b) All adults.

(c) All (married and single) men and all married women.

(d) All women and children.

(e) All mothers and children.

(f) The history of all Christian marriages for all adults.

**Fig. 6.5** Example relationship types for a closed community of people.

Firstly, the application comprises four entity types:

- *Parcels*: each entity instance of this type is described by the following simple and single-valued attributes: parcel identifier (ID as a key), parcel use, built factor, and address (as a single string).
- *Owners* (physical persons): each entity instance of this type is described by the following simple and single-valued attributes: Social Insurance Number (SIN as a key), surname, name, and date of birth. A composite attribute is used to describe the address as a compound of: street name, number, city, province, zip

(postal) code, and country. The derived attribute: age, is used to report the owner's age, while the multivalued attribute: telephones, is used to record the owner's multiple telephone numbers.

- *Polygons*: each entity instance of this type is described by the following simple and single-valued attributes: polygon identifier (PLID as a key), parcel area, and perimeter. Notice that the values of area and perimeter can be derived by the coordinates of the polygon vertices (stored in entity type: Points), however, in this diagram they are handled are stored attributes.
- *Points*: each entity instance of this type is described by the following simple and single-valued attributes: point identifier (PTID as a key), a triplet of coordinates: X (abscissa), Y (ordinate), and Z (height) in a common spatial reference system.

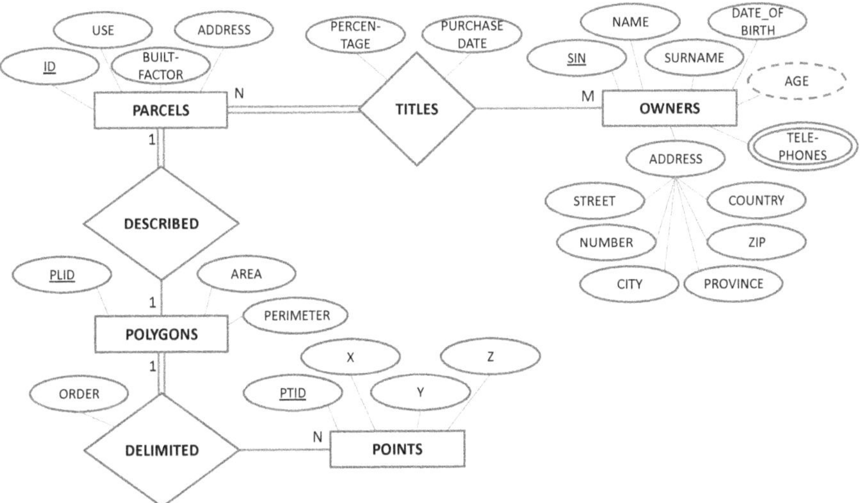


**Fig. 6.6** An ER-diagram example for cadastre.

Secondly, the application comprises the following three relationship types:

- *Titles*: this relationship type of degree two relates the entity types: Parcels and Owners. The cardinality ratio is N:M, with total participation of the entity type Parcels and partial participation of the entity type Owners. This means that each parcel should have at least one or more owners, while each owner may own none, one, or many parcels. The relationship type has assigned two stored attributes: the percentage of ownership and the date of purchase. Both refer to the relationship of a single parcel with a single owner.
- *Described*: this relationship type of degree two relates the entity types: Parcels and Polygons. The cardinality ratio is 1:1, with total participation of the entity type Parcels and partial participation of the entity type Polygons. This means that each parcel should have one polygon description, while some (not all) of

the polygons stored in the database may describe parcels (a polygon may describe a building block or a lake instead).

- *Delimited*: this relationship type of degree two relates the entity types: Polygons and Points. The cardinality ratio is 1:N, with total participation of the entity type Polygons and partial participation of the entity type Points. This means that each polygon is delimited by a set of vertices (points), while some (not all) of the points stored in the database may describe polygons (a point may describe the location of a hydrant or a traffic lights post instead). Additionally, the cardinality ratio 1:N defines that each polygon vertex is assigned to a single polygon and the vertices of adjacent polygons must be duplicated (Figure 6.7b). This could have been avoided if a cardinality ration N:M was chosen instead (Figure 6.7a). Finally, the relationship type Delimited has assigned one stored attribute: order of vertices. Notice that the order of vertices is part of the polygon definition (Figure 6.7c).

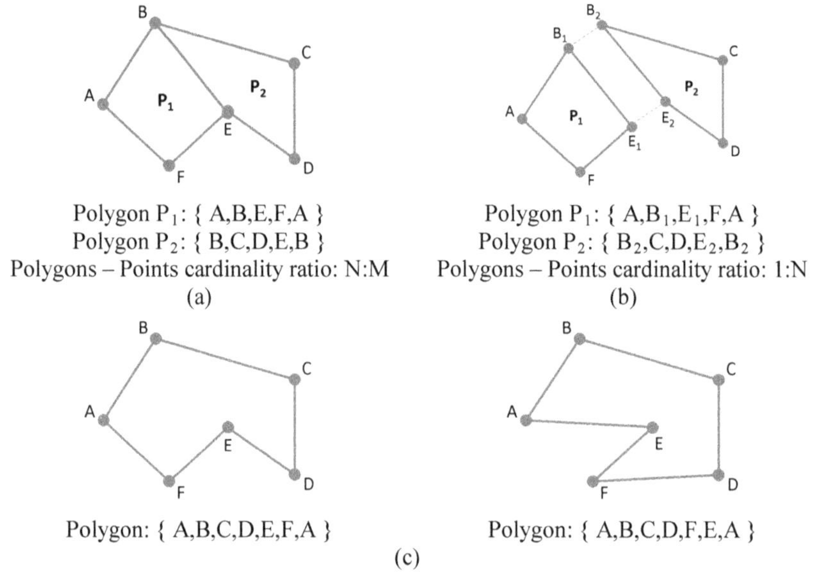

Polygon $P_1$: { A,B,E,F,A }
Polygon $P_2$: { B,C,D,E,B }
Polygons – Points cardinality ratio: N:M
(a)

Polygon $P_1$: { A,$B_1$,$E_1$,F,A }
Polygon $P_2$: { $B_2$,C,D,$E_2$,$B_2$ }
Polygons – Points cardinality ratio: 1:N
(b)

Polygon: { A,B,C,D,E,F,A }

Polygon: { A,B,C,D,F,E,A }

(c)

**Fig. 6.7** (a,b) Alternative cardinality ratios for the relationship type Delimited; (c) the order of vertices is part of a polygon description (different orders result into different polygons).


## 6.5 Logical Schema Design

The conceptual schema eliminates any vagueness related to the user and application requirements. However, it cannot be implemented directly in a computer system. For this to occur, the *conceptual schema* should be mapped to a *logical schema* (Figure 6.1). The logical schema can easily be implemented in a Database Management System (DBMS), while at the same time it hides the physical database storage details. The logical schema is dependent on the *data model* adopted by the underlying DBMS.

In the past, several (logical) data models have been proposed, such as the hierarchical, network, relational, object-oriented, and object-relational. The most widely used has been the *relational model*. Since its introduction in 1970, it has been adopted by the most popular commercial and open source DBMSs, e.g., Oracle, Informix, Ingres, and PostgreSQL. Next, the relational model is presented. The object-oriented and object-relational models are presented in Chapters 7 and 15, respectively.

## 6.5.1 Relational Model

The *relational model* was introduced by Codd in 1970. In this model, the database consists of a collection of *relations*. Each relation is similar to a *table* or *file of records* (Chapter 8). Each record in a relation is called *tuple* and accommodates the values for a set of attributes. Each column in a relation corresponds to an *attribute*, whose name is typed at the header line of the relation. Each attribute has a *data type* (e.g., integer, character string, date) and takes values from a specific *domain*. Figure 6.8 shows an example relation of the owners in the cadastral application.

OWNERS

| SIN | SURNAME | NAME | DoB | STREET | No | CITY | PROV | ZIP | COUNTRY |
|---|---|---|---|---|---|---|---|---|---|
| 464197351 | SMITH | JOHN | 08/15/1952 | GEORGE | 191 | FREDERICTON | NB | E3A5A3 | CANADA |
| 576129452 | COOPER | MARY | 11/23/1945 | QUEEN | 2581 | HALIFAX | NS | B3J2H8 | CANADA |
| ... | ... | ... | ... | ... | ... | | | ... | ... |

**Fig. 6.8** Example relation.

The relational model, by definition, adopts the concept of Sets from Mathematics. Each tuple in a relation is unique, i.e., it is not expected for two tuples with the same combination of values assigned to all of the attributes to appear in a relation. In addition, the minimum number of attributes that identify the relation tuples specify the *key attributes* of the relation, e.g., the attribute SIN in the owners' relation (Figure 6.8).

Each value in a relation is *atomic*, by the means that it is indivisible as far as the relational model is concerned. Hence, the relational model by definition cannot handle nested relations, composite, or multivalued attributes explicitly. Finally, the *ordering of tuples* in a relation is not part of the definition, but it occurs randomly or artificially on an on-screen or paper report (print-out).

## *6.5.2 Example: The Cadastral Application*

Figure 6.10 presents the logical schema of the cadastral application presented in Figure 6.6 (conceptual schema), while Figure 6.11 shows the corresponding instance for the snapshot in Figure 6.9.

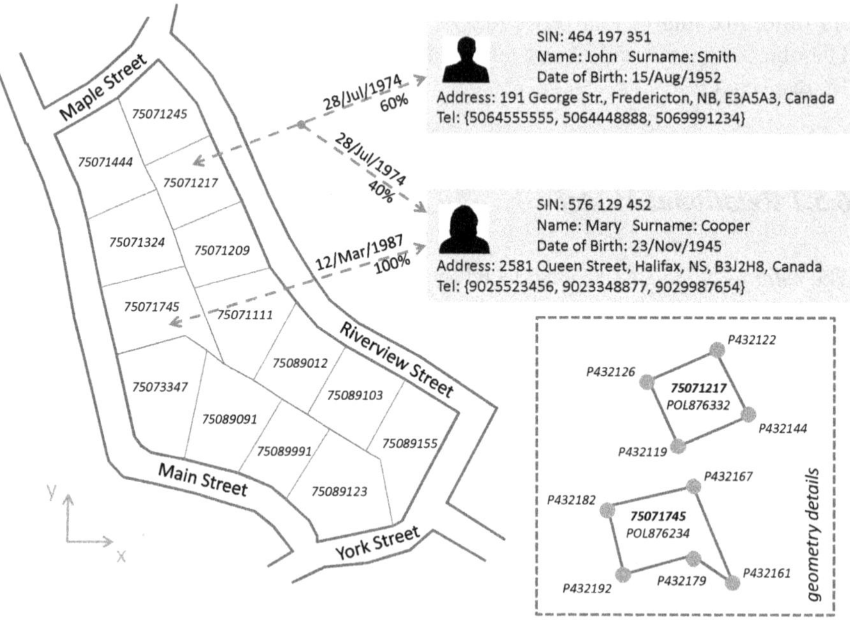


**Fig. 6.9** An example cadastral plan.

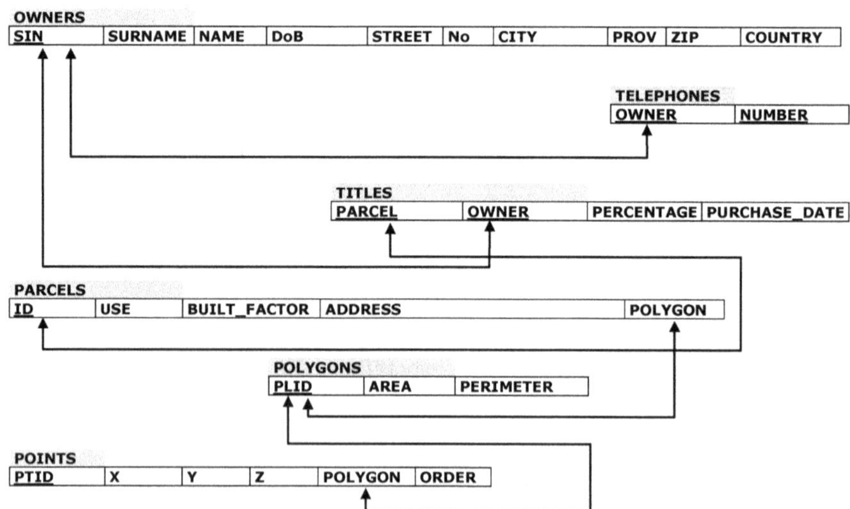


**Fig. 6.10** An example schema of the cadastral database.

The entities in the relational data model are represented through the tuples of a relation (e.g., OWNERS). The entity attributes are represented by the columns in that relation (e.g., in relation OWNERS the attributes are: SIN, NAME SURNAME, DoB, STREET, NUMBER, CITY, PROVINCE, ZIP, COUNTRY). The representation of relationships with cardinality ration N:M is also realized through the tuples in a relation (e.g., TITLES). On the other hand, the relationships with cardinality ration 1:1 or 1:N are represented through attributes of common domain in multiple relations; e.g., the arrow connecting PARCELS with POLYGONS in Figure 6.10. As for the relationship attributes they are represented by the columns in a relation (e.g., in relation TITLES the attributes are: PERCENTAGE and PURCHASE_DATE).

**OWNERS**

| SIN | SURNAME | NAME | DoB | STREET | No | CITY | PROV | ZIP | COUNTRY |
|---|---|---|---|---|---|---|---|---|---|
| 464197351 | SMITH | JOHN | 08/15/1952 | GEORGE | 191 | FREDERICTON | NB | E3A5A3 | CANADA |
| 576129452 | COOPER | MARY | 11/23/1945 | QUEEN | 2581 | HALIFAX | NS | B3J2H8 | CANADA |
| ... | ... | ... | ... | ... | ... | | | ... | ... |

**TELEPHONES**

| OWNER | NUMBER |
|---|---|
| 464197351 | 5064555555 |
| 464197351 | 5064448888 |
| 464197351 | 5069991234 |
| 576129452 | 9025523456 |
| 576129452 | 9023348877 |
| 576129452 | 9029987654 |
| ... | ... |

**TITLES**

| PARCEL | OWNER | PERCENTAGE | PURCHASE_DATE |
|---|---|---|---|
| 75071217 | 464197351 | 60% | 07/28/1974 |
| 75071217 | 576129452 | 40% | 07/28/1974 |
| 75071745 | 576129452 | 100% | 03/12/1987 |
| ... | ... | ... | ... |

**PARCELS**

| ID | USE | BUILT_FACTOR | ADDRESS | POLYGON |
|---|---|---|---|---|
| 75071217 | HOUSING | 1.40 | 542 RIVERVIEW STREET, FREDERICTON | POL876332 |
| 75071745 | PARKING | 1.20 | 323 MAIN STREET, FREDERICTON | POL876234 |
| ... | ... | ... | ... | ... |

**POLYGONS**

| PLID | AREA | PERIMETER |
|---|---|---|
| POL876332 | 1.235 | 112 |
| POL876234 | 1.440 | 169 |
| ... | ... | ... |

**POINTS**

| PTID | X | Y | Z | POLYGON | ORDER |
|---|---|---|---|---|---|
| P432122 | 45678.34 | 8938.89 | 34.20 | POL876332 | 1 |
| P432144 | 45705.56 | 8879.67 | 32.85 | POL876332 | 2 |
| P432119 | 45621.12 | 8845.87 | 31.97 | POL876332 | 3 |
| P432126 | 45592.56 | 8910.91 | 32.88 | POL876332 | 4 |
| P432167 | 45650.33 | 8813.12 | 30.71 | POL876234 | 1 |
| P432161 | 45692.11 | 8726.44 | 28.12 | POL876234 | 2 |
| P432179 | 45653.98 | 8749.92 | 28.65 | POL876234 | 3 |
| P432192 | 45550.19 | 8730.51 | 27.92 | POL876234 | 4 |
| P432182 | 45539.87 | 8802.01 | 29.33 | POL876234 | 5 |
| ... | ... | ... | ... | ... | ... |

**Fig. 6.11** An example instance of the cadastral database (for the snapshot in Figure 6.9).

### *6.5.3 Key Attributes*

In the relational model there are three types of *key attributes*:

- *Candidate key:* each attribute or combination of attributes that identifies the tuples in a relation. For instance, the tuples in the relation of owners (Figure 6.8) can be identified either through the SIN (candidate key 1) or the combination of attributes SURNAME-NAME-DoB (candidate key 2), assuming that there are no two owners in the database sharing the same combination of values in these three attributes. A candidate key is called *simple*, when it consists of a single attribute (e.g., the candidate key 1) or *composite*, when it comprises more than one attributes (e.g., the candidate key 2).
- *Primary key:* the candidate key chosen to identify the tuples in a relation. For practical reasons, it is the one with the fewest attributes. For example, the SIN (candidate key 1) is an ideal key for the relation of owners (Figure 6.8). The names of the primary key attributes are underlined in the relational schema and the relation header line.
- *Foreign key:* each attribute or combination of attributes in a relation sharing the same domain with the key attribute in another relation. For example, the relation of titles (Figure 6.10) has as attribute the owner identifier (OWNER). This attribute corresponds to the owner's SIN, which is the key attribute in the relation of owners. Hence, the attribute OWNER is a foreign key in the relation TITLES and can be used to combine (join) the content of two relations: TITLES and OWNERS. This is represented by arrows in the relational schema of Figure 6.10.

The concept of key attributes in the relational model forces the following *constraints*:

- *Key constraint:* the candidate key values are unique for each tuple in a relation.
- *Entity constraint:* the key value cannot be null.
- *Foreign key constraint:* the foreign key value must be present in the key attribute of the reference relation.

## 6.6 Mapping the Conceptual Schema to the Logical Schema

The mapping of the conceptual schema (ER-model) to the logical schema (relational model) is a straightforward process and attained through a series of standardized rules (steps). Rule 1 dictates how to map the entity types. Rules 2, 3, and 4 describe the mapping of relationship types with cardinality ratio one-to-one, one-to-many, and many-to-many, respectively. Rule 5 refers to the mapping of multi-valued attributes. These five rules are presented next.

## 6.6.1 Rule 1: Mapping the Entity Types

The first rule refers to the mapping of the entity types into the relational schema. This rule consists of the following steps:

i.    for each entity type form one relation with all simple attributes included as columns in this relation;

ii.    represent all composite attributes as simple attributes, while ignoring all derived and multivalued attributes;

iii.    select a candidate key as the primary key of the relation.

An example execution of these steps is shown in Figure 6.12. Obviously, each entity type in the ER-diagram is mapped into a separate relation in the relational schema.

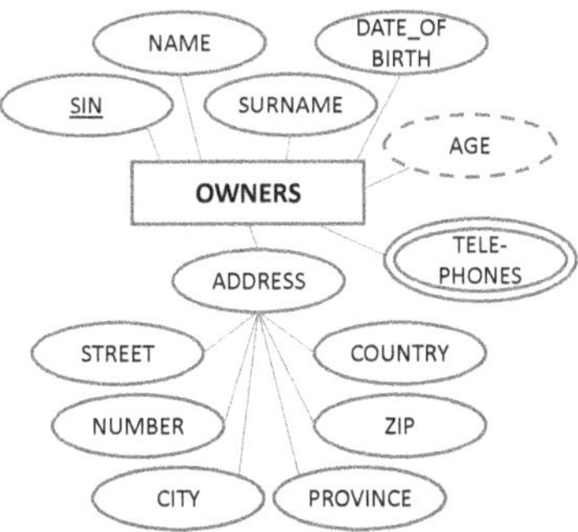


**Step (i):** for each entity type form one relation with all simple attributes included as columns in this relation;

**OWNERS**

| SIN | SURNAME | NAME | DoB |
|-----|---------|------|-----|
| | | | |

**Step (ii):** represent all composite attributes as simple attributes, while ignore all derived and multivalued attributes;

**OWNERS**

| SIN | SURNAME | NAME | DoB | STREET | No | CITY | PROV | ZIP | COUNTRY |
|-----|---------|------|-----|--------|-----|------|------|-----|---------|
| | | | | | | | | | |

**Step (iii):** select a candidate key as the primary key of the relation.

**OWNERS**

| SIN | SURNAME | NAME | DoB | STREET | No | CITY | PROV | ZIP | COUNTRY |
|-----|---------|------|-----|--------|-----|------|------|-----|---------|
| | | | | | | | | | |

**Fig. 6.12** Mapping the ER-diagram into the relational schema. Rule 1: Mapping the entity types.

## 6.6.2 Rule 2: Mapping the 1:1 Relationship Types

This rule refers to all relationship types of cardinality ratio 1:1. The following steps show how these relationship types are mapped into the relational schema:

iv.    for each 1:1 relationship type include (as a foreign key) in the relation which represents the entity type with total participation, the primary key of the other entity type (with partial participation) in the relationship (if none or both are in full participation, make a choice);

v.     include the attributes of the relationship type (if any) as attributes in the same relation.

An example execution of these steps is shown in Figure 6.13. The execution of Rule 1 has already generated two relations, one for each entity type participating in the 1:1 relationship type. For completeness of the discussion (i.e, for the presentation of step v), a new attribute has been attached to the relationship type: GDATE (i.e., the date when the polygon geometry description was attached to the parcel). Notice that this attribute is missing in Figure 6.6.

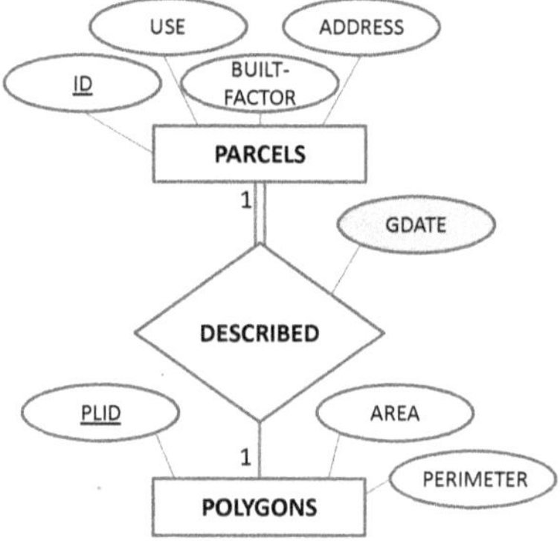


From execution of Rule 1 the following two relations have been generated:

**PARCELS**

| ID | USE | BUILT_FACTOR | ADDRESS |
|---|---|---|---|

**POLYGONS**

| PLID | AREA | PERIMETER |
|---|---|---|

**Step (vi):** for each 1:1 relationship type, include (as foreign key) in the relation, which represent the entity type with total participation, the primary key of the other entity type (with partial participation) in the relationship;

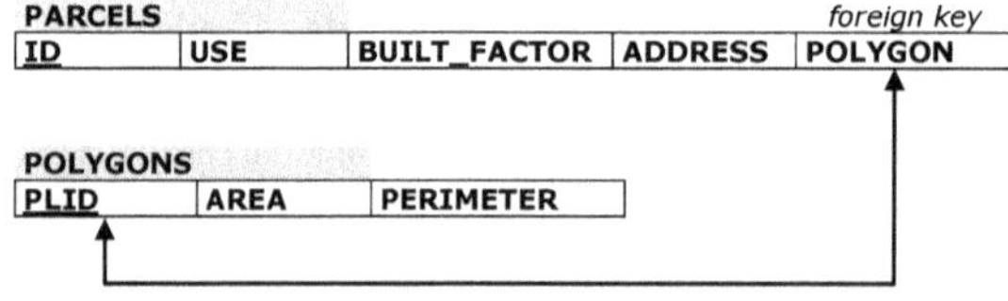

**PARCELS**

| | | | | *foreign key* |
|---|---|---|---|---|
| ID | USE | BUILT_FACTOR | ADDRESS | POLYGON |

**POLYGONS**

| PLID | AREA | PERIMETER |
|---|---|---|

**Step (v):** include in this same relation the attributes of the relationship type (if any) as attributes.

**PARCELS**

| | | | | *foreign key* | *rel. attribute* |
|---|---|---|---|---|---|
| ID | USE | BUILT_FACTOR | ADDRESS | POLYGON | GDATE |

**Fig. 6.13** Mapping the ER-diagram into the relational schema. Rule 2: Mapping the 1:1 relationship types.

## 6.6.3 *Rule 3: Mapping the 1:N Relationship Types*

This rule refers to all relationship types of cardinality ratio 1:N. The following steps show how these relationship types are mapped into the relational schema:

vi.    for each 1:N relationship type include (as foreign key) in the relation which represents the entity type with participation N, the primary key of the entity type with participation 1;

vii.    include the attributes of the relationship type (if any) as attributes in the same relation.

An example execution of these steps is shown in Figure 6.14. The execution of Rule 1 has already generated two relations, one for each entity type participating in the 1:N relationship type.

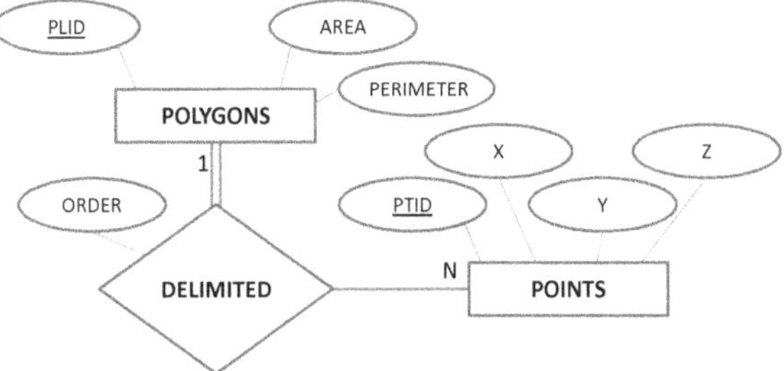


From execution of Rule 1 the following two relations have been generated:

**POLYGONS**

| PLID | AREA | PERIMETER |
|------|------|-----------|

**POINTS**

| PTID | X | Y | Z |
|------|---|---|---|

**Step (vi):** for each 1:N relationship type, include (as foreign key) in the relation which represents the entity type with participation N, the primary key of the entity type with participation 1;

**POLYGONS**

| PLID | AREA | PERIMETER |
|------|------|-----------|

**POINTS**

| PTID | X | Y | Z | *foreign key* POLYGON |
|------|---|---|---|-----------------------|

**Step (vii):** include the attributes of the relationship type (if any) as attributes in this same relation.

**POINTS**

| PTID | X | Y | Z | *foreign key* POLYGON | *rel. attribute* ORDER |
|------|---|---|---|-----------------------|------------------------|

**Fig. 6.14** Mapping the ER-diagram into the relational schema. Rule 3: Mapping the 1:N relationship types.

### *6.6.4 Rule 4: Mapping the N:M Relationship Types*

This rule refers to all relationship types of cardinality ratio N:M. The following steps show how these relationships types are mapped into the relational schema:

viii.   for each N:M relationship type create a new relation; and include (as foreign keys) in the relation the primary keys of the two entity types participating in the relationship type (combined together they will form the primary key of the new relation);

ix.   include the attributes of the relationship type (if any) as attributes in the same relation.

An example execution of these steps is shown in Figure 6.15. The execution of Rule 1 has already generated two relations, one for each entity type participating in the N:M relationship type. The relationship type is also represented in a new relation.

### *6.6.5 Rule 5: Mapping the Multivalued Attributes*

This rule refers to the multivalued attributes assigned to either the entity or relationship types in the ER-diagram. These attributes are being mapped into the relational schema as follows:

x.   for each multivalued attribute, create a new relation and insert into it the attribute and the primary key of the entity (or relationship) type (their combination forms the primary key of the new relation.

An example execution of these steps is shown in Figure 6.16.

By applying the five rules as described above, the ER-diagram is mapped into an efficient relational schema (Figure 6.10). The rules are documented further in the following Section.

## 6.7 Comments on the Mapping Rules

The scope of the aforementioned rules to map the conceptual schema (ER-diagram) into the logical schema (relational model) is to avoid redundant data (first priority), while at the same time organize the data items into as few relations as possible (second priority). This scope is further elaborated through a series of examples in the following subsections.

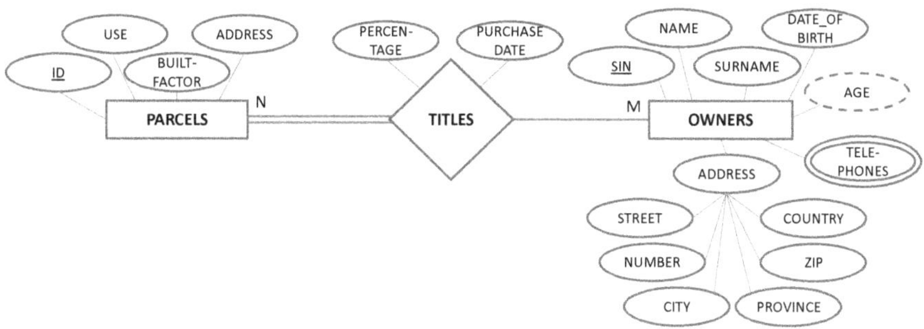

From execution of Rule 1 the following two relations have been generated:

**OWNERS**

| SIN | SURNAME | NAME | DoB | STREET | No | CITY | PROV | ZIP | COUNTRY |
|-----|---------|------|-----|--------|----|----|----|-----|---------|
| | | | | | | | | | |

**PARCELS**

| ID | USE | BUILT_FACTOR | ADDRESS | POLYGON |
|----|-----|--------------|---------|---------|
| | | | | |

**Step (viii):** for each N:M relationship type, create a new relation; and include (as foreign keys) in the relation the primary keys of the two entity types participating in the relationship type;

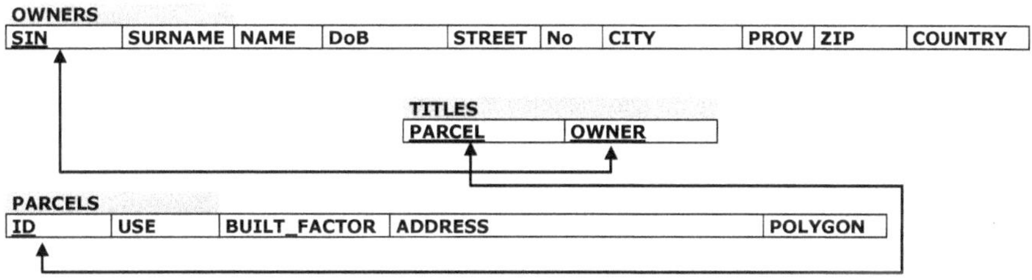

**Step (ix):** include the attributes of the relationship type (if any) as attributes in this same relation.

**TITLES**

| PARCEL | OWNER | PERCENTAGE | PURCHASE_DATE |
|--------|-------|------------|---------------|
| | | | |

**Fig. 6.15** Mapping the ER-diagram into the relational schema. Rule 4: Mapping the N:M relationship types.

## *6.7.1 Rule 1*

The first rule refers to the *mapping of an entity type* and its attributes into the relational schema. Obviously, all derived attributes are ignored, as they are computed by applying a function on the attribute values. Additionally, these attributes are usually dynamic (change values over time) and it is worthless to store their values. Example derived attributes are: the age of a person, the area of a (dynamic) polygon, or the number of employees in a company.

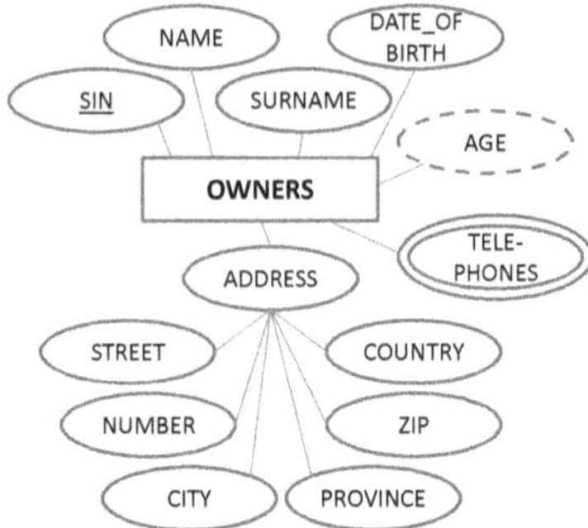

From execution of Rule 1 the following two relations have been generated:

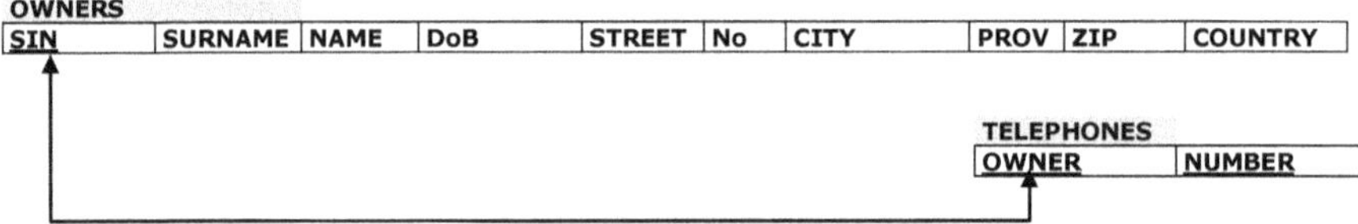

**OWNERS**

| SIN | SURNAME | NAME | DoB | STREET | No | CITY | PROV | ZIP | COUNTRY |
|-----|---------|------|-----|--------|----|------|------|-----|---------|

**Step (x):** for each multivalued attribute, create a new relation and insert into it the attribute and the primary key of the entity (or relationship) type.

**OWNERS**

| SIN | SURNAME | NAME | DoB | STREET | No | CITY | PROV | ZIP | COUNTRY |
|-----|---------|------|-----|--------|----|------|------|-----|---------|

**TELEPHONES**

| OWNER | NUMBER |
|-------|--------|

**Fig. 6.16** Mapping the ER-diagram into the relational schema. Rule 5: Mapping the multivalued attributes.

As for the age of a person, obviously it changes over time. Hence, if handled as a stored attribute, it must be updated every year. To avoid frequent updates, static attributes are chosen over dynamic ones, whenever possible. Such an attribute is the person's date of birth. The age is computed by subtracting this date from the current date.

All multivalued attributes are also ignored in this phase as they are handled in Rule 5.

## 6.7.2 Rule 2

The second rule is applied to *map all relationship types of cardinality ratio 1:1*. According to the steps of the rule, the relation with total participation (if any) is extended to accommodate the key attribute of the relation with partial participation. Why has this decision been taken and not the reverse, i.e., extend the relation with partial participation by adding the key attribute of the relation with total participation?

The snapshot in Figure 6.17 shows the justification, through an example. In the database, there are four polygon instances and only two of them represent parcels (partial participation). Obviously, if the second alternative is applied (i.e., insert the key of the relation with total participation as a foreign key to the relation with partial participation), empty cells in the relation of the entity type with partial participation are likely to appear (Figure 6.17b). The number of these empty cells depends on the application and may be very large. As a consequence there is waste of storage. Hence, the first alternative (as stated by Rule 2) qualifies instead (Figure 6.17a).

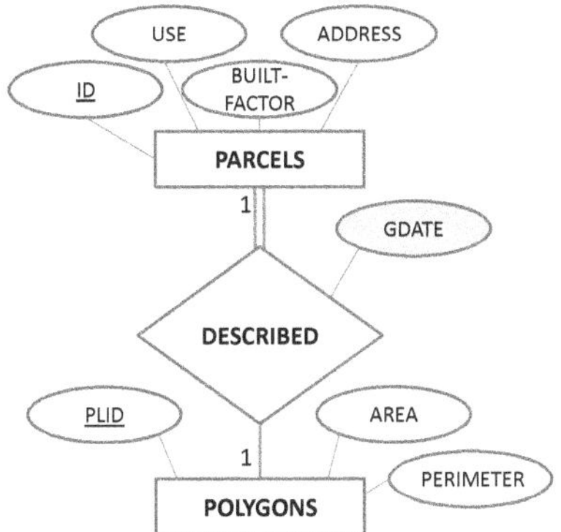

**PARCELS**

| | | | | foreign key | rel.attribute |
|---|---|---|---|---|---|
| **ID** | **USE** | **BUILT_FACTOR** | **ADDRESS** | **POLYGON** | **GDATE** |
| 75071217 | HOUSING | 1.40 | 542 RIVERVIEW STREET, FREDERICTON | POL876332 | 05/25/1969 |
| 75071745 | PARKING | 1.20 | 323 MAIN STREET, FREDERICTON | POL876234 | 07/01/2000 |

**POLYGONS**

| **PLID** | **AREA** | **PERIMETER** |
|---|---|---|
| POL876332 | 1.235 | 142 |
| POL876234 | 1.440 | 169 |
| POL555555 | 5.234 | 1345 |
| POL555556 | 12.341 | 4322 |

(a)

**PARCELS**

| **ID** | **USE** | **BUILT_FACTOR** | **ADDRESS** |
|---|---|---|---|
| 75071217 | HOUSING | 1.40 | 542 RIVERVIEW STREET, FREDERICTON |
| 75071745 | PARKING | 1.20 | 323 MAIN STREET, FREDERICTON |

**POLYGONS**

| | | | foreign key | rel.attribute |
|---|---|---|---|---|
| **PLID** | **AREA** | **PERIMETER** | **PARCEL** | **GDATE** |
| POL876332 | 1.235 | 142 | 75071217 | 05/25/1969 |
| POL876234 | 1.440 | 169 | 75071745 | 07/01/2000 |
| POL555555 | 5.234 | 1345 | | |
| POL555556 | 12.341 | 4322 | | |

(b)

**Fig. 6.17** Relationship type 1:1 and mapping to the relational model: (a) the key of the relation with partial participation in the relationship as a foreign key to the relation with total participation, (b) the key of the relation with total participation in the relationship as a foreign key to the relation with partial participation.

### *6.7.3 Rule 3*

The third rule is applied to *map all relationship types of cardinality ratio 1:N*. According to the steps of the rule, the relation with cardinality ratio N is extended to accommodate the key attribute of the relation with cardinality ratio 1. Why has this decision been taken and not the reverse, i.e., extend the relation with cardinality ratio 1 with the key attribute of the relation with cardinality ration N?

The snapshot in Figure 6.18 shows the justification, through an example. In the database there are two polygons which consist of four or five vertices each. Obviously, the second alternative causes the repetition of data items (area, perimeter) in the relation of polygons (Figure 6.18b). The first alternative (Figure 6.18a) is free of such a problem; hence it is the one dictated by Rule 3.

### *6.7.4 Rule 4*

The fourth rule is applied to *map all relationship types of cardinality ratio N:M*. According to the steps of the rule, a new relation must be created for the representation of the relationship type. Any attempt to link the relations of the two entity types by attaching to any of them the key of the other as a foreign key, leads to highly redundant data (repetitions of other attribute values; in shaded cells). Figure 6.19b,c shows these alternatives through an example.

### *6.7.5 Rule 5*

The fifth rule is applied to represent the *multivalued attributes* in the relational schema. The integration of a multivalued attribute to a relation leads to redundant data. An example is given in Figure 6.20a. This can be avoided by generating a new relation as Rule 5 dictates (Figure 6.20c). The alternative approach of using multiple columns in the relation, as shown in Figure 6.20b, can be applied only when the maximum number of values is known in advance. Even so, this approach might lead to multiple empty cells.

Alternatively, the values of the multivalued attribute can be stored all in a single column of type string (Figure 6.20d). In this case, the retrieval of a single value involves the parsing of the string, which is beyond the definition of the relational model that handles the attribute values as atomic values.

Notice that a multivalued attribute may be represented as a relationship type of cardinality ratio 1:N or N:M. An example for the owners' telephone numbers is shown in Figure 6.21. It is assumed that each telephone number is assigned to a single owner, while an owner may have one or more telephone numbers. This situation is mapped into the relational schema by applying the Rules 1 and 3, which leads to a result identical to the one obtained by Rule 5.

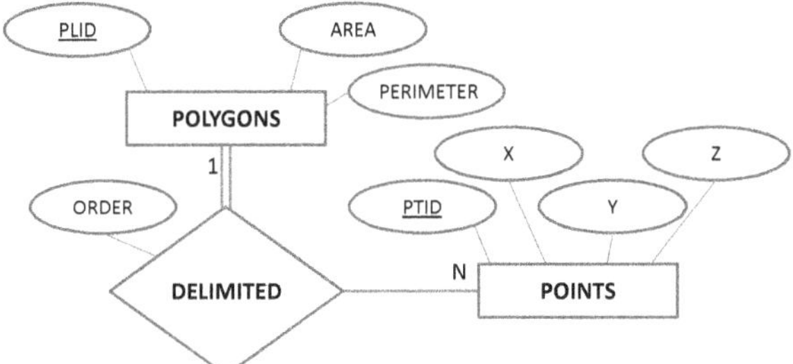


**POLYGONS**

| PLID | AREA | PERIMETER |
|------|------|-----------|
| POL876332 | 1.235 | 142 |
| POL876234 | 1.440 | 169 |

**POINTS**  *foreign key   rel.attribute*

| PTID | X | Y | Z | POLYGON | ORDER |
|------|---|---|---|---------|-------|
| P432122 | 45678.34 | 8938.89 | 34.20 | POL876332 | 1 |
| P432144 | 45705.56 | 8879.67 | 32.85 | POL876332 | 2 |
| P432119 | 45621.12 | 8845.87 | 31.97 | POL876332 | 3 |
| P432126 | 45592.56 | 8910.91 | 32.88 | POL876332 | 4 |
| P432167 | 45650.33 | 8813.12 | 30.71 | POL876234 | 1 |
| P432161 | 45692.11 | 8726.44 | 28.12 | POL876234 | 2 |
| P432179 | 45653.98 | 8749.92 | 28.65 | POL876234 | 3 |
| P432192 | 45550.19 | 8730.51 | 27.92 | POL876234 | 4 |
| P432182 | 45539.87 | 8802.01 | 29.33 | POL876234 | 5 |

(a)

**POLYGONS**  *foreign key   rel.attribute*

| PLID | AREA | PERIMETER | POINT | ORDER |
|------|------|-----------|-------|-------|
| POL876332 | 1.235 | 142 | P432122 | 1 |
| POL876332 | 1.235 | 142 | P432144 | 2 |
| POL876332 | 1.235 | 142 | P432119 | 3 |
| POL876332 | 1.235 | 142 | P432126 | 4 |
| POL876234 | 1.440 | 169 | P432167 | 1 |
| POL876234 | 1.440 | 169 | P432161 | 2 |
| POL876234 | 1.440 | 169 | P432179 | 3 |
| POL876234 | 1.440 | 169 | P432192 | 4 |
| POL876234 | 1.440 | 169 | P432182 | 5 |

**POINTS**

| PTID | X | Y | Z |
|------|---|---|---|
| P432122 | 45678.34 | 8938.89 | 34.20 |
| P432144 | 45705.56 | 8879.67 | 32.85 |
| P432119 | 45621.12 | 8845.87 | 31.97 |
| P432126 | 45592.56 | 8910.91 | 32.88 |
| P432167 | 45650.33 | 8813.12 | 30.71 |
| P432161 | 45692.11 | 8726.44 | 28.12 |
| P432179 | 45653.98 | 8749.92 | 28.65 |
| P432192 | 45550.19 | 8730.51 | 27.92 |
| P432182 | 45539.87 | 8802.01 | 29.33 |

(b)

**Fig. 6.18** Relationship type 1:N and mapping to the relational model: (a) the key of the relation with ratio 1 as a foreign key to the relation with ratio N, (b) the key of the relation with ratio N as a foreign key to the relation with ratio 1.

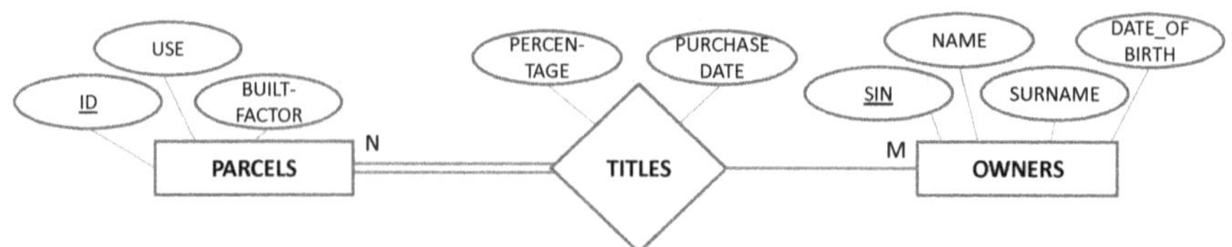

**OWNERS**

| SIN | SURNAME | NAME | DoB |
|---|---|---|---|
| 464197351 | SMITH | JOHN | 08/15/1952 |
| 576129452 | COOPER | MARY | 11/23/1945 |

**TITLES**

| PARCEL | OWNER | PERCENTAGE | PURCHASE_DATE |
|---|---|---|---|
| 75071217 | 464197351 | 60% | 07/28/1974 |
| 75071217 | 576129452 | 40% | 07/28/1974 |
| 75071745 | 576129452 | 100% | 03/12/1987 |

**PARCELS**

| ID | USE | BUILT_FACTOR |
|---|---|---|
| 75071217 | HOUSING | 1.40 |
| 75071745 | PARKING | 1.20 |

(a)

**OWNERS**

| SIN | SURNAME | NAME | DoB |
|---|---|---|---|
| 464197351 | SMITH | JOHN | 08/15/1952 |
| 576129452 | COOPER | MARY | 11/23/1945 |

**PARCELS**

| | | | foreign key | rel.attribute | rel.attribute |
|---|---|---|---|---|---|
| ID | USE | BUILT_FACTOR | OWNER | PERCENTAGE | PURCHASE_DATE |
| 75071217 | HOUSING | 1.40 | 464197351 | 60% | 07/28/1974 |
| 75071217 | HOUSING | 1.40 | 576129452 | 40% | 07/28/1974 |
| 75071745 | PARKING | 1.20 | 576129452 | 100% | 03/12/1987 |

(b)

**OWNERS**

| | | | | foreign key | rel.attribute | rel.attribute |
|---|---|---|---|---|---|---|
| SIN | SURNAME | NAME | DoB | PARCEL | PERCENTAGE | PURCHASE_DATE |
| 464197351 | SMITH | JOHN | 08/15/1952 | 75071217 | 60% | 07/28/1974 |
| 576129452 | COOPER | MARY | 11/23/1945 | 75071217 | 40% | 07/28/1974 |
| 576129452 | COOPER | MARY | 11/23/1945 | 75071745 | 100% | 03/12/1987 |

**PARCELS**

| ID | USE | BUILT_FACTOR |
|---|---|---|
| 75071217 | HOUSING | 1.40 |
| 75071745 | PARKING | 1.20 |

(c)

**Fig. 6.19** Relationship type N:M and mapping to the relational model: (a) generation of a new relation; (b) the key of the relation with ratio N as a foreign key to the relation with ratio M, (b) the key of the relation with ratio M as a foreign key to the relation with ratio N.

**OWNERS**

| SIN | SURNAME | NAME | DoB | TELEPHONE |
|---|---|---|---|---|
| 464197351 | SMITH | JOHN | 08/15/1952 | 5064555555 |
| 464197351 | SMITH | JOHN | 08/15/1952 | 5064448888 |
| 464197351 | SMITH | JOHN | 08/15/1952 | 5069991234 |
| 576129452 | COOPER | MARY | 11/23/1945 | 9025523456 |
| 576129452 | COOPER | MARY | 11/23/1945 | 9023348877 |
| 576129452 | COOPER | MARY | 11/23/1945 | 9029987654 |

(a)

**OWNERS**

| SIN | SURNAME | NAME | DoB | NUMBER_1 | NUMBER_2 | NUMBER_3 | NUMBER_4 |
|---|---|---|---|---|---|---|---|
| 464197351 | SMITH | JOHN | 08/15/1952 | 5064555555 | 5064448888 | 5069991234 | |
| 576129452 | COOPER | MARY | 11/23/1945 | 9025523456 | 9023348877 | 9029987654 | |

(b)

**OWNERS**

| SIN | SURNAME | NAME | DoB |
|---|---|---|---|
| 464197351 | SMITH | JOHN | 08/15/1952 |
| 576129452 | COOPER | MARY | 11/23/1945 |

**TELEPHONES**

| OWNER | NUMBER |
|---|---|
| 464197351 | 5064555555 |
| 464197351 | 5064448888 |
| 464197351 | 5069991234 |
| 576129452 | 9025523456 |
| 576129452 | 9023348877 |
| 576129452 | 9029987654 |

(c)

**OWNERS**

| SIN | SURNAME | NAME | DoB | TELEPHONES |
|---|---|---|---|---|
| 464197351 | SMITH | JOHN | 08/15/1952 | 5064555555; 5064448888; 5069991234 |
| 576129452 | COOPER | MARY | 11/23/1945 | 9025523456; 9023348877; 9029987654 |

(d)

**Fig. 6.20** Representation of a multivalued attribute in the relational model: (a) integration as an attribute in the entity type relation; (b) use of multiple columns, (c) generation of a new relation, and (d) use of a long single attribute of type string.

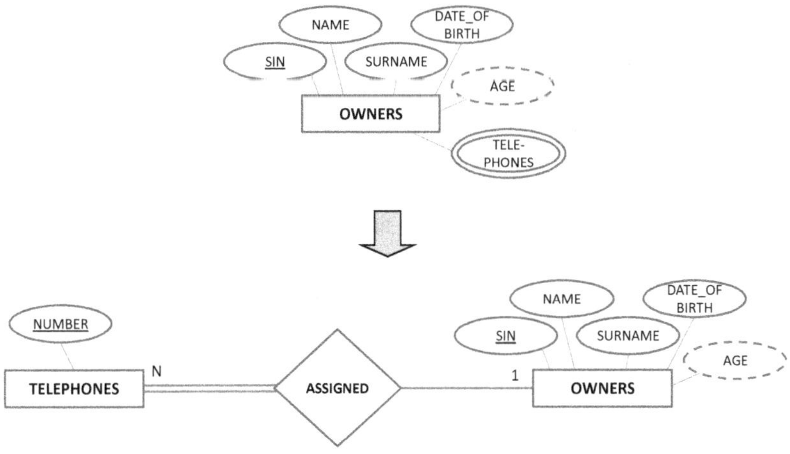


**Fig. 6.21** Representation of a multivalued attribute with a relationship type of cardinality ratio 1:N.

## 6.8 Normalization Theory

The relations in the relational data model are subject of a series of forms. These forms, when satisfied, result in both an efficient database design and effective organization of data items in the database are obtained. As stated already, the relational database design scope is twofold: to avoid redundant data and organize the data items in as few tables as possible.

There are seven forms to be met during the relational database design process. These forms are called *normal forms* and the relational schema that meets them is called *normalized*. Notice that the five rules presented in Section 6.6, when applied to map the ER-diagram to the relational model, they generate a normalized schema.

In the following subsections, three normal forms are presented. The other four forms refer to specialized cases and are not considered here.

### *6.8.1 First Normal Form*

The *first normal form (1NF)* is derived from the definition of the relational data model and specifies the following: (a) the domain of each attribute in a relation consists of atomic values, (b) no multivalued or composite attributes are allowed, and (c) nested relations are not recognized.

Figure 6.22a presents a relation that violates the first normal form. Although, this relation is compliant with the structuring of data items in spreadsheets and can easily be read by humans, it is not a valid relation in the relational model. Figure 6.22b shows a valid relation with the same content.

### *6.8.2 Second Normal Form*

The *second normal form (2NF)* introduces the following constraint: the non-key attributes in a relation must be dependent on the full composite key.

Figure 6.23 shows this form through an example. The relation TITLES has a composite primary key as a combination of the parcel id and the owner's SIN. Some of the attributes in the relation in Figure 6.23a are dependent on the *full* primary key, while others are dependent on a *part* of the primary key. Specifically, the attributes PERCENTAGE and PURCHASE_DATE are dependent on the full composite key. On the other hand, the attributes SURNAME, NAME, and DATE_OF_BIRTH describe the owner and hence depend partially on the composite key, i.e., solely on the SIN.

The presence of the latter attributes in the relation causes a data repetition in the corresponding instances. An example is given in Figure 6.23c. Obviously, each tuple describing a title for an owner should also host the surname, name and date of birth for this owner, along with the SIN (e.g., tuples 2 and 3; according to the 1NF).

The problem can be avoided if the relation is split into two new relations, which comply with the second normal form. These new relations are shown in Figure 6.23b. Figure 6.23d shows the normalized instance of Figure 6.23c, where redundant data items have been eliminated.

**POLYGONS**

| PLID | AREA | PERIMETER | POINT | ORDER |
|---|---|---|---|---|
| POL876332 | 1.235 | 142 | P432122 | 1 |
|  |  |  | P432144 | 2 |
|  |  |  | P432119 | 3 |
|  |  |  | P432126 | 4 |
| POL876234 | 1.440 | 169 | P432167 | 1 |
|  |  |  | P432161 | 2 |
|  |  |  | P432179 | 3 |
|  |  |  | P432192 | 4 |
|  |  |  | P432182 | 5 |

(a)

**POLYGONS**

| PLID | AREA | PERIMETER | POINT | ORDER |
|---|---|---|---|---|
| POL876332 | 1.235 | 142 | P432122 | 1 |
| POL876332 | 1.235 | 142 | P432144 | 2 |
| POL876332 | 1.235 | 142 | P432119 | 3 |
| POL876332 | 1.235 | 142 | P432126 | 4 |
| POL876234 | 1.440 | 169 | P432167 | 1 |
| POL876234 | 1.440 | 169 | P432161 | 2 |
| POL876234 | 1.440 | 169 | P432179 | 3 |
| POL876234 | 1.440 | 169 | P432192 | 4 |
| POL876234 | 1.440 | 169 | P432182 | 5 |

(b)

**Fig. 6.22** 1NF: The organization of data items in a spread sheet (a) and in a relation (b).

## 6.8.3 Third Normal From

The *third normal form (3NF)* introduces the following constraint: the non-key attributes in a relation must not be transitive dependent on the primary key.

Figure 6.24 shows this form through an example. Relation PARCELS has as primary key the ID (the parcel identifier). All other attributes are dependent on the key. However, this dependency can be characterized as *direct* or *transitive*. Specifically, the attributes USE, BUILT_FACTOR ADDRESS and POLYGON depend directly on the ID. Whereas, the attribute VERTEX depends transitively on the ID (the parcel identified), as it describes the POLYGON, which depends directly on the primary key (ID) (Figure 6.24a).

This transitive dependency causes redundancy in the data storage (repeated attribute values), as shown in Figure 6.24c, and for each polygon vertex, all attributes describing the parcel must be repeated (according to the 1NF). This problem can be avoided by splitting the relation into two new relations that meet the third normal form. Figure 6.24b shows the result of the normalization, which is free of redundant data.

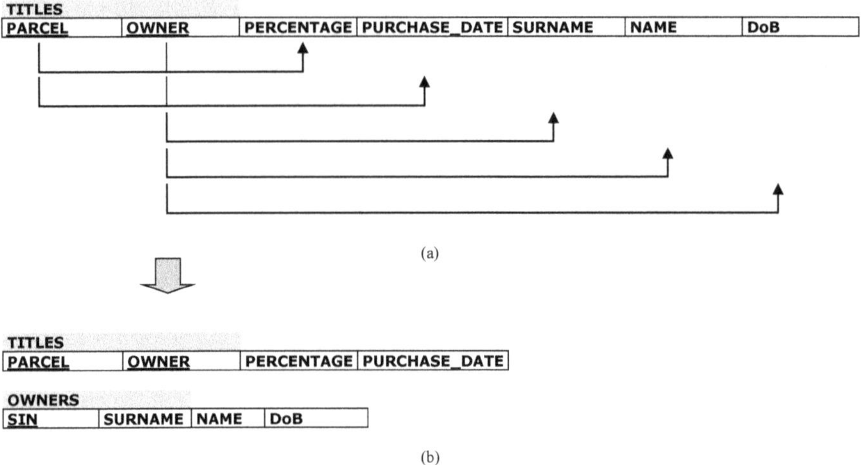

Fig. 6.23 2NF: violated in (a) and (c); and met in (b) and (d).

## 6.9 Relational Algebra

A data model should be accompanied by an algebra, which defines a compact set of operations for the management of data. The relational model uses the *relational algebra*, that inherits many concepts from the Set Theory in Mathematics. This Section briefly presents the operations offered by the relational algebra. The subsequent Section describes the implementation of this algebra into a *language* for relational database management systems (SQL).

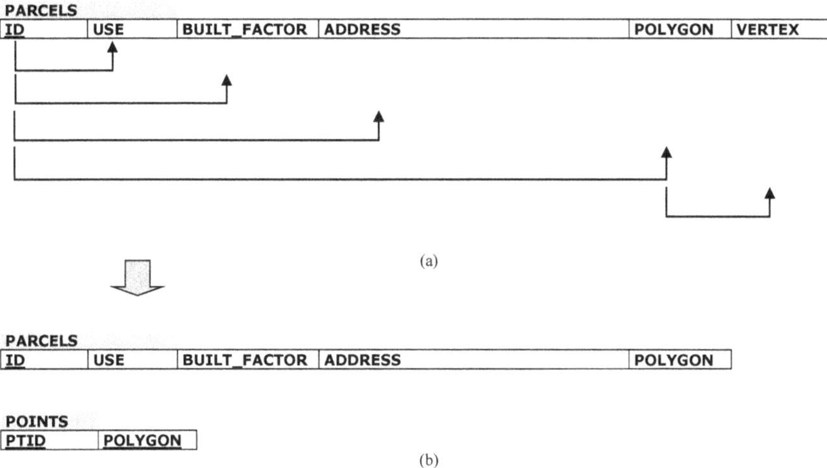

Fig. 6.24 3NF: violated in (a) and (c); and met in (b) and (d).

The relational algebra defines the operations for handling the relations as well as the data that resides into them. These operations are classified into four categories (Figure 6.25): (a) *relation operations*, (b) *set operations*, (c) *computational operations*, and (d) *update operations*.

The *relation operations* include the data selection from a relation as well as the combination of data residing into two relations through common attributes. The most important operations in this category are:

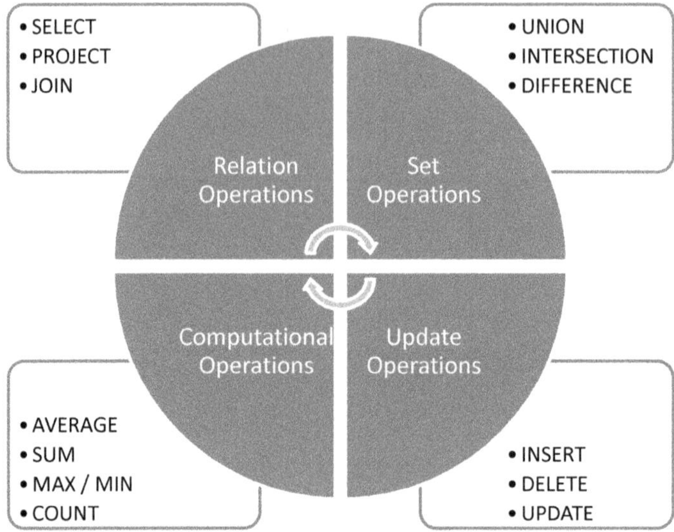


**Fig. 6.25** Relational algebra operations.

- SELECT: selection of tuples from a relation.
- PROJECT: selection of columns (projection of attributes) from a relation.
- JOIN: combination (join) of relative tuples from two relations into single tuples.

The *set operations* are adopted from the set theory in mathematics and include the basic set operations among relations with common attributes. The basic operations in this category are:

- UNION: the union of the tuples from two relations into a single relation.
- INTERSECTION: the intersection of the tuples from two relations (only the common tuples are reported).
- DIFFERENCE: the difference of the tuples from two relations (the tuples in one relation that are not present in the second relation are reported).

The third category of the relational algebra includes the *computational operations*. These operations apply basic mathematical functions on the values that reside in the cells (tuples and columns) of a relation. Some of the most commonly used functions are:

- AVERAGE: computes the average value for an attribute in a relation.
- SUM: computes the sum of the values for an attribute in a relation.
- MAX and MIN: computes the maximum and minimum value for an attribute in a relation.
- COUNT: computes the number of tuples in a relation.

The last category includes the *update operations*. These operations update the content of a relation. The most important operations in this category are:

- INSERT: inserts new tuples in a relation.
- DELETE: deletes tuples from a relation.
- UPDATE: updates the values in the tuples of a relation.

## 6.10 The Structured Query Language (SQL)

The *Structured Query Language (SQL)* is a language implementation for managing data in relational database management systems (RDBMS). It is based on the relational algebra and is a *declarative* (non-procedural) language. This means that the user describes what s/he wants and not how the content will be retrieved from the database. The DBMS *optimizer* is responsible for the execution plan (Chapter 14).

SQL was initially developed by IBM in the early 1970s; it became a standard of the American National Standards Institute (ANSI) in 1986, and of the International Organization for Standards (ISO) in 1987. Since then, the standard has been enhanced several times. Nowadays, all commercial and open source RDBMS offer interfaces to SQL. In 1999, SQL3 was introduced. This extension takes into consideration the concepts of the object-relational technology (Chapter 15).

SQL consists of six components: (a) the data definition language (DDL), (b) the data manipulation language (DML), (c) the embedded and dynamic SQL, (d) data security, (e) transaction management, and (f) client-server services. In the next subsections the first two components are described briefly. Then, a simplified cadastral database is defined, constructed (populated), and manipulated using SQL.

### *6.10.1 Data Definition Language (DDL)*

The *data definition language (DDL)* comprises a set of commands to create relations and indices over relation fields (attributes). The most common commands are:

- CREATE TABLE
- CREATE INDEX
- CREATE VIEW
- DROP TABLE
- DROP INDEX
- DROP VIEW
- MODIFY  (modify the definition of existing tables, indices or views)

CREATE TABLE command defines a *new relation* in the database. The syntax of this command is as follows:

```
CREATE TABLE relation_name (column_name data_type, ...,
PRIMARY KEY (column_name, ...) );
```

   e.g.,

```
CREATE TABLE OWNERS (
  SIN CHAR(9) NOT NULL, SURNAME VARCHAR(15) NOT NULL,
  NAME VARCHAR(10) NOT NULL,
  DoB DATE, ...
PRIMARY KEY (SIN) );
```

CREATE INDEX command defines a new index on top of a column in a relation. The syntax of this command is as follows:

```
CREATE INDEX index_name ON relation (column_name);
```

   e.g.,

```
CREATE INDEX IND_OWNERS ON OWNERS (SURNAME);
```


## *6.10.2 Data Manipulation Language (DML)*

*Data manipulation language (DML)* consists of a set of commands to insert, delete, or modify the values in database relations. The most common commands are:

* SELECT (retrieve data from a relation)
* INSERT (populate a relation)
* DELETE (delete data from a relation)
* UPDATE (modify data values in a relation)
* COMMIT WORK (termination of a transaction)
* ROLLBACK WORK (cancel a transaction and rollback to the state of the last COMMIT)

SELECT command (query) *retrieves the tuples* from one or more relations. This command has three main *clauses*: (a) SELECT, (b) FROM, and (c) WHERE. In the FROM clause the relations involved in the query are stated. In the SELECT clause the attributes to be reported in the result are listed. Finally, in the WHERE clause both the query criteria and the relation join conditions (if any) are typed. The syntax of the SELECT command is as follows:

```
SELECT column_name, ...
FROM relation_name, ...
WHERE condition AND/OR condition ...;
```

   Three example SELECT queries are given next:

```
SELECT SURNAME, NAME
FROM OWNERS;
```

```
SELECT ID, BUILT_FACTOR
FROM PARCELS
WHERE USE = 'HOUSING';

SELECT OWNERS.SURNAME, OWNERS.NAME
FROM OWNERS, TITLES
WHERE TITLES.PURCHASE_DATE > 1/1/1972
AND OWNERS.SIN = TITLES.OWNER;
```

**INSERT** command *inserts new tuples* in a relation. The syntax of this command is as follows:

```
INSERT INTO relation_name [ (column_name, ...) ]   VALUES
(value, ...);
```

e.g.,

```
INSERT INTO OWNERS VALUES
  ('464197351', 'SMITH', 'JOHN', '08/15/1952');
```

**DELETE** command *deletes the tuples* in a table that satisfy a condition. The syntax of this command is as follows:

```
DELETE FROM relation_name WHERE condition;
```

e.g.,

```
DELETE FROM OWNERS WHERE SIN = '464197351';
```

Finally, **UPDATE** command *updates the values* stored in the relation tuples according to a condition. The syntax of this command is as follows:

```
UPDATE relation_name SET column_name = value WHERE condition;
```

e.g.,

```
UPDATE PARCELS SET BUILT_FACTOR = 2.2
WHERE ID = '75071745';
```

### *6.10.3 Example Cadastral Scenario*

Instead of a detailed description of the SQL syntax, the language is presented through a simplified example. Readers, who wish further detail in the language, are referred to the SQL specification (check the references at the end of this Chapter).

A simplified version of the cadastral scenario described in the previous Sections is considered next. As in the Figure 6.26a,b the database schema consists of three relations: PARCELS, OWNERS and TITLES. The corresponding data instances are shown in Figure 6.26c.

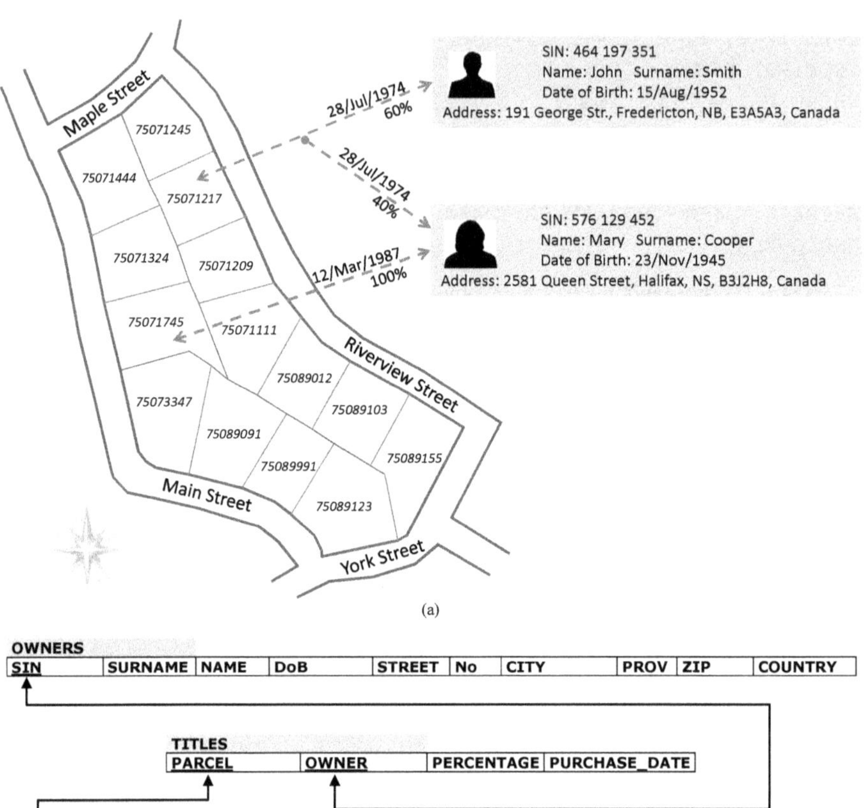


(a)

**OWNERS**

| SIN | SURNAME | NAME | DoB | STREET | No | CITY | PROV | ZIP | COUNTRY |
|---|---|---|---|---|---|---|---|---|---|

**TITLES**

| PARCEL | OWNER | PERCENTAGE | PURCHASE_DATE |
|---|---|---|---|

**PARCELS**

| ID | USE | BUILT_FACTOR | ADDRESS |
|---|---|---|---|

(b)

**OWNERS**

| SIN | SURNAME | NAME | DoB | STREET | No | CITY | PROV | ZIP | COUNTRY |
|---|---|---|---|---|---|---|---|---|---|
| 464197351 | SMITH | JOHN | 08/15/1952 | GEORGE | 191 | FREDERICTON | NB | E3A5A3 | CANADA |
| 576129452 | COOPER | MARY | 11/23/1945 | QUEEN | 2581 | HALIFAX | NS | B3J2H8 | CANADA |
| ... | ... | ... | ... | ... | ... | | | ... | ... |

**TITLES**

| PARCEL | OWNER | PERCENTAGE | PURCHASE_DATE |
|---|---|---|---|
| 75071217 | 464197351 | 60% | 07/28/1974 |
| 75071217 | 576129452 | 40% | 07/28/1974 |
| 75071745 | 576129452 | 100% | 03/12/1987 |
| ... | ... | ... | ... |

**PARCELS**

| ID | USE | BUILT_FACTOR | ADDRESS |
|---|---|---|---|
| 75071217 | HOUSING | 1.40 | 542 RIVERVIEW STREET, FREDERICTON |
| 75071745 | PARKING | 1.20 | 323 MAIN STREET, FREDERICTON |
| ... | ... | ... | ... |

(c)

**Fig. 6.26** A simplified cadastral scenario (a); the database schema (b); and a snapshot of the data instances (c).

The following three DDL commands *define* the database relations. Specifically, the next SQL command creates (defines) the relation of owners:

```
CREATE TABLE OWNERS (
   SIN CHAR(9) NOT NULL,
   SURNAME VARCHAR(15) NOT NULL,
   NAME VARCHAR(10) NOT NULL,
   DoB DATE,
   STREET VARCHAR(15),
   No VARCHAR(5),
   CITY VARCHAR(12),
   PROV CHAR(2),
   ZIP CHAR(6),
   COUNTRY VARCHAR(12),
PRIMARY KEY (SIN) );
```

The following SQL command creates (defines) the relation of parcels:

```
CREATE TABLE PARCELS (
   ID CHAR(10) NOT NULL,
   USE VARCHAR(20),
   BUILT_FACTOR REAL,
   ADDRESS VARCHAR(40),
PRIMARY KEY (ID) );
```

The following SQL command creates (defines) the relation of titles. The last two lines define the foreign keys and implement the join of the three relations as of Figure 6.26b. The third line from the bottom defines a constraint on the value of the percentage of ownership (unsigned integer less than or equal to 100).

```
CREATE TABLE TITLES (
   PARCEL CHAR(10) NOT NULL,
   OWNER CHAR(9) NOT NULL,
   PERCENTAGE INT,
   PURCHASE_DATE DATE,
PRIMARY KEY (PARCEL, OWNER),
CONSTRAINT Perc_chk (PERCENTAGE>=0 AND PERCENTAGE<=100),
FOREIGN KEY (PARCEL) REFERENCES PARCELS(ID),
FOREIGN KEY (OWNER) REFERENCES OWNERS(SIN) );
```

Figure 6.27 shows the *relationship diagram* after the definition of the three relations.

The relations can be *populated* with data instances through a series of DML commands. Specifically, the following two commands insert two tuples in the relation of owners (as of Figure 6.26c):

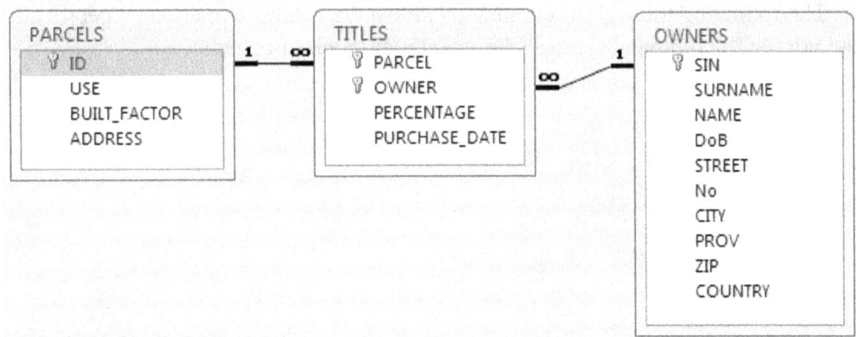

**Fig. 6.27** The relationship diagram of the three relations in Figure 6.26 as drawn by MS Access.

```
INSERT INTO OWNERS
   VALUES ('464197351', 'SMITH', 'JOHN', '08/15/1952',
   'GEORGE', '191', 'FREDERICTON', 'NB', 'E3A5A3', 'CANADA');

INSERT INTO OWNERS
   VALUES ('576129452', 'COOPER', 'MARY', '11/23/1945',
   'QUEEN', '2581', 'HALIFAX', 'NS', 'B3J2H8', 'CANADA');
```

In a similar way, data instances are inserted into the other two relations of the cadastral database to end up with the snapshot of Figure 6.26c:

```
INSERT INTO PARCELS
   VALUES ('75071217', 'HOUSING', '1.40', '542 RIVERVIEW
   STREET, FREDERICTON');

INSERT INTO PARCELS
   VALUES ('75071745', 'PARKING', '1.20', '323 MAIN STREET,
FREDERICTON');

INSERT INTO TITLES
   VALUES ('75071217', '464197351', '60', '07/28/1974');

INSERT INTO TITLES
   VALUES ('75071217', '576129452', '40', '07/28/1974');

INSERT INTO TITLES
   VALUES ('75071745', '576129452', '100', '03/12/1987');
```

The result of these commands is illustrated in the snapshot of Figure 6.26c. Figure 6.28 shows a screenshot of the relations as created and populated.

In the following example, a series of SELECT queries (Queries 1-9) are listed to *retrieve data* from the database, which satisfy various criteria. After these queries a few operations to *modify* the database content are also given (Queries 10-12).

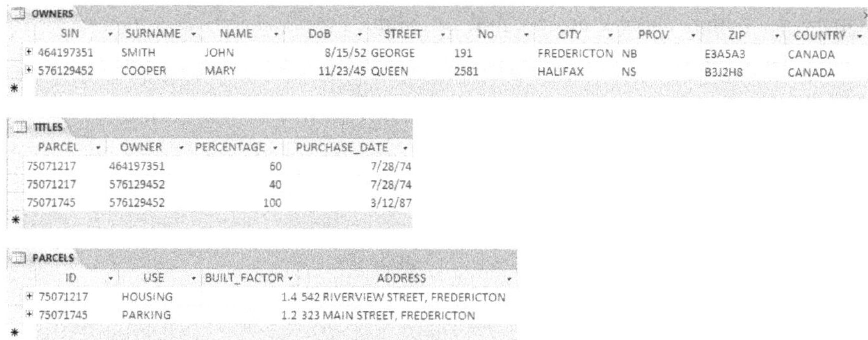

**Fig. 6.28** The relations as created and populated in MS Access (screenshot).

<u>Query 1:</u> How many parcels are stored in the database?

SQL Statement:
```
SELECT count(*) AS NUMBER_OF_PARCELS
FROM PARCELS;
```

Result:

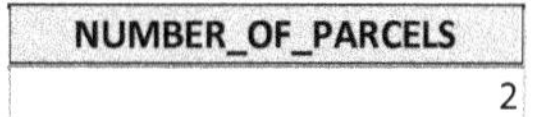

| NUMBER_OF_PARCELS |
| --- |
| 2 |

<u>Query 2:</u> Which is the land use of the parcels with a built factor greater than 1.3? The records in the report are grouped based on the use value (GROUP BY clause). As a result, every use will appear once in the report.

SQL Statement:
```
SELECT USE
FROM PARCELS
WHERE BUILT_FACTOR > 1.3
GROUP BY USE;
```

Result:

| USE |
| --- |
| HOUSING |

<u>Query 3:</u> List the name and surname of all owners who live in Fredericton. The owners will be ordered based on their surname (ORDER BY clause).

SQL Statement:
```
SELECT SURNAME, NAME
FROM OWNERS
WHERE CITY = 'FREDERICTON'
ORDER BY SURNAME;
```

Result:

| SURNAME | NAME |
| --- | --- |
| SMITH | JOHN |

<u>Query 4</u>: Which parcels are located on Main street? (LIKE is a partial string matching operator).

SQL Statement:
```
SELECT *
FROM PARCELS
WHERE ADDRESS LIKE '*MAIN STREET*'
```

Result:

| ID | USE | BUILT_FACTOR | ADDRESS |
|---|---|---|---|
| 75071745 | PARKING | 1.2 | 323 MAIN STREET, FREDERICTON |

<u>Query 5:</u> Report the name and surnames of all owners in the database along with the ids of the parcels they own and the percentage of ownership (two tables involved; the join condition in the WHERE clause).

SQL Statement:
```
SELECT OWNERS.SURNAME, OWNERS.NAME,
       TITLES.PARCEL, TITLES.PERCENTAGE
FROM OWNERS, TITLES
WHERE OWNERS.SIN = TITLES.OWNER
ORDER BY OWNERS.SURNAME;
```

Result:

| SURNAME | NAME | PARCEL | PERCENTAGE |
|---|---|---|---|
| COOPER | MARY | 75071745 | 100 |
| COOPER | MARY | 75071217 | 40 |
| SMITH | JOHN | 75071217 | 60 |

<u>Query 6:</u> In Query 5, also list the address of each parcel (three tables involved; the two join conditions in the WHERE clause).

SQL Statement:
```
SELECT OWNERS.SURNAME, OWNERS.NAME,
       TITLES.PARCEL, TITLES.PERCENTAGE,
       PARCELS.ADDRESS
FROM OWNERS, TITLES, PARCELS
WHERE OWNERS.SIN = TITLES.OWNER
   AND TITLES.PARCEL = PARCELS.ID
ORDER BY OWNERS.SURNAME;
```

Result:

| SURNAME | NAME | PARCEL | PERCENTAGE | ADDRESS |
|---|---|---|---|---|
| COOPER | MARY | 75071745 | 100 | 323 MAIN STREET, FREDERICTON |
| COOPER | MARY | 75071217 | 40 | 542 RIVERVIEW STREET, FREDERICTON |
| SMITH | JOHN | 75071217 | 60 | 542 RIVERVIEW STREET, FREDERICTON |

<u>Query 7:</u> Compute the sum of percentage of ownership for all parcels in the database (consistency control).

SQL Statement:
```
SELECT PARCEL, SUM(PERCENTAGE) AS TOTAL_PERC
FROM TITLES
GROUP BY PARCEL;
```

Result:

| PARCEL | TOTAL_PERC |
|---|---|
| 75071217 | 100 |
| 75071745 | 100 |

Query 8: Which was the age (derived attribute) of the owners in the database in 2012?

SQL Statement:
```
SELECT SIN, SURNAME, NAME, DoB AS BIRTH_DATE,
       (2012-YEAR(BIRTH_DATE)) AS AGE
FROM OWNERS;
```

Result:

| SIN | SURNAME | NAME | BIRTH_DATE | AGE |
|---|---|---|---|---|
| 464197351 | SMITH | JOHN | 8/15/52 | 60 |
| 576129452 | COOPER | MARY | 11/23/45 | 67 |

Query 9: Report the parcels with use 'housing' and built factor greater than 1 as well as those with land use 'parking' and built factor smaller than 2.

SQL Statement:
```
SELECT ID, USE, BUILT_FACTOR
FROM PARCELS
WHERE (USE='HOUSING' AND BUILT_FACTOR > 1)
   OR (USE='PARKING' AND BUILT_FACTOR < 2);
```

Result:

| ID | USE | BUILT_FACTOR |
|---|---|---|
| 75071217 | HOUSING | 1.4 |
| 75071745 | PARKING | 1.2 |

Query 10: Mrs. Cooper has sold her title (40%) over the parcel with ID= '75071217' to Mr. Smith on November 15, 2011. The database can be updated accordingly after executing the following two commands:

Statement 1: Delete the title of Mrs. Cooper over the parcel 75071217
```
DELETE FROM TITLES
WHERE PARCEL = '75071217'
AND OWNER = (SELECT OWNERS.SIN
             FROM OWNERS
             WHERE OWNERS.SURNAME ='COOPER' );
```

Statement 2: Update the title of Mr. Smith over parcel 75071217 to 100%. Also update the date for the new ownership.

```
UPDATE TITLES
SET PERCENTAGE = 100,
    PURCHASE_DATE ='15/11/2011'
WHERE PARCEL = '75071217'
AND OWNER = (SELECT OWNERS.SIN
                FROM OWNERS
                WHERE OWNERS.SURNAME ='SMITH' );
```

Result: The relation of titles after the update.

| TITLES | | | |
|---|---|---|---|
| PARCEL | OWNER | PERCENTAGE | PURCHASE_DATE |
| 75071217 | 464197351 | 100 | 11/15/11 |
| 75071745 | 576129452 | 100 | 3/12/87 |

Query 11: The Government of New Brunswick in Canada has decided to increase all built factors by 10%.

SQL Statement:
```
UPDATE PARCELS
SET BUILT_FACTOR = BUILT_FACTOR*1.10;
```

Result: The relation of parcels after the update.

| PARCELS | | | |
|---|---|---|---|
| ID | USE | BUILT_FACTOR | ADDRESS |
| 75071217 | HOUSING | 1.54 | 542 RIVERVIEW STREET, FREDERICTON |
| 75071745 | PARKING | 1.32 | 323 MAIN STREET, FREDERICTON |

Query 12: Create an index over the surnames of owners. Obviously, an index is useless for so few records in a table. The statement is given in here for educational purposes only.

SQL Statement:
```
CREATE INDEX OWN_INDEX
ON OWNERS (SURNAME);
```

The readers may get acquainted with the SQL language by practicing the above commands in a commercial or an open source relational DBMS, such as MS-Access, Oracle, or PostgreSQL.


## References and Further Reading


ANSI, 1986. *The Database Language SQL*. Document ANSI X3.135. American National Standards Institute.

Date, C.J., 2003. *An Introduction to Database Systems*. Addison-Wesley.

Elmasri, R., and Navathe, S.B., 2000. *Fundamentals of Database Systems*. Addison-Wesley.

Garcia Molina, H., Ullman, J.D., and Widom, J., 2000. *Database Systems Implementation*. Prentice Hall.

Melton, J., and Simon, A.R., 1993. *Understanding the New SQL: A Complete Guide*. Morgan-Kauffmann.
Ramakrishnan, R, and Gehrke, J., 2002. *Database Management Systems*. McGraw-Hill.
Ullman, J.D., Widom, J., 2001. *A First Course in Database Systems*. Prentice Hall.

# 7 Advanced Topics in Database Systems

## 7.1 Introduction

This Chapter is a follow up on the discussion in Chapter 6 and presents some advanced topics in database system design. Specifically, the discussion is organized as follows. In the next Section, some special cases in conceptual design are highlighted using the entity relationship (ER-) diagrams. Then, an enriched model, the enhanced entity relationship (EER-) model (and diagram) is introduced. The EER-model provides more semantics and can better represent complex reality.

The discussion proceeds to the object-oriented database systems, which are based in the object-oriented model and overcome many of the database design and implementation issues. Finally, the Chapter concludes by briefly presenting the Unified Markup Language (UML) class diagrams in conceptual database design.

## 7.2 Special Cases in the ER-diagrams

In Chapter 6 the basic constructs and concepts of the entity-relationship diagrams were presented. The theory of these diagrams also covers some special cases in conceptual modeling. The following Subsections briefly discuss three such cases.

### *7.2.1 Entity Roles and Recursive Relationship Types*

A relationship type of degree two connects two entity types. For example, the relationship type of TITLES (Figures 6.4, 6.6) relates the instances in the entity type OWNERS with those in the entity type PARCELS. The entity types participate in the relationship with specific *roles*. Each instance in the entity type OWNERS has the role of *owner*, while each instance in the entity type PARCELS has got the role of *parcel*. In a typical ER-diagram these roles are not annotated, because the name of the entity type usually conveys the corresponding role. An exemption is with recursive relationship types, where the roles are clearly annotated.

A relationship type is called *recursive* when a single entity type participates in the relationship type more than once. Should this occur, the entity type participates in the recursive relationship type with multiple roles. An example of a recursive relationship type is shown in Figure 7.1. The relationship type represents the topological relationship of "contains" between polygon geometries. The role describes if the instance participates in the relationship type as an external (bounding) or internal (bounded) polygon. In the cadastral plan of Figure 6.9, the polygon representing the building block (bounded by the roads: Main, Maple, Riverview, and York) has the role of the external polygon, while the individual parcels receive the role of the internal polygons. Notice that the degree of a recursive relationship type is one.

Figure 7.2b represents the recursive relationship type CONTAINS for the polygons in Figure 7.2a. The lines connecting the instances between the entity type and the relationship type are directed (in arrows). The direction as chosen in Figure 7.2b is from the external to the internal polygon. This means that the entity instance setting off an arrow refers to an external polygon, while the one ending an arrow refers to an internal polygon.

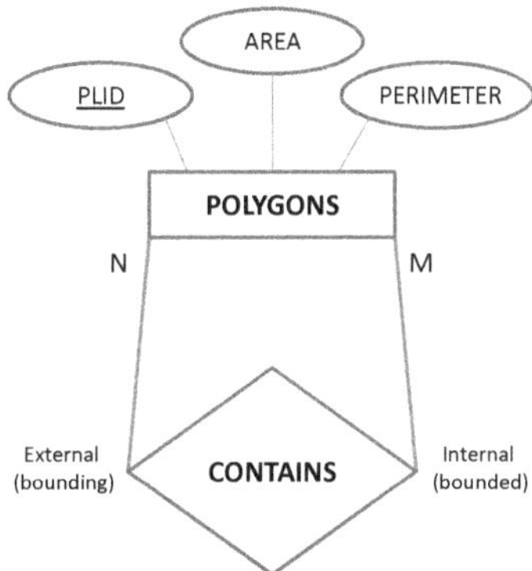

**Fig. 7.1** An example of a recursive relationship type.

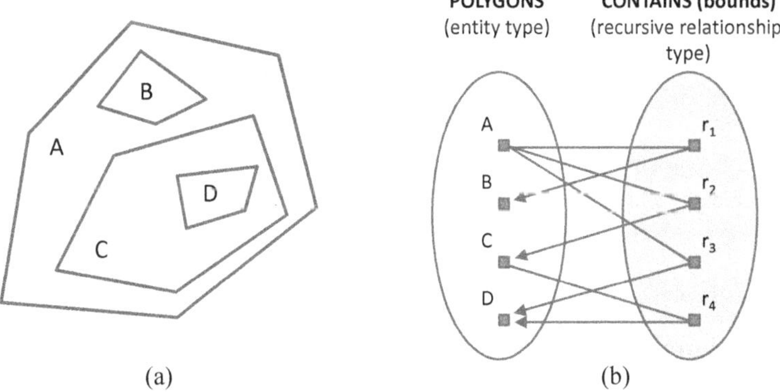

(a)(b)

**Fig. 7.2** Representation of the recursive relationship type contains (b) for a set of polygons (a).

## 7.2.2 Weak Entity Type and Weak Relationship Type

An entity type is called *weak* (or dependent) when the instances are not autonomous and their existence depends on their participation in a relationship with the instances of another entity type. The latter entity type is called *strong* entity type, while the relationship type connecting the weak with the strong entity type is called *weak* (or *dependent*) *relationship type.*

An example of a weak entity type in the cadastral example of Figure 6.9 could be the dependents (e.g., family members) of an owner. The presence of the weak entity instances (dependents) in the database is meaningful if the corresponding strong entity instance (owner) is also present. Figure 7.3b shows the weak relationship type PROTECTS between the owners (strong entity type: OWNERS) and their dependents (weak entity type: DEPENDENTS).

The notation of the weak entity and relationship types in an ER-diagram is through the symbols shown in Figure 7.3a. Notice that the weak entity type always has a full participation on the weak relationship type, provided that each entity instance must be related to an instance in the strong entity type.

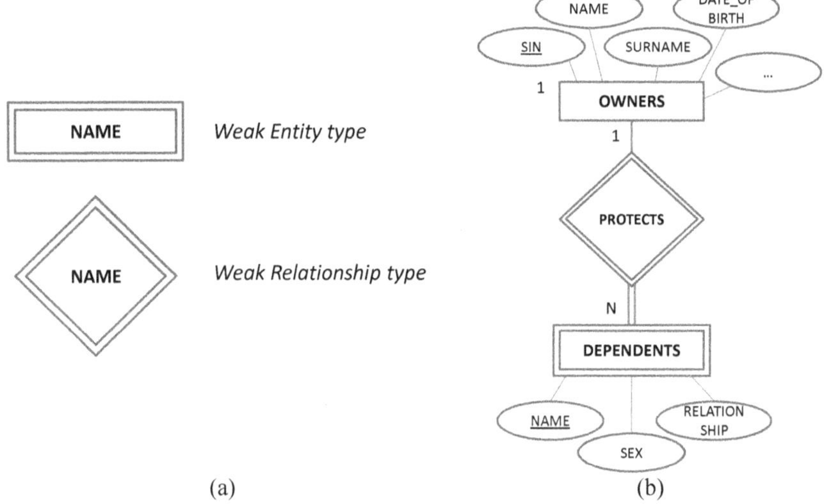


**Fig. 7.3** Weak entity and relationship types: their notation in an ER-diagram (a) and an example diagram (b).

### 7.2.3 Relationship Types of Degree greater than two

In Chapter 6, only relationship types of degree two were considered, i.e., relationship types that either connect two entity types or one entity type with two roles (in the recursive relationship types). In the real world, there are also relationship types of degree three or higher.

One example of a relationship type of degree three is shown in Figure 7.4a. The relationship type MORTGAGE connects the parcels (PARCELS) owned by owners (OWNERS) and mortgaged by banks (BANKS). The relationship type also has two attributes: the mortgage reference number (M_REF), and the date the mortgage expires (M_DATE).

Figure 7.4b presents three entity types: PARCELS, OWNERS, BANKS related to each other with three relation types of degree two: MORTGAGE, TITLES, CLIENT. The relationship type of degree three (Figure 7.4a) represents more details than that of the three relationship types of degree two (Figure 7.4b). Specifically, the situation that owner o gets a mortgage for parcel p from bank b, i.e., the triple instance (o,p,b), cannot be represented by the three relationship types of degree two, provided that the next three pair instances (o,p), (p,b), (o,b) do not necessarily correspond to the triplet (o,p,b).

Proof for the last sentence is given in Figure 7.5 through the following example. Figure 7.5a presents the relationship type (of degree 3) between the entity types in Figure 7.4a, while Figure 7.5b presents the alternative three relationship types (of degree 2) between the entity types in Figure 7.4b. In this example, there are two owners ($o_1$, $o_2$), two parcels ($p_1$, $p_2$) and two banks ($b_1$, $b_2$). According to the Figure 7.5a the following three triplets are valid:

$$(o_1, p_1, b_1), (o_2, p_1, b_2), (o_2, p_2, b_1)$$

On the other hand, Figure 7.5b illustrates the following nine pairs of instances, as derived from the relationship instances in Figure 7.5a:

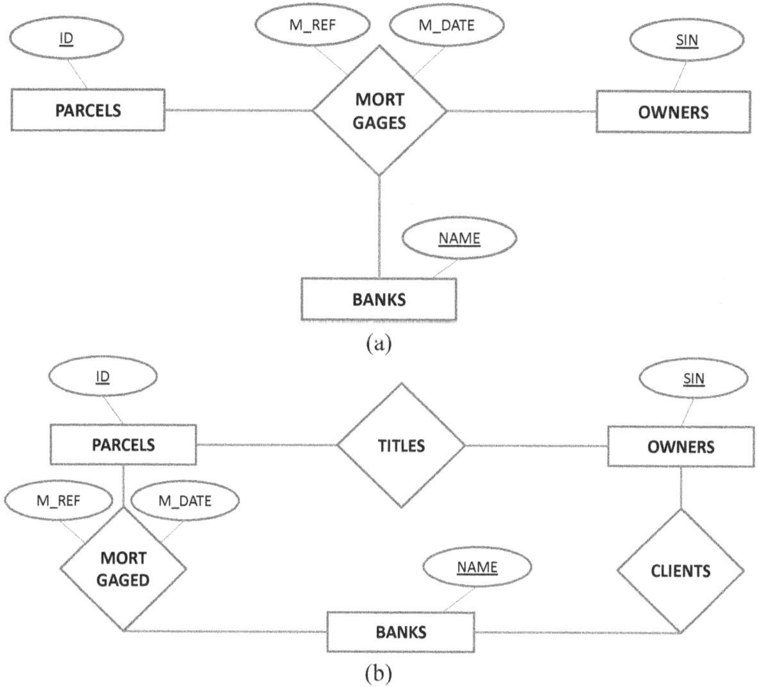


**Fig. 7.4** Example of a relationship type of degree 3 (a); and three relationship types of degree 2 (b).

$(o_1, p_1), (o_2, p_1), (o_2, p_2)$
$(o_1, b_1), (o_2, b_2), (o_2, b_1)$
$(p_1, b_1), (p_1, b_2), (p_2, b_1)$

Notice that Figure 7.5b does not allow ruling that owner $o_2$ has mortgaged parcel $p_1$ in bank $b_2$ (as Figure 7.5a does, through the triplet $o_2, p_1, b_2$). As of Figure 7.5b, $o_2$ is the owner of parcel $p_1$ $(o_2, p_1)$, $o_2$ is a client of bank $b_2$ $(o_2, b_2)$, and parcel $p_1$ is mortgaged in bank $b_2$ $(p_1, b_2)$. Additionally, $o_2$ is also a client of bank $b_1$ $(o_2, b_1)$ in which parcel $p_1$ is mortgaged $(p_1, b_1)$. Hence, a conclusion that owner $o_2$ has mortgaged parcel $p_1$ in bank $b_1$ $(o_2, p_1, b_1)$ could be derived, which is wrong according to Figure 7.5a. The mortgage of parcel $p_1$ from bank $b_1$ is through owner $o_1$ $(o_1, p_1, b_1)$. In other words, it is not possible to derive from Figure 7.5b which owner $o_1$ or $o_2$ has mortgaged his/her title on the parcel $p_1$ to the bank $b_2$.

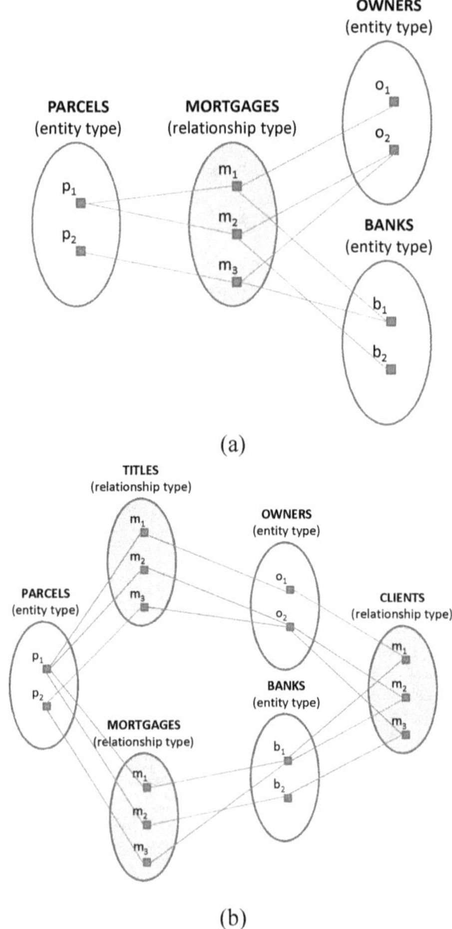

(a)

(b)

**Fig. 7.5** A snapshot of the instances in the entity and relationship types shown in Figure 7.4.

Obviously, replacing the relationship type of degree three with three relationship types of degree two is not possible. An equivalent representation of the relationship type of degree three with relationship types of degree two can be obtained if the relationship type of degree three is replaced by a weak entity type, which will be in turn connected to the initial entity types with three weak relationship types. An example of this configuration is shown in Figure 7.6. Mortgages are handled as weak entity types and the relationship type of degree three is mapped into three weak relationship types of degree two.

The mapping of a relationship type of degree three to relationship types of degree two might be required, because existing database design tools usually support only relationship types of degree two in the ER-model (and ER-diagram; Chapter 6). More details can be found in the references at the end of this Chapter.

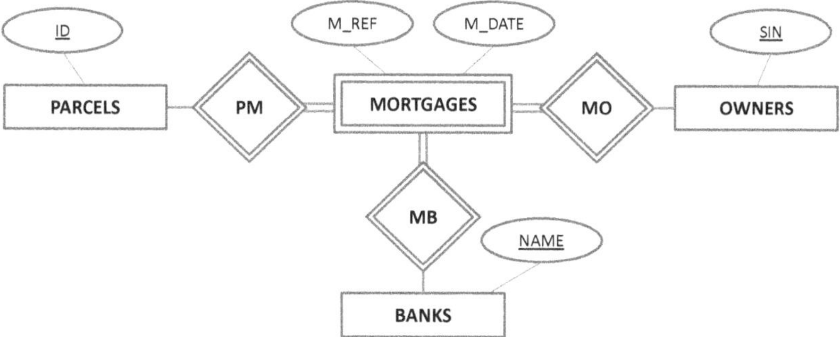


**Fig. 7.6** An equivalent representation of a relationship type of degree 3 (Figure 7.4a) with one weak entity type and three weak relationship types of degree 2.


## 7.3 Enhanced Entity-Relationship Model

As stated already, the Entity-Relationship model offers a set of concepts to conceptually represent the reality. However, these concepts are not adequate to meet the complex requirements introduced by modern technologies, e.g., mechanical design, computer graphics, multimedia, cartography, GIS, etc.

The need to enrich the ER-model, to support these new technologies led to the development of conceptual models with enhanced expressive power. One of these models, which extends the concepts of the ER-model, is the *Enhanced Entity Relationship (EER-) model*.

The *new concepts* introduced by the EER-model are: (a) superclasses and subclasses, (b) class inheritance, (c) specialization and generalization, and (d) inheritance. These concepts are described in the following Sections.

### 7.3.1 Superclasses and Subclasses

The instances of an entity type, e.g., A, can usually be grouped further into other entity types, e.g., B, C and D (Figure 7.7). The broader entity type (A) is called *superclass* of the other types (B, C, D). The latter types (B,C,D) are called *subclasses* of the superclass (A). Some or all instances of the superclass A may also appear in a subclass (B, C, or D). In the example of Figure 7.7, the instances $a_4$ and $a_5$ do not fall under any subclass, while the others do (e.g., instances $a_1$, $a_2$, $a_3$ fall under the subclass B). Another case is to have one instance in A falling under more than one subclasses, e.g., $a_{12}$.

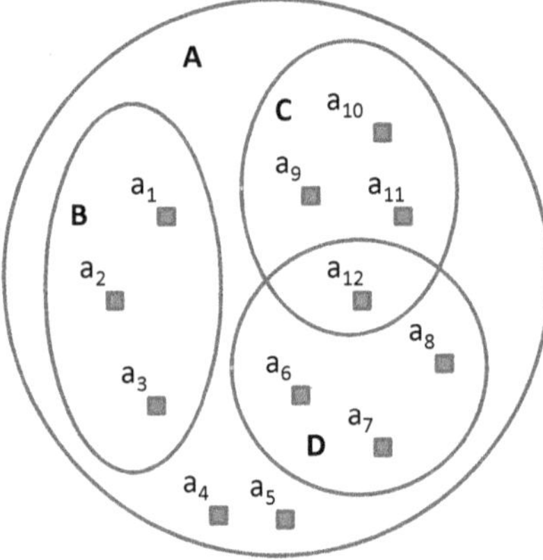


**Fig. 7.7** Example of a superclass (A) and three subclasses (B,C,D).

In the simplified cadastral application examined in Chapter 6, the instances of the entity type representing the owners (superclass: OWNERS), may be clustered into two new types (subclasses): (a) the physical persons (subclass: PERSONS), which accommodates people as owners of parcels, and (b) the legal entities (subclass: LEGAL_ENTITIES), which accommodates the companies, institutions, municipalities, etc., as owners of parcels.

The relationship type which connects the broader entity type (e.g., OWNERS) with each of the narrower entity types (e.g., PERSONS) is called a *superclass-subclass* or simply a *class-subclass* relationship type. Notice that a class-subclass relationship type relates one instance of the superclass with one instance in the subclass. However, there is a significant difference between this type and a relationship type of cardinality ratio 1:1. In a class-subclass relationship type, the instances in the superclass and the subclass refer to the same entity. On the flip side, in a relationship type of ratio 1:1, the instances refer to different entities.

Figure 7.8 highlights the difference. In Figure 7.8a a class-subclass relationship type is shown. Mary Cooper, represented as $o_1$, is a person and owner. Hence, she appears in both the superclass (OWNERS) and subclass (PERSONS) as an instance. The same applies for the City of Fredericton, represented as $o_3$ in the same Figure. On the other hand, Figure 7.8b shows a relationship type of ratio 1:1. This relationship refers to the spouses of owners. Each instance in this relationship type (MARRIAGE) connects one instance from the entity type of owners (PERSONS) with another (discrete) instance from a group of people (SPOUSES). The assumption that the spouse is not an owner is made here.

Obviously, each entity as an instance in a subclass must also appear as an instance in the corresponding superclass, e.g., Mary Cooper as an instance in the class PERSONS (i.e., persons as owners) must also appear as an instance in the superclass OWNERS. The opposite is not necessarily true. For instance, a legal entity (e.g., the City of Fredericton) as an instance in the superclass OWNERS will not appear as an instance in the subclass PERSONS. It may or may not appear in another subclass. Here it does in the subclass LEGAL_ENTITIES.

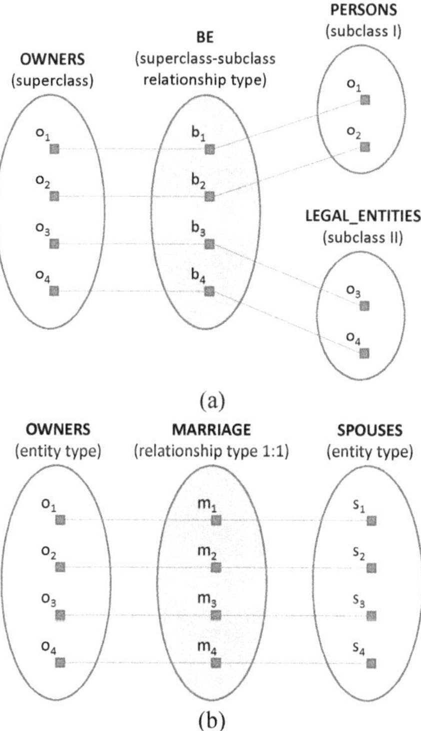

(a)

(b)

**Fig. 7.8** Representation of a superclass-subclass relationship (a) and of a relationship type of cardinality ratio 1:1 (b).

## 7.3.2 Inheritance

Another significant concept that is closely related with the class-subclass definition is *inheritance*. According to this, an entity instance, which is a member of a subclass, inherits all attributes and relationships of the corresponding entity in the superclass. This is a result of the fact that the instances in a superclass and its sub-classes correspond to the same entities.

In the cadastral example, the owners (superclass: OWNERS) are grouped into persons and legal entities (subclasses: PERSONS, LEGAL_ENTITIES) (Figure 7.8a). Figure 7.9 shows a class-subclass relationship type in an EER-diagram. The relationship type is denoted with the mathematical symbol of subsets, i.e., $\subset$. The notations of the circle with the letter d and the double line from the entity type of OWNERS to the circle are described in the next Section.

Mrs. Mary Cooper, as a person, she is an instance in the subclass: PERSONS and inherits all the attributes of the superclass OWNERS, e.g., the owner's identi-fier, registration date in the database, telephone numbers, etc. She also inherits the relationship instances (if any) with the relationship types that the superclass OWNERS participates in, e.g., the relationship type TITLES with the entity type PARCELS.

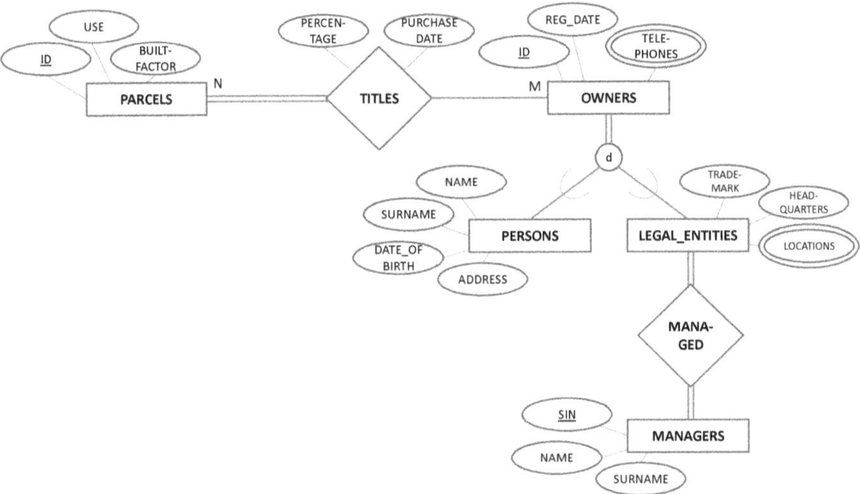


**Fig. 7.9** An example class-subclass relationship type for the domain of cadastre.

One reason for introducing a relationship type of class-subclass in a data model is that some attributes are applicable to some (and not all) the instances of an enti-ty type (the superclass). For example, as shown in the Figure 7.9, an owner, either person or legal entity, has an identifier, a registration date in the database, and a set of telephone numbers. The persons also have a name, surname, data of birth, address, etc. On the other hand, the legal entities have a trademark, headquarters location, branch locations, etc.

Another reason of introducing a class-subclass relationship type in a data model is that some relationship types may apply to the instances of a subclass only. As shown in the example of Figure 7.9, the subclass of the legal entities is connected with the entity type MANAGERS through the relationship type MANAGED. On the other hand, the instances of the subclass PERSONS do not participate in the relationship type MANAGED.


## *7.3.3 Specialization and Generalization*

Both *specialization* and *generalization* constitute two basic concepts of the EER-model, which are closely related to the concept of inheritance. *Specialization* is the process of defining the subclasses of an entity type. The definition of the subclasses is based on the attributes and relationships of the entity instances in the superclass. For example, the subclasses PERSONS and LEGAL_ENTITIES in the example of Figure 7.9 are a specialization of the class OWNERS and result from the specific attributes and relationship types of the instances in these subclasses. *Generalization* is the reverse process of specialization.


### 7.3.3.1 Definition of a Specialization

The definition of a specialization is a *three-step process*: (1) the subclasses are defined, (2) attributes are assigned to the superclass and each of the subclasses, and (3) specific relationship types for each subclass with other entity types are defined.

These steps for the specialization in the example of Figure 7.9 {superclass: OWNERS, subclasses: PERSONS, LEGAL_ENTITIES} are as follows:

- Step (1): The entity type OWNERS consists of physical persons (citizens) and legal entities (companies, institutions, municipalities, etc.). Hence, two subclasses are defined: {PERSONS, LEGAL_ENTITIES}.
- Step (2): All entity instances of the type OWNERS, regardless their nature (persons or legal entities) have the following attributes: identifier (for physical persons the SIN, and for legal entities the taxation number), date of first registration in the database, and a set of telephone numbers as contact information. All common attributes are assigned to the superclass OWNERS. If the entity instance is a physical person, it is also shown up in the subclass PERSONS. This subclass has assigned some additional attributes that describe a person: name, surname, date of birth, address, etc. On the other hand, if the entity type is a legal entity, it is also present in the subclass LEGAL_ENTITIES. This subclass has assigned some additional attributes that correspond to legal entities: trademark, headquarters location, branch locations, etc.
- Step (3): The superclass participates in the relationship type TITLES, because all owners regardless their nature are related to parcels. Additionally, a legal entity is managed by one or more managers. Hence, there is a relationship type connecting the instances of the subclass LEGAL_ENTITIES with the entity type of managers. For simplicity, it is assumed that these managers are not owners themselves. Hence, the subclass LEGAL_ENTITIES participates in the

relationship type MAGAGED and is connected to the entity type MANAGERS (Figure 7.9).

Notice that a subclass may be the superclass of another specialization. For example, the classes {PUBLIC, PRIVATE} can be two subclasses of the class LEGAL_ENTITIES, to further divide the instances into the set of public and another set of private legal entities (Figure 7.10), respectively. This approach results to a *hierarchy of specializations*.

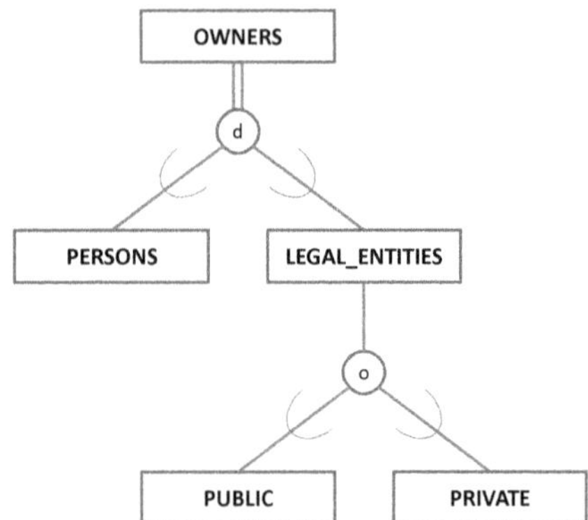


**Fig. 7.10** An example hierarchy of specializations.


### 7.7.3.2 Types of Specialization

A specialization is either *total* or *partial*. In addition, a specialization is either *disjoint* or *overlapping*. These new terms are described in the following paragraphs.

A specialization is *total*, if all instances in the superclass are also present in one or more subclasses. For example, in Figure 7.10, all instances of the class OWNERS are either physical persons or legal entities and obviously are present in one subclass {PERSONS, LEGAL_ENTITIES}. The total specialization is denoted by the double line coming out the superclass rectangle to the specialization circle.

A specialization is *partial*, if there are instances in the superclass, which do not show up in any subclass. For example, the bottom specialization in Figure 7.10 is a partial specialization. An entity in the superclass LEGAL_ENTITIES may be public or private and also appear in either of the corresponding subclasses {PUBLIC, PRIVATE}. However, there are LEGAL_ENTITIES which do not fall within any of these two subclasses (such as an NGO, monastery, farmer union, political party, etc.). The partial specialization is denoted by the single line coming out the superclass rectangle to the specialization circle.

A specialization is overlapping when it is allowed for the instances of the superclass to be members of more than one subclass. For example (Figure 7.10), a legal entity may have a shared scheme in the public and private sectors. Hence, it will be a member of both subclasses {PUBLIC, PRIVATE} of the class

LEGAL_ENTITIES. An overlapping specialization is denoted by the letter (o) written inside the specialization circle.

Finally, a specialization is disjoint when the subclasses of a class are disjoint, i.e., they do not share any entity instance. The top specialization in Figure 7.10 is an example of a disjoint specialization. An owner is either a person or a legal entity. Hence, the entity instances of the superclass fall into one and only subclass {PERSONS, LEGAL_ENTITIES}. A disjoint specialization is denoted by the letter (d) written inside the specialization circle.

The combination of the specialization types in pairs (total/partial and overlapping/disjoint) forms four distinct specialization categories: (a) *partial and overlapping*, (b) *partial and disjoint*, (c) *total and overlapping*, and (d) *total and disjoint* specialization. Figure 7.11 shows the four categories of specializations in the EER-diagram notation.

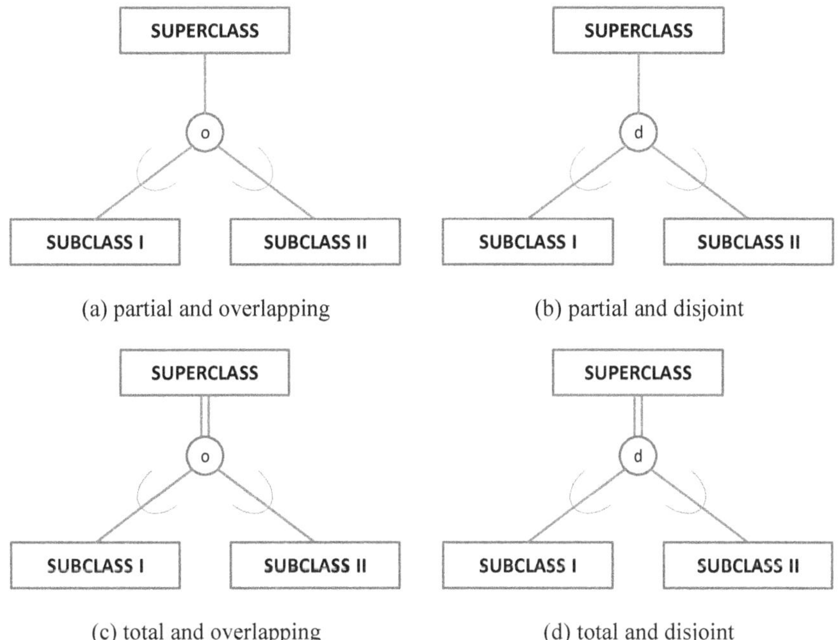

(a) partial and overlapping       (b) partial and disjoint

(c) total and overlapping       (d) total and disjoint

**Fig. 7.11** Four categories of specialization (EER-diagram notation).

### 7.7.3.3 Generalization

*Generalization* is the reverse process of specialization. Specifically, given two or more entity types, based on the common attributes and/or relationship types a superclass can be formed out of them. The common attributes and relationship types are then eliminated from the initial entity types, which take the role of the subclasses of the new superclass.

Figure 7.12 presents an example generalization process for the cadastral scenario. The entity types PARCELS and BUILDINGS (Figure 7.12a) represent the land parcels and the buildings in a cadastral database, respectively. These types have some common attributes. In addition, they both participate in the relationship type TITLES, which relates them to the entity type of owners (OWNERS). All these common attributes and relationships are collected to define a new (generalized) entity type with name REAL_ESTATE (Figure 7.12b). This new type takes the role of the superclass over the initial types PARCELS and BUILDINGS. Specifically, the superclass accommodates the common attributes (id, address) and participates in the common relationship type (TITLES) as the initial types.

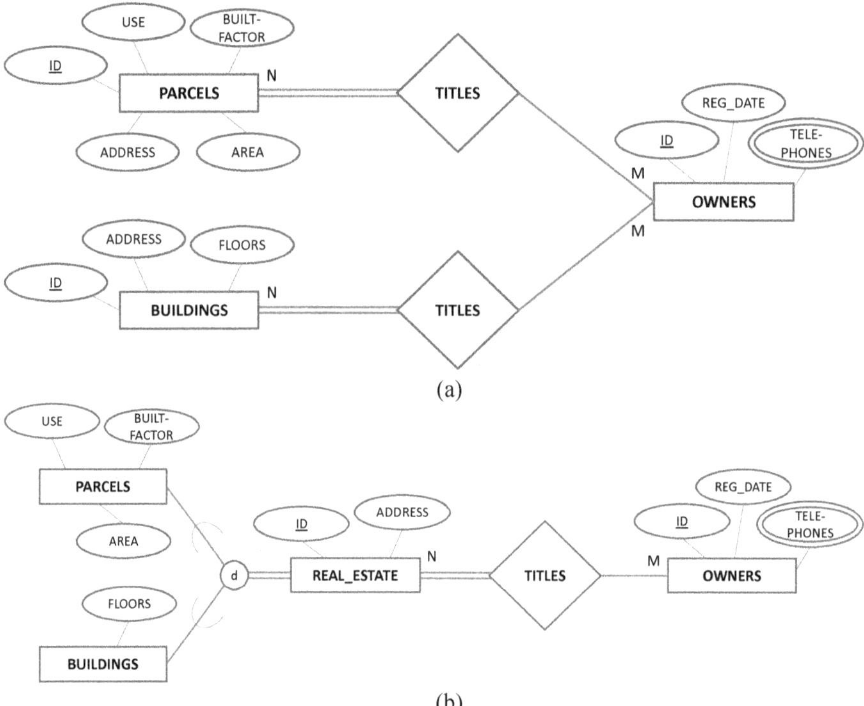


**Fig. 7.12** Generalization process example. The initial state (a) and the result of the generalization (b).

### 7.3.3.4 Specialization Networks

All previous examples refer to *specialization hierarchies* (e.g., Figure 7.10). This means that each entity type is a subclass to a single superclass. The EER-model also allows for the formation of *specialization networks*. In a specialization network an entity type may be in the role of the subclass in more than one superclass. According to the concept of inheritance, the subclass in a specialization network inherits the attributes and relationships of all its superclasses. This is the concept of *multiple inheritance.*

Figure 7.13 shows an example of a specialization hierarchy in the simplified cadastral application. Two specializations are present in this example for the entity type REAL_ESTATE: {PARCELS, BUILDINGS} and {COASTAL, IN_LAND}. The subclass PARCELS is further specialized into {BUILDABLE, NON-BUILDABLE}, while the non-buildable parcels are specialized into {FOREST_LANDS, ARCHAEOLOGICAL_SITES}. Finally, an entity type which corresponds to the coastal archaeological lands (and named COASTAL_ARCHAEOLOGIC_LANDS) is defined as shown in Figure 7.13. This type is the subclass of two specializations. This results in a specialization network.

Due to the multiple inheritance, a coastal archaeological land has the following attributes: identifier, address (from the class REAL_ESTATE), area (from PARCELS), law of characterization as a non-buildable land (from NON-BUILDABLE), name, number of visitors per year (from ARCHAEO_SITES), aspect and distance from the sea (from COASTAL). It also has an attribute: distance from the nearest town. Additionally, the coastal archaeological lands inherit all the relationships of the superclasses in the network, which are omitted from the Figure 7.13.

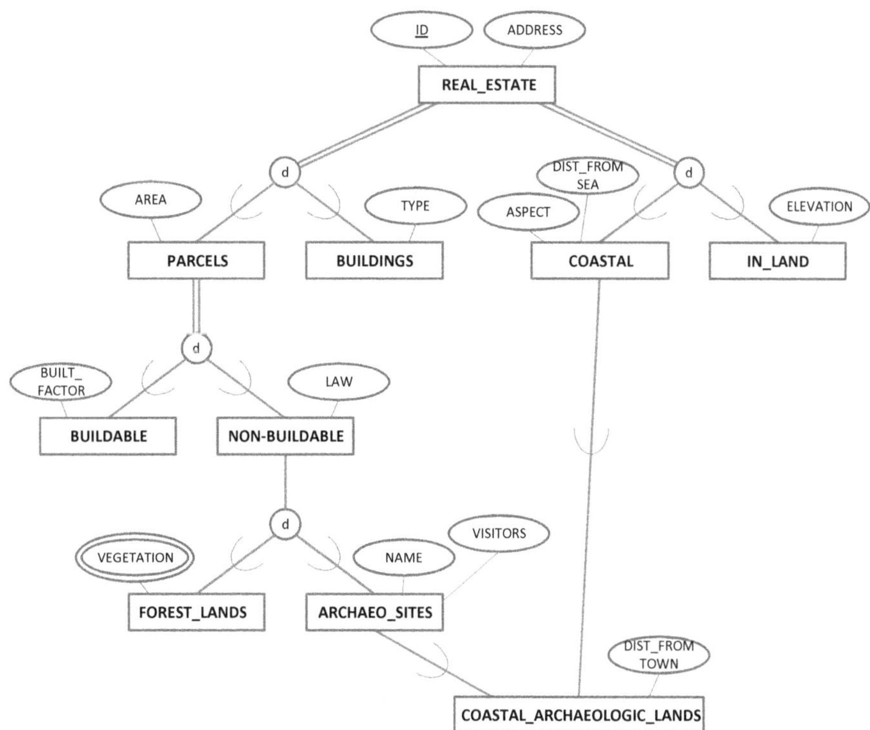


**Fig. 7.13** An example specialization network.

## *7.3.4 Categories*

For a more complete representation of the reality the EER-model also allows for class-subclass relationship types with more than one entity type in the role of the superclass. In this case, the subclass is a result of merging two or more superclasses and the relationship type is called *category*.

Notice that the superclasses in a category represent different entity types, while each instance in the subclass inherits the attributes and relationships of the superclass. In addition, the superclasses in a category may have the same or different key attributes.

Figure 7.14 shows two example categories. Letter (c) written inside the circle denotes the category in an EER-diagram. Figure 7.14a refers to the category of non-buildable parcels (NON-BUILDABLE_PARCELS), which are derived by merging the forest land parcels (FOREST_LANDS) and the archaeological sites (ARCHAEO_SITES). Figure 7.14b refers to the category of coastal real estate (COASTAL_ESTATE), which is derived by merging the coastal buildings and the coastal parcels.

As coastal buildings and parcels belong to the broader entity types of BUILDINGS and PARCELS respectively, two *predicates* are applied for the selection of the instances to appear in the category COASTAL_ESTATE.

A category is either *total* or *partial*. The non-buildable parcels (NON-BUILDABLE_PARCELS) in Figure 7.14a is a total category, as all instances of the superclasses {FOREST_LANDS, ARCHAEO_SITES} are merged to form the subclass (category). The total category is denoted by the double line from the circle to the subclass rectangle. On the other hand, the category of the coastal real estate (COASTAL_ESTATES) is a partial category (Figure 7.14b), as only some of the instances in the superclasses {BUILDINGS, PARCELS} are merged to form the subclass (category). The partial category is denoted by the single line from the circle to the subclass rectangle.

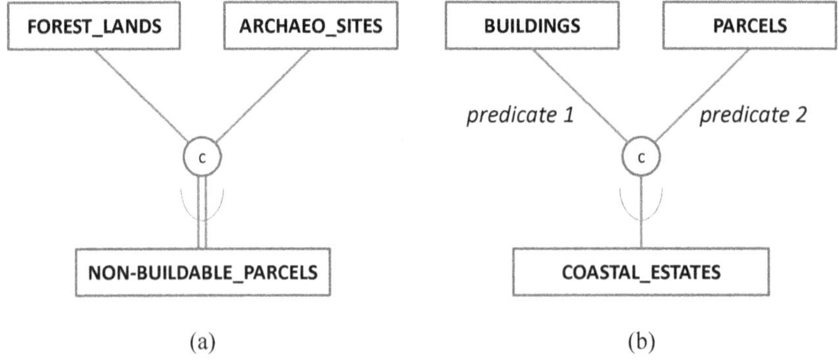


**Fig. 7.14** Examples of categories: total category (a) and partial category (b).

In case of a total category (Figure 7.15a), all the instances of the superclasses are merged to form the subclass. This means that a total category may be *reversed* to an equivalent specialization, as shown in Figure 7.15b. Obviously, the reversal is not valid in case of a partial category, provided that in a specialization the entities in the subclasses must all be present in the superclass.

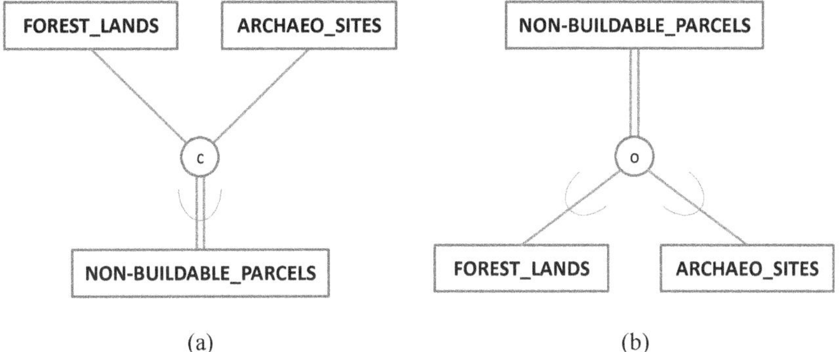


**Fig. 7.15** An example of a total category (a) and its reverse specialization (b).

Notice that, the connection of the subclass NON-BULDABLE_PARCELS with the specialization circle is through a single line (partial specialization) in Figure 7.13, while – in contrast – a double line (total specialization) is used in Figure 7.15b. The reason for this is that in Figure 7.13 the assumption that there are other types of non-buildable parcels is made (e.g., a beach). On the other hand, the superclass in Figure 7.15b is the result of merging two sets: the forest lands and the archaeological sites. Hence there are no other types of non-buildable parcels included in the superclass.

## 7.4 Mapping the EER-Diagram into a Relational Schema

An EER-diagram is an extended ER-diagram. Hence, the five rules for mapping an ER-diagram into the relational schema (Chapter 6) can still be applied to map all ER-constructs that are present in an EER-diagram. Two additional rules are described next to map the additional constructs of the EER-diagram into the relational schema.

### 7.4.1 Rule 6: Mapping the Classes and their Subclasses

This rule refers to the mapping of a *class* and its *subclasses* into the relational schema. This rule consists of the following steps:

xi.    create a relation for the superclass and one for each subclass according to the first rule (Rule 1; Chapter 6),

xii.    insert in the relations of the subclasses the key attribute of the superclass as a key attribute.

An example of these steps is given in Figure 7.16.

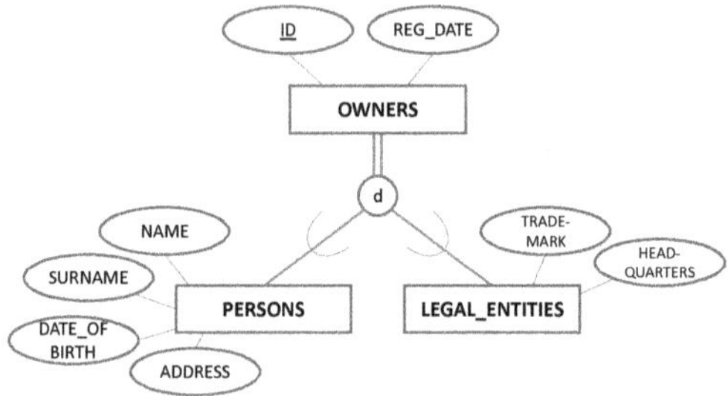


**Step (xi):** Create one relation for the superclass and each subclass.

OWNERS

| ID | REG_DATE |
|---|---|

PERSONS

| NAME | SURNAME | DoB | ADDRESS |
|---|---|---|---|

LEGAL_ENTITIES

| TRADEMARK | HEADQUARTERS |
|---|---|

**Step (xii):** Insert the key of the superclass as a key attribute to each relation of the subclasses.

OWNERS

| ID | REG_DATE |
|---|---|

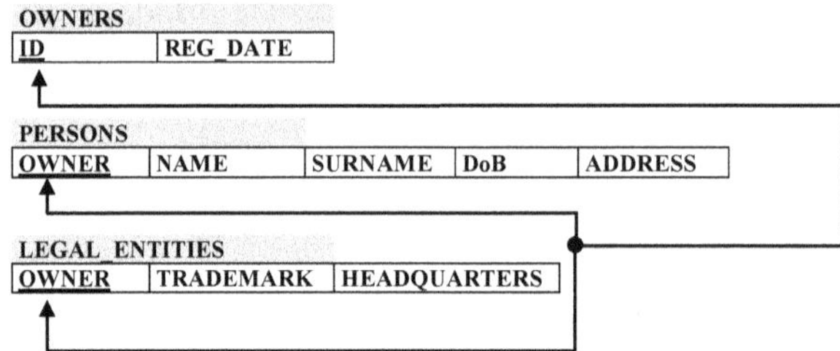

PERSONS

| OWNER | NAME | SURNAME | DoB | ADDRESS |
|---|---|---|---|---|

LEGAL_ENTITIES

| OWNER | TRADEMARK | HEADQUARTERS |
|---|---|---|

**Fig. 7.16** An example mapping of a specialization from the EER-diagram to the relational schema.

## 7.4.2 Rule 7: Mapping the Categories

This rule refers to the mapping of a *category* to the relational schema. This rule consists of the following steps:

xiii.    create a relation for each superclass of the category and one for the category itself (the subclass) according to the first rule (Rule 1; Chapter 6),

xiv.    make the following distinction: (a) if the superclasses have the same key attribute(s), insert it (them) as a key to the relation of the category; (b) if the superclasses have different key attributes, create a new key attribute and insert it as a foreign key in all relations of the superclasses and the category.

An example of these steps is given in Figure 7.17. Figure 7.17a shows a category with superclasses of the same key attribute. Figure 7.17b shows a category with superclasses of different key attributes.

## 7.5 Object-Oriented Database Systems

The object-oriented database systems were first introduced in 1980, as a merging of two technologies: (a) the database systems, and (b) the object-oriented programming. This merging fetched multiple benefits to the database technology, in both the design and implementation levels. The object model has many advantages: it is expressive, extensible, portable, and easy to maintain. This Section briefly presents the object-oriented database system technology and highlights its advantages over the traditional relational database systems.

The discussion is organized as follows. First, the weaknesses of the relational model are presented. Then, the basic concepts of the object-oriented database systems and the advantages of the object model over the relational one are highlighted. The basic steps of the object-oriented database design are elaborated after along with a set of rules to map an EER-schema into the object schema.

### *7.5.1 Weaknesses of the Relational Model*

The relational model organizes the entities and relationships into a set of relations or tables. The design of the relations is based on the normalization theory which aims at (a) the elimination of redundant data, and (b) the minimization of the number of relations (Chapter 6).

On one hand, composite entities are being split into their atomic elements (simple entities) before being stored in database relations (First Normal Form: e.g., a line is decomposed into the set of vertices, and a vertex into its coordinate values). On the other hand, the information describing an entity can be decayed and stored in multiple relations, which need to be merged back to reconstruct the entity (Second and Third Normal Forms: e.g., information regarding the ownership of a parcel is split into multiple relations: owners, titles, and parcels).

Although the relational model has been widely adopted in traditional database systems, its deficiencies render it inadequate to support geographic databases. These deficiencies are highlighted next.

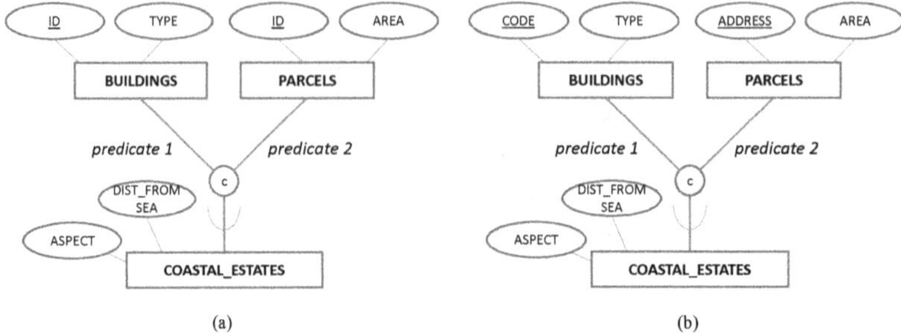

(a)                                     (b)

**Step (xiii):** Create a relation for the category and each superclass.

Relations for the EER-diagram in (a): The superclasses {BUILDINGS, PARCELS} have a common key attribute (an ID generated according to the cadastral encoding).

**BUILDINGS**

| ID | TYPE |
|----|------|

**PARCELS**

| ID | AREA |
|----|------|

**COASTAL_ESTATES**

| ASPECT | DIST_FROM_SEA |
|--------|---------------|

Relations for the EER-diagram in (b): The superclasses {BUILDINGS, PARCELS} have different key attributes, an code for the buildings (CODE) and the address for the parcels (ADDRESS).

**BUILDINGS**

| CODE | TYPE |
|------|------|

**PARCELS**

| ADDRESS | AREA |
|---------|------|

**COASTAL_ESTATES**

| ASPECT | DIST_FROM_SEA |
|--------|---------------|

**Step (xiv-a):** insert the common key attribute (ID) as a key in the relation of the category (COASTAL_ESTATES).

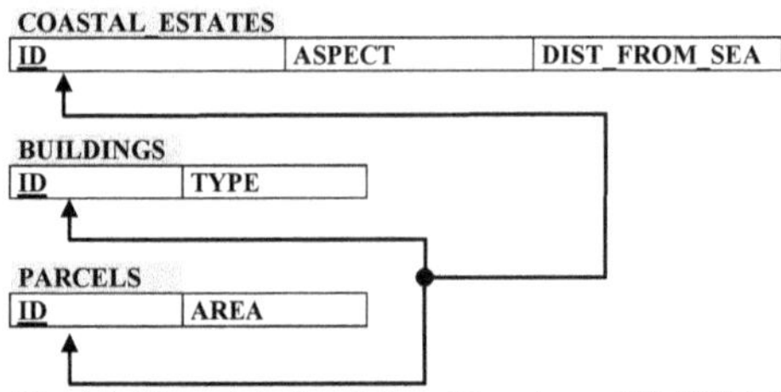

**Step (xiv-b):** create in the relation of the category (COASTAL_ESTATES) a new key attribute (C_ID) and insert it as a foreign key in the relations of the superclasses (BUILDINGS, PARCELS).

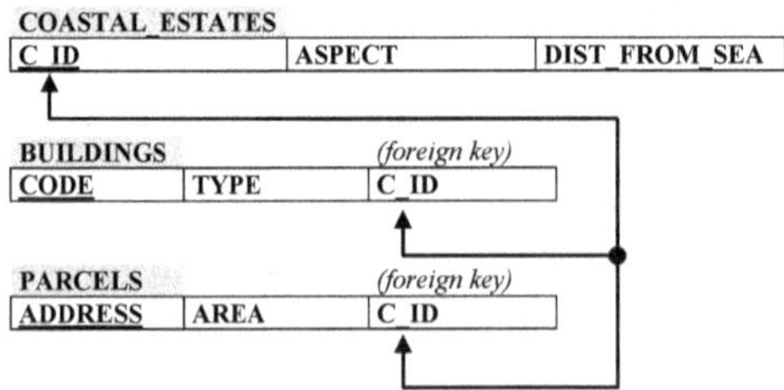

**Fig. 7.17** An example mapping of a category from the EER-diagram to the relational schema.

### 7.5.1.1 Limited Efficiency

The *efficiency* of a database management system refers to the how fast the database content can be managed. Relational database systems are usually slow when they are applied to support applications involving non-traditional data, such as geometries.

As described previously, normalization theory (Chapter 6) usually forces the data describing an entity to reside into multiple relations. For example, the telephone numbers of the owners in the cadastral application were stored in a separate table (Figure 6.11). Hence the queries posed by the user might require merging back the data from multiple relations (*entity reconstruction*). Merging is usually a Cartesian product (join operation) of two large sets of records and this is a very slow process. In large databases involving geographic data, the response time is usually inacceptable.

### 7.5.1.2 Poor Semantics

*Semantics* describe the expressive power of a model and give meaning to the entities, relationships, and their attributes. The relational model lacks semantics. Although, it can connect the data residing into multiple relations, the relational model cannot describe the meaning of these connections. For example, the relationship TITLES describes a relationship between the parcels and owners in a cadastral application, however, only the user is aware of this relationship.

A data model provides rich semantics when it is capable of distinguishing entities from relationships as well as different types of relationships that apply between the database entities. This is significant for a system to assure the consistency of data and a user friendly environment for the general user.

Three kinds of relationship types are of high importance in a geographic database system: (a) *association*, (b) *generalization*, and (c) *aggregation*. Association refers to the generic relationship type between two or more entity types. Generalization refers to the class-subclasses relationship type. Finally, the aggregation refers to the relationship between the whole (e.g., an entity) and its components (parts). The relational model cannot distinguish between these three types of relationships.

Another weakness of the relational model is that a column in a relation may represent either an entity attribute or the relationship between entities (*lack of semantics*). In addition, the model provides two alternative ways to represent an association between two entities (*semantic overflow*): (i) inside a relation (relationship between columns) through a column, or (ii) between relations (join process) through foreign keys.

### 7.5.1.3 Degraded Model Extension

A data model to be flexible must provide a wide variety of data types, while allowing the user for creating *new data types*. The relational model offers the most commonly used data types (e.g., integer, real number, character sets, date, etc.) as *built-in* types. By combining those types the user is able to define relations (tables) as complex data types (record types).

However, these types (the relations) cannot be used as constructs to create other types, as a true data type does. According to the First Normal Form (Chapter 6), the relational model does not allow nested relations (i.e., the nesting of a relation inside a cell of another relation).

### 7.5.1.4 Ambiguous Identification Scheme

Database entities must be assigned a unique identifier, which will remain unchanged during any update and restructuring of the database. The *identification scheme* is closely related to the database model. The relational model uses unique key values, which are derived from the attribute values assigned to the entities. For example, the SIN is a common attribute to identify physical persons (e.g., the owners) in a relational database.

The approach of mixing the attribute values with the identification of entities has been proven prone to inconsistencies in the database. For instance, a change of a person's SIN (e.g., as a result of a change in her immigration status) would cause inconsistencies in the database.

### 7.5.1.5 Inefficient Interfaces with External Programs

The DBMS user can be either a person or an external program interacting with the database (Figure 7.18). In the latter case, the interaction is supported by an interface which takes into consideration the database model. Modern software programs are written in an object-oriented language. Any differences in the data models used by the external program and the database itself may hamper the smooth interaction.

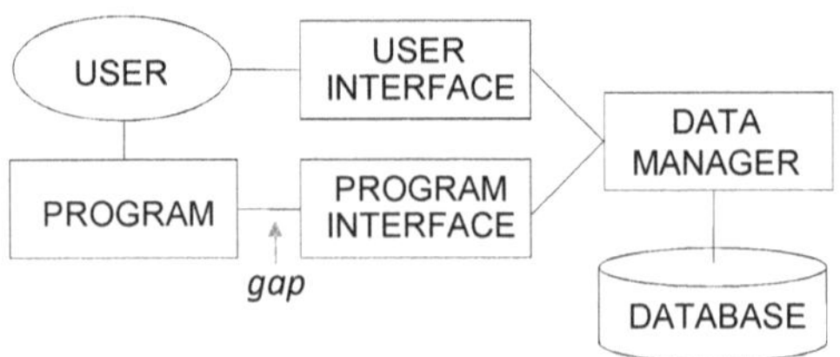


**Fig. 7.18** The gap between the data models and types used by the external program and the DBMS hampers the interaction.

Specifically, when an object-oriented program accesses a relational database there is a gap in the interaction: (a) the languages used by the two sides are not of comparable power, and (b) the data types supported by the two sides do not match.

## *7.5.2 Basic Concepts of the Object-Oriented Databases*

The basic concept of an object-oriented database (OO-DB) and the underlying object model is the *object*. An object (e.g., a parcel) has three components (Figure 7.19): (a) *identity*, which is used to identify the object with a unique code, (b) *state*, that refers to the values assigned to the object's attributes (e.g., parcel's area = 1500 m$^2$), and (c) *behavior*, which refers to the operations that can be applied to the object (e.g., divide a parcel into two).

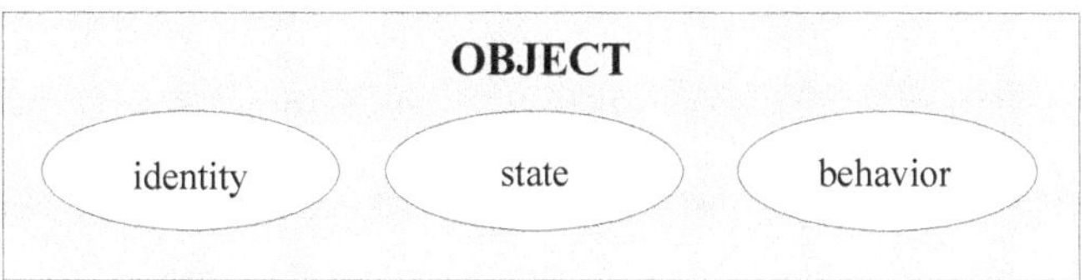


**Fig. 7.19** The components of an object in the object model.

### 7.5.2.1 Object Identity

In an OO-DB, each object has a unique *identity*, which is generated automatically by the system when the object has first registered in the database. The object identity (or oid) is an internal system code that is not visible to the system users.

The system is responsible to assure the *uniqueness* of the identity for each object in the database. Even if the object is deleted from the database, its identity will never be assigned to another object. In addition, the system is responsible to maintain the identity of an object *unchangeable* throughout its life.

### 7.5.2.2 Object State

An object is either *simple* or *composite* (Figure 7.20). In the former case, the object state is described by an atomic value (atom) of a specific data type (integer, real, date, character string, etc.). In the latter case, the object state comprises a set of other simple or composite objects.

A composite object type can be either *tuple* or *collection*. The state of an object of type tuple consists of pointers to the identities of other simple or composite objects of the same or different data types. On the other hand, the state of an object of type collection consists of pointers to the identities of a set of objects of the same data type. This set can be ordered or not. The state of an ordered collection is of type *list* or *array*, while the state of an unordered collection is of type *set* or *bag* (these types are explained later).

The object model allows an indefinite nesting of state types to the generation of composite objects according to the domain needs.

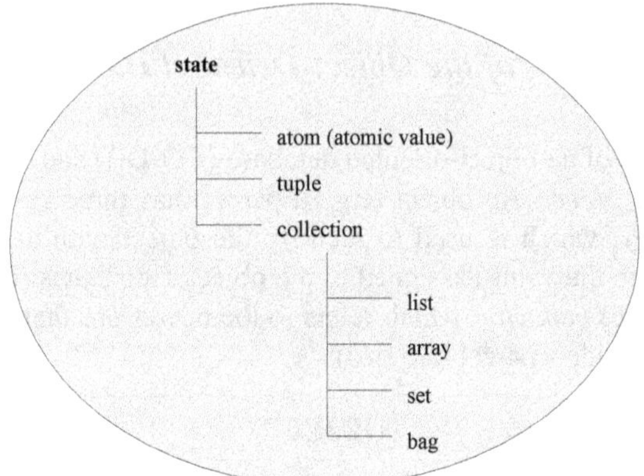


**Fig. 7.20** State types of an object.

The representation of an object in an object-oriented database is realized through the *triplet* (oid, type, value), which defines the object identity, the object type, and the current object value, respectively. Hence, a simple object (*atom*) is represented as: (i, atom, v), where i is the identity, and v the atomic value from the domain of values V. For instance, the triple:

```
(i, atom, 'SMITH')
```

refers to an atom and represents the surname of a person.

A composite object of type *tuple* is represented as: (i, tuple,v), where i is the identity, and v a tuple of n elements of type $<a_1:i_1, a_2:i_2,..., a_n:i_n>$, where $a_k$ is the name of attribute k and $i_k$ is the identity of the object k (for k = 1, ... n).

Notice that the objects of a tuple may be of either the same or different type. For instance the object:

```
(i, tuple, <SIN:i₁, SURNAME:i₂, NAME:i₃, DoB:i₄>)
```

with pointers to the identities of the objects:

```
(i₁, atom, '464197351')      (i₂, atom, 'SMITH')
(i₃, atom, 'JOHN')           (i₄, atom, 08/15/1952)
```

constitutes a composite object-tuple to describe a physical person (owner).

A composite object of state *collection* is represented as: (i, c, v), where i the object identity, c the collection type, i.e., a value form the set {list, array, set, bag}, and v a set of n objects of the same type $<i_1, i_2,..., i_n>$, where $i_k$ the identity of object k (where k = 1, ... n). For instance the object:

```
(i, list, <i₁, i₂, i₃>)
```

with pointers for the identities of objects:

```
(i₁, atom, 'SMITH')            (i₂, atom, 'COOPER')
(i₃, atom, 'THOMPSON')
```

constitutes a composite object of type list, and represent a collection of three person surnames. Notice that the objects $i_1$, $i_2$, $i_3$, ... may be composite (e.g., objects of type tuple) but of the same type.

These objects may also comprise a composite collection of type array, which is represented as follows:

```
(i, array, <i₁, i₂, i₃>)
```

The objects in a collection of type *list* or *array* are ordered. An object of type list is implemented in a linked-list structure, while an object of type array is implemented in a table structure (see Chapter 8). Based on the underlying data structures, the former type does not impose a predefined maximum number of objects in the collection, while the latter does. An example of an ordered collection is the collection of vertices in a polyline (Figure 5.6). The value of either the list or array is an ordered collection of points that comprise the polyline.

When the ordering is not necessary, the collection can be of type *set* or *bag*. The definition of the collection of surnames using these types would be as follows:

```
(i, set, <i₁, i₂, i₃>)
(i, bag, <i₁, i₂, i₃>)
```

Obviously, because the notion of ordering does not apply to collections of type set or bag, the following collections are equivalent by row:

```
(i, set, <i₁, i₂, i₃>)         (j, set, <i₂, i₃, i₁>)
(i, bag, <i₁, i₂, i₃>)           (j, bag, <i₂, i₃, i₁>)
```

On the other hand, the following collections are not equivalent by row:

```
(i, list, <i₁, i₂, i₃>)  (j, list, <i₂, i₃, i₁>)
(i, array, <i₁, i₂, i₃>) (j, array, <i₂, i₃, i₁>)
```

The difference between the alternative types of non-ordered collections, i.e., sets or bags, is that sets do not allow for duplicates, while bags do. In other words, an object may appear only once in a set. On the other hand, it may appear multiple times in a bag, e.g., object $i_2$ appears twice in the following bag:

```
(i, bag, <i₁, i₂, i₃, i₂>)
```

Figure 7.21 presents the state of a composite object, an owner for the simplified cadastral application (Figure 6.9) using the notation of the triplet. As shown in Figure 7.21, all values which refer to the personal data of a person have been stored as atomic objects (atoms). The person's telephones are handled as a collection of type set, while the address as a composite object of type tuple. Finally, the attributes of the person are also represented in a tuple.

The state of the object can also be represented in a graph, as shown in Figure 7.23, using the notation of Figure 7.22.

```
(i₁,atom,'464197351')
(i₂,atom,'SMITH')
(i₃,atom,'JOHN')
(i₄,atom,'08/15/1952')
(i₅,atom,'GEORGE')
(i₆,atom,'191')
(i₇,atom,'FREDERICTON')
(i₈,atom,'NB')
(i₉,atom,'E3A5A3')
(i₁₀,atom,'CANADA')
(i₁₁,atom,'5064555555')
(i₁₂,atom,'5064448888')
(i₁₃,atom,'5069991234')
(i₁₄,tuple,<STREET:i₅, NUMBER:i₆, CITY:i₇, PROVINCE:i₈,
          ZIP:i₉, COUNTRY:i₁₀>)
(i₁₅,set,<i₁₁, i₁₂, i₁₃>)
(i₁₆,tuple,<SIN:i₁, SURNAME:i₂, NAME:i₃, DoB:i₄,
          ADDRESS:i₁₄, TELEPHONES:i₁₅>)
```

**Fig. 7.21** Example object state: A person-owner in the cadastral database.

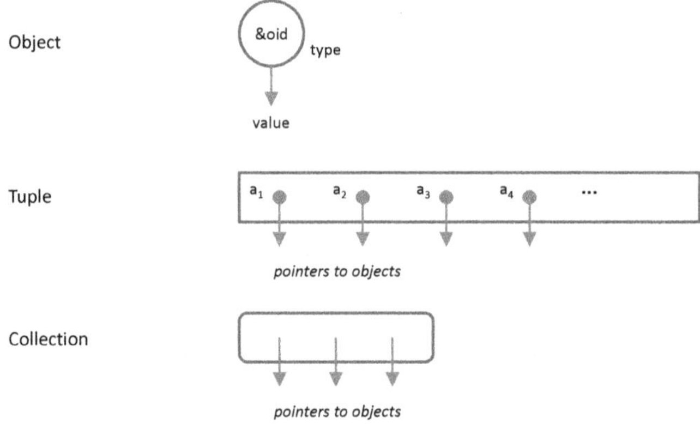


**Fig. 7.22** Notation for representing an object state in a graph.

### 7.5.2.3 Object Behavior

The *behavior* of an object has to do with the operations that can be performed on the object. These operations can result in the creation or deletion of an object. They can also access, process or modify the state of an existing object.

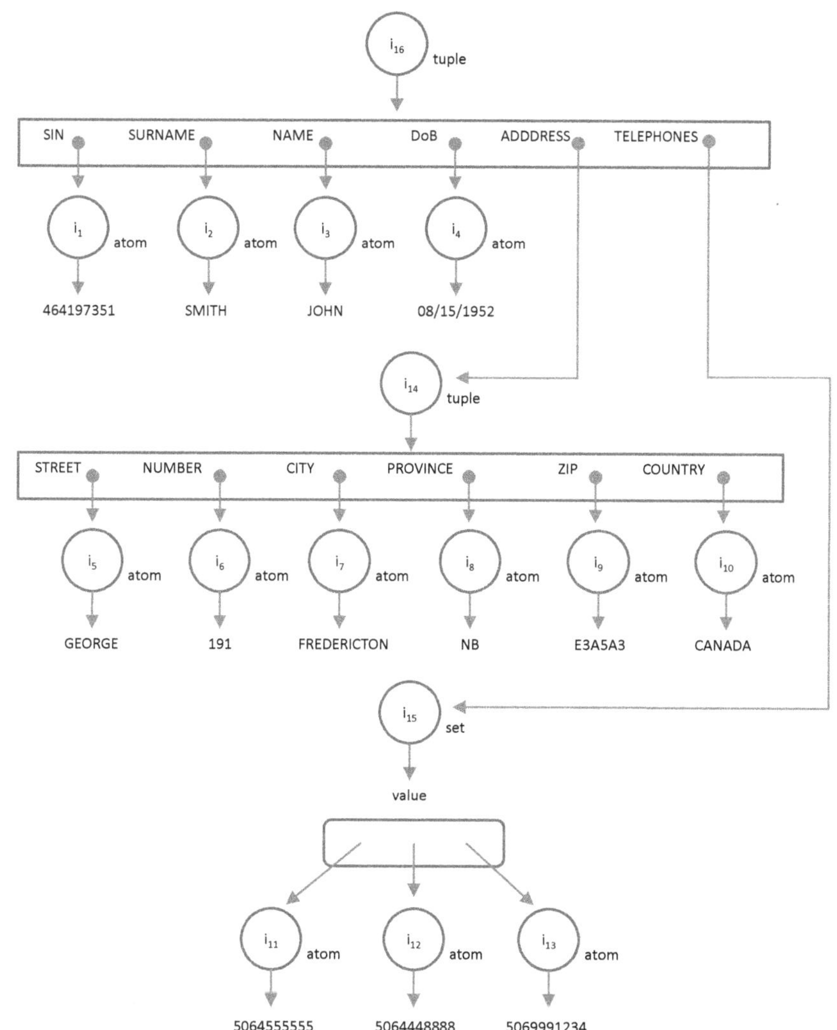

**Fig. 7.23** An object state (Figure 7.21) represented in a graph form.

The names and parameters of the operations of an object (as shown to the users) form the object *interface*. The interface part for an operation is called *signature* of that operation. The implementation of an operation in a programming language, such as C++ or Java, is called *method*. A method is activated when a *message* is sent to the object.

The behavior is an integral part of the object definition. The integration of the methods in an object is called *encapsulation*. The encapsulation is either total or partial. In a *total* encapsulation, the user can access the object state (i.e., its attributes) only through predefined operations, designed for this purpose. On the other hand, in a *partial* encapsulation, the user is still allowed to access some attributes even when the methods are not available. These attributes are called *visible attributes*. Attributes which cannot be accessed unless an appropriate method is available are called *hidden attributes*. An attribute is characterized as visible or hidden in the definition of the object state.


### 7.5.2.4 Object Classes and Relationships

*Class* is any object type in its entirety, i.e., the state (attributes) and the behavior (operations). For example, the object type in Figure 7.24 is an object class for the owners in the simplified cadastral application. The state of the class is the one shown in Figure 7.23.

The behavior refers to the operations to create a new owner (newOwner; returns true if successfully created or false otherwise), remove an owner (remOwner; returns true if successfully removed or false otherwise), change an owner's address (modOwnAddr; returns true if successfully modified or false otherwise), count the number of owners (numOfOwn; returns the number of owners in the database), derive the age of an owner (ageOwner; returns the owner's age), retrieve the telephone numbers of an owner (telOwner; returns a set of telephone numbers). Notice that the notation of the class in Figure 7.24 does not follow any standardization.

```
Object Class Owners:
   State
      tuple (
         sin            string
         surname        string
         name           string
         dob            date
         address        tuple (
                           street        string
                           number        integer
                           city          string
                           province      string
                           zip           string
                           country       string )
         telephones  set (
                           telephone     integer )
      )
   Behavior
      newOwner       boolean
      remOwner       boolean
      modOwnAddr     boolean
      numOfOwn       integer
      ageOwner       integer
      telOwner       set(integer)
```

**Fig. 7.24** Example object class: Owner.


Figure 7.25 presents the class of parcels in the cadastral application. The state of a parcel consists of four atomic attributes (Figure 6.6): id (a unique identifier), use (the land use), built_factor (specifies the construction rules), and address (a string for the parcel indirect location). The behavior of the parcel includes the following operations: newParcel (creates a new parcel), remParcel (removes a parcel from the database), useParcel (returns the land use of the parcel).

```
Object Class Parcels:
   State
      tuple (
         id              string
         use             string
         built_factor    decimal
         address         string
      )
   Behavior
      newParcel          boolean
      remParcel          boolean
      useParcel          string
```

**Fig. 7.25** Example object class: Parcel.

As described in Chapter 6, there is a relationship type (TITLES) between the two classes defined previously (Figures 7.24 and 7.25). A parcel belongs to one or more owners and an owner may have one or more parcels.

The representation of a relationship between two classes is realized by inserting the *reference attributes* from one class to the other, as shown in Figure 7.26. The values assigned to these attributes in one class are used to identify specific objects in the other class. A relationship type of degree two (Chapter 6) can be represented as a *single-directional* (Figure 7.26a,b) or *bi-directional* (Figure 7.26c) relationship. In the latter case, the reverse references must be consistent. A bi-directional relationship is declared using the keyword *inverse*, as shown in Figure 7.26c. Notice that the relationship type in Figure 7.26 has a cardinality ratio N:M. This is the reason of using the collection type set.

Should the relationship type have attributes, as occurs in the type of TITLES (percentage and purchase date; see Figure 6.6), these attributes characterize the corresponding reference. Figure 7.27 presents the definition of the reference attribute of Figure 7.26a along with the relationship type attributes (as a tuple).

### 7.5.2.5 Inheritance

In object-oriented databases new object classes can be defined based on existing classes by simply adding new attributes or operations. This practice results in the generation of *class hierarchies*, similar to the entity hierarchies described in the Enhanced Entity-Relationship model earlier in this Chapter. In a class hierarchy, the concept of *inheritance* for both the state (attributes) and behavior (operations) from the superclass to the subclasses also apply. Each subclass can also have additional attributes and methods.

A class in the object-model may inherit the state and behavior from more than one superclasses. This is the concept of *multiple inheritance*, which extends the class hierarchies to *class networks*.

In a class network, it is not always clear how the state and/or behavior of the superclasses are inherited by the subclasses. For instance, a subclass may inherit from two discrete superclasses, two distinct attributes or methods with the same name (*attribute/method synonymy*). This type of problem can be resolved after an appropriate action is taken by the database user.

To solve the problem of synonymy, the user may decide to change the attribute or method name at the subclass level. Notice that the object model allows the *renaming* of the attributes and methods that are inherited from another classes to better reflect their meaning (enriched semantics).

Closing this Section, it is worthy to mention the concept of *selective inheritance* in a class hierarchy. According to this, a subclass may inherit part of the state and/or the behavior of its superclass.

```
Object Class Owners:
   State
      tuple (
          sin          string
          ...
          owns         set(Parcels) )
   Behavior
      ...

Object Class Parcels:
   State
      tuple (
          id           string
          ...   )
   Behavior
      ...

            (a)

Object Class Owners:
   State
      tuple (
          sin          string
          ...   )
   Behavior
      ...

Object Class Parcels:
   State
      tuple (
          id           string
          ...
          ownedBy      set(Owners) )
   Behavior
      ...

            (b)

Object Class Owners:
   State
      tuple (
          sin          string
          ...
          owns         set(Parcels) inverse Parcels:ownedBy )
   Behavior
      ...

Object Class Parcels:
   State
      tuple (
             id         string
             ...
             ownedBy    set(Owners) inverse Owners:own )
   Behavior
      ...
            (c)
```

**Fig. 7.26** Alternative representations of a relationship type of degree two with single-directional (a), (b) and bi-directional references (c).

```
Object Class Owners:
   State
      tuple (
         sin    string
         ...
         owns   set(
                   tuple (
                      parcel         Parcels
                      percentage     integer
                      parchase_date  date ) )
      Behavior
         ...


Object Class Parcels:
   State
      tuple (
            id        string
            ...
      Behavior
         ...
```

**Fig. 7.27** Integration of the relationship type attributes in the reference.


## 7.5.2.6 Advantages of the Object-Model

The object-model eliminates or diminishes the deficiencies of the relational model that were described earlier in this Chapter. Specifically, with respect to *efficiency*, the object model may accelerate the processing, due to the encapsulation of the methods in the object classes and the implementation of relationships through direct references. As shown in Figure 7.28, the data related to an object (entity) is clustered or linked through direct references, as opposite to the relational model, where it is usually split into multiple relations that need to be re-joined. The retrieval of a composite object in the object model is a matter of posting a single message.

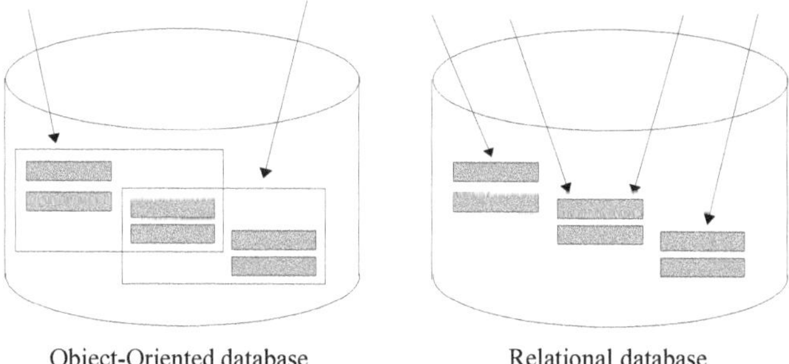


**Fig. 7.28** Schematic representation of object retrieval in the object and the relational model (after Lee).

As for the *semantics*, the object-model, in contrast to the relational model, has a clear distinction between the three types of relationships: association, generalization and aggregation. Additionally, the object-model is *extensible*, provided that the object classes can be used as constructs to other classes, similarly to the *abstract data types (ADT)* in programming.

Each object (instance) in an object-oriented database system has a *unique identity*. The identity is created by the system and it is independent on the physical address or the values attached to the object attributes. Hence, there is not any risk of losing the identity when an update or restructuring takes place in the database. Finally, the object model adopted in object-oriented databases shares the same concepts with the model applied in object-oriented programming. Therefore, the *communication/interface* between an object-oriented database system and any external program is smooth and of equivalent power.

On the other hand, the object-model has its own deficiencies. Among them are the lack of widely accepted specifications, the limitations imposed by the definition of the object behavior, the complexity of the model (complicated concepts difficult to be understood by non-programmers), etc. For these reasons, modern database systems adopt a hybrid model that combines concepts from both the relational and object models. This hybrid model is called *object-relational model* and will be discussed in Chapter 15.

### 7.5.2.7 Object-Oriented Database Design

As described in Chapter 6, relational databases are based on the concepts of relational algebra and make use of the SQL language to define and manipulate the content. A similar set of specifications for object-oriented databases has been introduced by Object Data Management Group (ODMG), a working group of object-oriented developers. This group introduced the first version of the ODMG standard for object-oriented database in 1993. Two new versions were introduced later, the ODMG 2.0 (1997) and the ODMG3.0 (2000). In 2001 the group disbanded.

The *ODMG specification* consists of three parts: (a) the *object model*, (b) the *object definition language*, and (c) the *object query language*. The following Sections briefly present these parts. More details can be found in the ODMG 3.0 specification.

7.5.2.7.1 The Object Model

The basic constructs of the object model are: (a) *objects*, and (b) *literals*. An *object* is described by: (i) *identity*, (ii) *name*, (iii) *duration*, and (iv) *structure*. The *identity* is system generated. The *name* is an optional identifier generated by the user (person or external program). The *duration* denotes whether the object is stored in the database (permanent object) or it just shows up during the execution of a program and disappears after its completion (non-permanent object). Finally, the *structure* describes the object state and behavior.

A *literal*, on the other hand, is a state (value), which has not any identity attached to it. The state of a literal can be: (i) *atomic* (atomic literal), (ii) *tuple* (tuple literal), or (iii) *collection* (collection literal).

The ODMG model uses the term *interface* to describe the attributes, the relationships and the methods of an object class. The interfaces define a series of methods which can be inherited from the object classes defined by the users and referring to the application domain. These classes are denoted with the term *class* in the ODMS model.

### Built-In Interfaces

The ODMG object model has a set of *built-in interfaces*. The very basic interface, which is inherited by all objects, is the *interface Object*, which is defined (part of it) in Figure 7.29. Hence, all objects inherit the following methods from this interface: (a) copy, which creates a replica of the object, (b) delete, which removes the object, etc.

```
interface Object {
    ...
    Object    copy();
    void      delete();
}
```

**Fig. 7.29** Definition of the interface Object (part of).

The ODMG object model has a set of *built-in interfaces*, which are described in the corresponding specifications. Figure 7.30 presents the hierarchy that applies to some built-in interfaces of the object model.

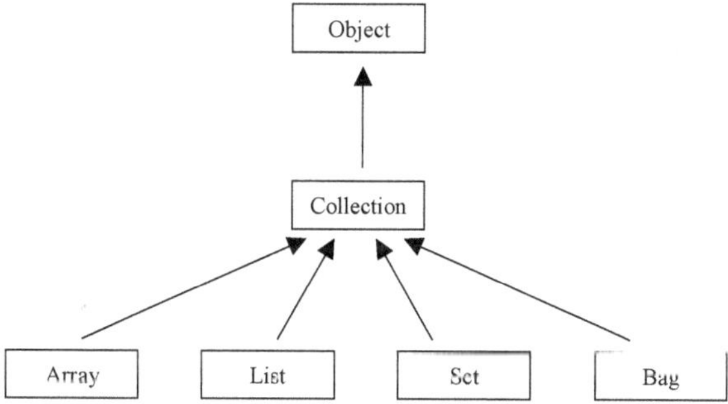


**Fig. 7.30** Hierarchy of some of the built-in interfaces in the object model.

Figure 7.31 presents a part of the interfaces *Object, Collection,* and *Set.* The *colon* ':' denotes a relationship of type *class-subclass* in the ODMG model. Specifically, as shown in Figure 7.30, the interface Set is a subclass of Collection, which is in turn a subclass of Object. The *period* '.' denotes the execution of a *method* in the object. For instance, given an object $o_1$, the notation: $o_2 = o_1.copy()$, creates a replica of $o_1$, the object $o_2$.

The interface of *collection* (Collection) includes an *exception* (Figure 7.31). The exception is used in the ODMG model to denote an error or violation during the execution of a method. Specifically, the method remove_element, which is used to remove an object instance from a collection, raises an exception whenever the instance is missing from the collection.

The interface of collection also includes the *methods*: is_empty (which returns true if the collection is empty and false otherwise), contains_element (which takes as an argument an object instance and returns true if the object is present in the collection or false otherwise), insert_element (which takes as an argument an object instance and inserts it in the collection). The interface of collection also *inherits* the methods from the interface Object (i.e., copy, delete, etc.).

The methods of the interfaces Object and Collection are inherited by the *interface of set* (Set), which in turn introduces its own methods. Figure 7.31 presents two of them: create_union (which integrates in a set the instances of another set), and is_subset_of (which takes as argument a set and returns true if the set is a subset of the initial set or false otherwise).

```
interface Object {
   ...
   Object   copy();
   Void     delete();
};

interface Collection : Object {
   ...
   exception   ElementNotFound(any element);
   boolean     is_empty();
   boolean     contains_element(in any element);
   void        insert_element(in any element);
   void        remove_element(in any element) raises(ElementNotFound);
};

interface Set : Collection {
   ...
   Set         create_union(in Set other_set);
   boolean     is_subset_of(in Set other_set);
};
```

**Fig. 7.31** Definition (part of) of three built-in interfaces in ODMG model.


*User-defined Objects – Classes*

As stated previously, any user-defined object type in the ODMG model is denoted as *class*. Each *class* has three components: (i) *attributes*, (ii) *relationships*, and (iii) *methods*. The attributes accommodate the values that describe the object and they are literals of simple or composite structure. The attributes of a class are denoted with the keyword *attributes*. The relationships between the objects in the database are denoted with the keyword *relationship*. The ODMG model represents explicitly only binary relationships (relationships of degree two) and this is achieved through pairs of reverse references. The maintenance of the integrity of references is a responsibility of the system and denoted with the keyword *inverse*. The operations that can take place over an object are denoted with the keyword *methods*. The signature of each method includes its name, its arguments (input parameters), and its values (output after execution).

Figure 7.32 presents the classes of owners and parcels described previously (Figures 7.24, 7.25, 7.26c, and 7.27) as defined in the ODMG model using the *Object Definition Language (ODL)*.

The definition of a *class* in ODL consists of three parts: (i) *class head*, (ii) *attributes and relationships*, and (iii) *methods*. In the *class head* the *name* of the class, its *subclasses* (if any), its *extensions* (if any), and its *key attributes* are listed.

Each class in the ODMG model is assigned an object, which behaves as a whole and includes all objects in the class. This object is denoted as *extent* and has a name. In the example of Figure 7.32, the extents of the two classes are named as all_owners and all_parcels. These two extents represent all object instances in the database. The *key attribute* of a class consists of one or more attributes, whose values identify the object instances. This attribute is also called *external or user-defined key* and identifies the database instances as of the user. In the classes Owners and Parcels the external keys are the attributes sin and parcel id, respectively. As stated previously, each object instance is assigned an *internal or system-key* (oid) when it is created. This key is not accessible by the users.

```
class Owners
( extent      all_owners
  key         sin )
{
  attribute   string   sin;
  attribute   string   surname;
  attribute   string   name;
  attribute   date     date_of_birth;
  attribute   struct   address(string street, string number, string city, string province,
                               string zip, string country);
  attribute   set<integer> telephones;
  relationship set<struct titles(Parcels parcel, integer percentage, date purchase_date)> owns
                          inverse Parcels::ownedBy;
  boolean     newOwner(in Owners);
  boolean     remOwner(in string sin)
                 raises(sin_not_valid);
  boolean     modAddress(in string street, in string number, in string city, in string province,
                         in string zip, in string country);
  integer     numOwvers();
  integer     ageOwner(in string sin);
  set<integer> telOwner(in string sin);
};

class Parcels
( extent      all_parcels
  key         id )
{
  attribute   string   id;
  attribute   decimal  built_factor;
  attribute   string   use;
  attribute   string   address;
  relationship set<struct titles(Owners owner, integer percentage, date purchase_date)> ownedBy
                          inverse Owners::owns;
  boolean     newParcel(in Parcels);
  boolean     remParcel(in string id) raises(id_not_valid);
  string      useParcel(in string id);
};
```

**Fig. 7.32** The classes of owners and parcels for the cadastral application as defined in ODMG (using ODL).

In the part of *class attributes and relationships*, the data types of the class attributes are defined along with the class relationships. The keyword *inverse* denotes the reverse relationship references. As shown in Figure 7.32, the reference owns is defined from the class Owners to the class Parcels, while the inverse reference ownedBy is defined from the class Parcels to the class Owners. These references are accompanied by the *relationship attributes* (percentage and purchase_date of the title).

The last part includes the *class methods*, their arguments as well as the names and types of the output values. Figure 7.32 presents some method definitions for the classes of Owners and Parcels.

*Inheritance*

The ODMG model distinguishes two types of inheritance: (a) *inheritance of be-havior from interfaces*, and (b) *inheritance of state and behavior from classes*. The first type has been described previously with the use of notation of colon ':'. Notice that the inheritance from an interface to another interface or class refers to the methods only and not the state of the interface.

On the other hand, the second type of inheritance offers the option to create a class hierarchy, where all three parts of the superclass (attributes, relationships, and methods) can be inherited by the subclass. The keyword *extends* is used to denote an inheritance from classes.

Figure 7.33b presents an example inheritance of behavior from an interface, while Figure 7.33c shows an example of inheritance of state and behavior from a class. The notation used is given in Figure 7.33a.

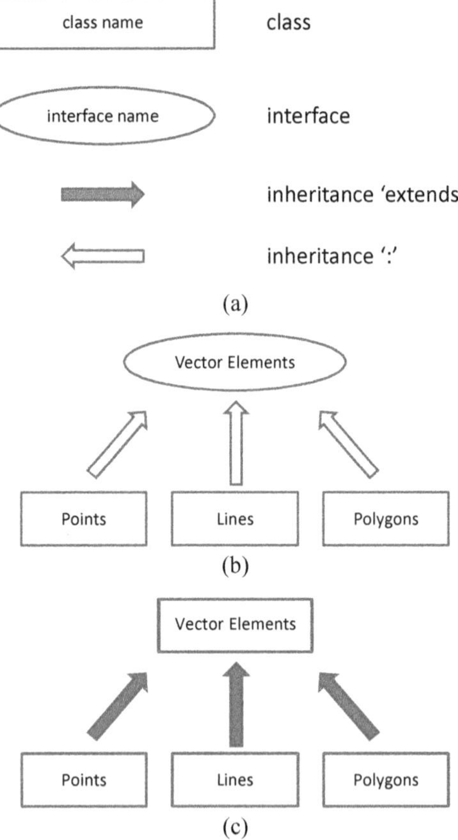


**Fig. 7.33** Diagrammatic representation of the two inheritance types.

Figure 7.34 presents the ODMG definitions (in ODL) for the first scenario (Figure 7.33b). Three classes *Points*, *Lines*, and *Polygons*, which represent the basic geometric types on a plane surface, inherit the behavior of the interface *Vectors*. The inheritance from the interface is denoted with the colon ':' at the head of the classes.

Figure 7.35 shows an alternative representation of the base geometric types with classes only and the definition of a class-subclass relationship type (i.e., inheritance of state and behavior) denoted by 'extends'.

Notice that the ODMG model does not allow for multiple inheritance through extends. On the other hand, it allows for multiple inheritance through ':'. This means that a class can participate in one only class-subclass relationship in the role of the subclass. However, it can inherit the behavior form multiple interfaces.

```
interface VectorElements
{
   attribute    enum elements(Points, Lines, Polygons);
   attribute    struct origin(integer x, integer y);
   attribute    struct azimuth(integer x, integer y);
   void         transition(in integer transition_X in integer transition_Y);
   void         rotation(in float rotationAngle_X, in float rotationAngle_Y);
   void         scale(in float scaleFactor_X, in float scaleFactor_Y);
};

class Points:VectorElements
( extent        all_points    )
{
   attribute    struct origin(integer x, integer y);
   attribute    struct azimuth(integer x, integer y);
   attribute    integer  x;
   attribute    integer  y;
}

class Lines:VectorElements
( extent        all_lines    )
{
   attribute    struct origin(integer x, integer y);
   attribute    struct azimuth(integer x, integer y);
   attribute    integer  x1;
   attribute    integer  y1;
   attribute    integer  x2;
   attribute    integer  y2;
   float        length();
}

class Polygons:VectorElements
( extent        all_polygons    )
{
   attribute    struct origin(integer x, integer y);
   attribute    struct azimuth(integer x, integer y);
   attribute    list<struct vertex(integer x, integer y)>;
   float        perimeter();
   float        area();
}
```

**Fig. 7.34** The definition of interfaces and classes (Figure 7.33b) in ODMG (using ODL).

### 7.5.2.7.2 Object Definition Language (ODL)

The *object definition language (ODL)* is the second component of the ODMG model. ODL has been designed to support the definition of objects according to the ODMG model and is independent of any programming language. Its basic functionality has to do with the definition of interfaces and classes. The examples of the previous Section are compliant to the ODL specification. More details can be found in the ODL specification.

### 7.5.2.7.3 Object Query language (OQL)

The *object query language (OQL)* is the third component of the ODMG model. OQL is compliant with the object-oriented programming languages, such as C++, Java, and Smalltalk. An OQL query can explicitly be embedded into the code of external programs and offer direct access and manipulation of the object-oriented database content.

The structure of OQL statements is similar to that in SQL, i.e., it consists of the three main clauses: SELECT... FROM... WHERE...; with the appropriate extensions to include other ODMG concepts (like methods, inheritance, etc.).

Next, some basic features of the OQL are presented through simple examples. The queries refer to the object classes Owners and Parcels, as they were defined in Figure 7.32. For an extended description of OQL the reader may refer to the OQL specification.

```
class VectorElements
(  extent      all_vectorelements
   key         id )
{
   attribute   string   id;
   attribute   enum element(Points, Lines, Polygons);
   attribute   struct origin(integer x, integer y);
   attribute   struct azimuth(integer x, integer y);
   void        transition(in integer transition_X in integer transition_Y);
   void        rotation(in rotationAnge_X, in float  rotationAngle_Y);
   void        scale(in float scaleFactor_X, in float scaleFactor_Y);
};

class Points extends VectorElements
(  extent      all_points )
{
   attribute   integer x;
   attribute   integer y;
}

class Lines extends VectorElements
(  extent      all_lines )
{
   attribute   integer  x1;
   attribute   integer  y1;
   attribute   integer  x2;
   attribute   integer  y2;
   float       length();
}

class Polygons extends VectorElements
(  extent      all_polygons   )
{
   attribute   list<struct vertex(integer x, integer y}>;
   float       perimeter();
   float       area();
}
```

**Fig. 7.35** The definition of classes (Figure 7.33c) in ODMG (using ODL).


The simplest type of queries in OQL language is the retrieval of attribute values from an object class. For instance, the following query retrieves the sin, surname, and name of all owners born before 1960.

```
SELECT struct<d.sin, d.surname, d.name>
FROM d in all_owners
WHERE d.DoB < 01/01/1960;
```

In this query a search will apply to all owner instances stored in the database. According to the class definitions (Figure 7.32), the set of owners is represented in the collection all_owners of the class Owners. The variable d scans the instances of the collection in a sequence and this is defined in the FROM clause of the query. In the SELECT clause, the attributes values to be reported are listed, while in the WHERE clause the conditions (filters) of the query are stated.

Notice that the result of a general OQL query is usually of type bag<tuple>. Specifically, the query above returns a set<tuple>, as the sin is the key attribute of the class Owners (no duplicates are expected).

Each OQL query requires the presence of a *persistent object* with a name. This object can be of any type and is called *entry point* to the database. In the previous query, the entry point to the database is the collection all_owners. Apart from the typical structure SELECT... FROM... WHERE..., OQL can also handle any expression that includes an entry point. For instance, the following expression is an OQL query:

```
all_owners;
```

This query, when executed, a reference to the collection of all persistent objects in the class of owners is returned. The type of the result is set<Owners>.

The following expressions are also OQL queries:

```
464197351.surname;
464197351.address.city;
464197351.telephones;
464197351.titles.owns.parcel;
```

464197351 is the name (owner's SIN) of an object in the object class Owners. Specifically, as shown in Figure 6.9, it is the SIN of Mr. Smith. Hence, the expressions above return the following, respectively: (a) the value of the attribute surname, i.e., 'SMITH' (type: <string>), (b) the value of the attribute city in the address tuple attached to the object, i.e., 'FREDERICTON' (type: <string>), (c) the values in the set of telephones, i.e., {'5064555555', '5064448888','5069991234'} (type: <set>), and (d) the parcels owned by the owner, i.e., {75071217} (type: <set>; in this example, Mr. Smith owns only one parcel).

As shown in the previous examples, the period '.' is used to represent the *nesting* of attributes into composite objects. Each expression of this type is called *path* and is used to indicate the appropriate entry point in an OQL query.

The Section concludes by presenting two more query expressions, which return the same result, i.e., the set of parcels owned by Mr. Smith with a built factor greater than 1.0:

```
SELECT d.id
FROM d in 464197351.titles.owns.parcels
WHERE d.built_factor > 1.0
ORDER BY d.id;
```

```
SELECT d.id
FROM d in all_parcels
WHERE d.titles.ownedBy.owners.sin = '464197351'
AND d.built_factor > 1.0
ORDER BY d.id;
```

## 7.5.3 Mapping the EER-Diagram to the Object Schema

The enhanced entity-relationship (EER-) model describes the database entities and relationships at the conceptual level. As a model, it adopts many concepts from the object model, such as those of class and inheritance. In a previous Section, the rules to map an EER-diagram to the relational schema were discussed. A set of rules can also be applied to map the EER-diagram to the object schema. These rules are briefly presented next. For simplicity, the cases of categories, relationship types of degree higher than two, and weak entities are not considered. For these cases, the reader may refer to the literature listed at the end of this Chapter.

Notice that an EER-schema describes only the state and ignore the behavior of the database objects. Hence, the rules for mapping an EER-diagram to the object schema only consider the state of the objects. The resulting schema must then be enriched with the methods that describe the behavior of the objects in the database.

Figure 7.36 shows a part of the EER-diagram for the cadastral application discussed in the previous Sections. This diagram can be easily mapped into an object schema by applying the following rules.

### 7.5.3.1 Rule 1: Mapping the Entity Types

The first rule has to do with the *mapping of the entity types* in the EER-diagram to the object schema. This rule consists of the following steps:

i.    for each entity type create an object class with all simple attributes represented as atomic attributes in the object schema

ii.    represent the multivalued attributes with the appropriate collection type, i.e., set, bag or list, depending on whether the values need to be ordered (list), non-ordered (set) or non-ordered with repetitions (bag) (this information is not recorded explicitly in the EER-diagram).

iii.    represent all composite attributes as records using the composite type tuple (or struct)

Figure 7.37 shows an example of these steps for the EER-diagram in Figure 7.36. Four object classes have been created to represent the four entity types in the EER-diagram.

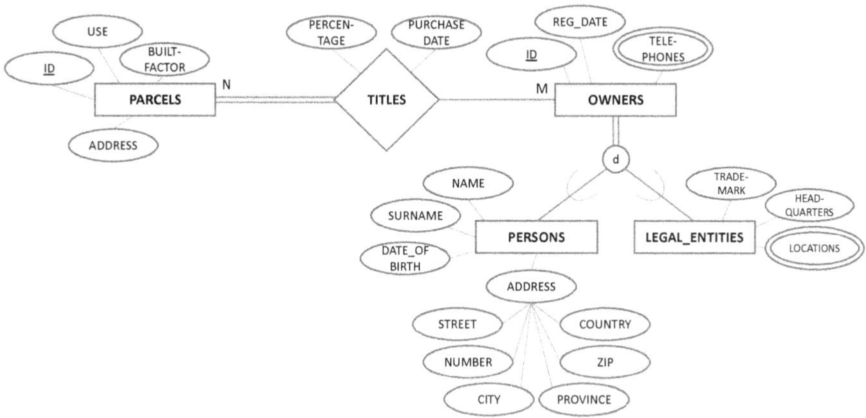



**Fig. 7.36** A snapshot of the EER-diagram for the cadastral application example.

```
class Parcels
( extent      all_parcels
  key         id)
{
  attribute   string      id;
  attribute   decimal     built_factor;
  attribute   string      use;
  attribute   string      address;
};

class Owners
( extent      all_owners
  key         sin)
{
  attribute   string      id;
  attribute   date        reg_date;
  attribute   set<integer> telephones;
};

class Persons
( extent      all_persons   )
{
  attribute   string      surname;
  attribute   string      name;
  attribute   date        dob;
  attribute   struct      address(string street, string number, string city, string province,
                              string zip, string country);
};

class Legal_Entities
( extent      all_legal_entities   )
{
  attribute   string      trademark;
  attribute   string      headquarters;
  attribute   set<string> locations;
};
```

**Fig. 7.37** The 1st Rule for mapping the EER-diagram (Figure 7.36) to the object schema.


## 7.5.3.2 Rule 2: Mapping the Relationship Types of Degree Two

This rule has to do with the mapping of *relationship types of degree two* in the EER-diagram to the object schema. This is achieved by adding the corresponding reference attributes (denoted by the keyword: relationship) to the object classes participating in a relationship type. The reference attributes may be either single or bi-directional. This rule consists of the following steps:

iv.     add the reference attributes to the object classes (both or one of them) that represent entity types and participate in relationship types. Depending on the cardinality ratio of the relationship type, the corresponding reference attribute must be of type: (a) atomic for relationships 1:1, (b) collection (set or list) for relationships N:M, or (c) composite for relationships 1:N, with atomic type at the N-side, and collection type at the opposite side (1).

v.     add the attributes of the relationship type to the object schema. If there are attributes attached to the relationship type, use a tuple with the following structure: <reference attribute, {relationship attributes}>. The bi-directional representation of the relationship types causes redundancies, as the relationship attributes are stored twice.

Figure 7.38 presents the result after executing step (iv) for the EER-diagram in Figure 7.36. Specifically, the classes Owners and Parcels (Figure 7.37) have been modified, so that their relationship is represented. A bi-directional relationship is applied. Figure 7.39 presents the same classes with the relationship attributes integrated (step v).

### 7.5.3.3 Rule 3: Encapsulation of Methods

This rule corresponds to the *encapsulation of methods* in the object classes created from the first rule. These methods are not part of the EER-diagram, but they should be defined during the analysis of the system functionality (Figure 6.1: processing requirements). This rule consists of the following steps:

vi.     add the appropriate methods to each class.

vii.     write and compile the program code to implement these methods.

viii.     embed in the program code all modules that assure the database integrity, e.g., in the creation or deletion of an object.

```
class Parcels
(  extent          all_parcels
   key             id )
{
   attribute       string      id;
   attribute       decimal     built_factor;
   attribute       string      use;
   attribute       string      address;
   relationship    set<Owners> ownedBy inverse Owners::owns;
};

class Owners
(  extent          all_owners
   key             id )
{
   attribute       string      id;
   attribute       date        reg_date;
   attribute       set<integer> telephones;
   relationship    set<Parcels> owns inverse Parcels::ownedBy;
};
```

**Fig. 7.38** Rules 1 and 2 in mapping the EER-diagram (Figure 7.36) to the object schema. Modification of the classes Owners and Parcels (Figure 7.37).

Figure 7.40 presents the methods that have been embedded in the class Owners. These methods have already been described in the previous Section.

```
class Parcels
( extent          all_parcels
  key             id)
{
  attribute       string      id;
  attribute       decimal     built_factor;
  attribute       string      use;
  attribute       string      address;
  relationship    set<struct  titles(Owners owner, integer percentage, date purchase_date)>
                  ownedBy inverse Owners::titles;
};

class Owners
( extent          all_owners
  key             id)
{
  attribute       string      id;
  attribute       date        reg_date;
  attribute       set<integer> telephones;
  relationship    set<struct  titles(Parcels parcel, integer percentage, date purchase_date)>
                  owns inverse Parcels::ownedBy;
};
```

**Fig. 7.39** The 2$^{nd}$ step of rule 2 in mapping the EER-diagram (Figure 7.36) to the object schema. Modification of the classes Owners and Parcels (Figure 7.38).

```
class Owners
( extent          all_owners
  key             id)
{
  attribute       string      id;
  attribute       date        reg_date;
  attribute       set<integer> telephones;
  relationship    set<struct  titles(Parcels parcel, integer percentage, date purchase_date)>
                  owns inverse Parcels::ownedBy;
  boolean         newOwner(in Owners);
  boolean         remOwner(in string id)  raises(sin_not_valid);
  integer         numOwners();
  set<integer>    telephones(in string id);
};
```

**Fig. 7.40** Rule 3 in mapping the EER-diagram (Figure 7.36) to the object schema. Modification of the class Owners by adding the methods (behavior) (Figure 7.39).


### 7.5.3.4 Rule 4: Mapping the Superclass-Subclass Relationships

This rule refers to the *mapping of the superclass-subclass relationships* of the EER-diagram to the object schema. This rule consists of the following step:

ix.    define each superclass-subclass relationship of the EER-diagram to the object schema through the relationship extends in the corresponding classes defined by the first rule.

Figure 7.41 presents the two subclasses of the EER-diagram in Figure 7.36 in the object schema.

```
class Persons extends Owners
( extent          all_persons )
{
   attribute    string       surname;
   attribute    string       name;
   attribute    date         dob;
   attribute    struct       address(string street, string number, string city, string province,
                                     string zip, string country);
};

class Legal_Entities extends Owners
( extent       all_legal_entities   )
{
   attribute    string       trademark;
   attribute    string       headquarters;
   attribute    set<string> locations;
};
```

**Fig. 7.41** Rule 4 in mapping the EER-diagram (Figure 7.36) to the object schema. The classes Persons and Legal_Entities are denoted as extends Owners (Figure 7.39).

## 7.6 Conceptual Database Design using UML

The *Unified Modeling Language (UML)* is a powerful and widely used specification for designing object-oriented software. UML provides many tools that are also valuable in the conceptual database design. The following subsections give a brief introduction to the *UML class diagrams*. A series of examples for the cadastral application are also presented. The Section is concluded by a comparison between the UML class diagrams and EER-diagrams.

### *7.6.1 UML Class Diagrams*

UML class diagrams were designed to support the object modeling. For this purpose, they include the following concepts: (a) *class*, (b) *attribute*, (c) *method*, (d) *inheritance*, and (e) *relationship*.

*Class* in a UML class diagram is a collection of objects in a domain sharing the same properties. These properties refer to the object attributes, the methods, and the relationships of the object with other classes (collections of objects) in the domain. In a UML class diagram, a class is represented with a frame of three sections. The top section denotes the class name. The middle section denotes the attributes of the objects in the class. The bottom section lists the methods (operations) that can be applied to the objects in the class. Figure 7.42 shows a representation of the class Owners in an EER-diagram and in a UML class diagram.

The *class attributes* describe the objects in the class. Example attributes in the class Owners (Figure 7.42b) are the owner's name and address. Each attribute has a *level of accessibility*, which defines whether the attribute is accessible (visible) from the class itself or other classes in the domain. There are three levels, which characterize an attribute as: (a) *public*, (b) *private*, and (c) *secured*.

An attribute is *public* when it is accessible from any class in the database. An attribute is *private* when it is accessible from its own class only. An attribute is *secured* when it is accessible from the subclasses of its class only. In a UML class diagram a special notation is applied to denote the level of accessibility. Specifically, the symbols '+', '−', and '#' are used to denote the public, private and secured attributes, respectively. Figure 7.43 shows an example of the class Owners with the level of accessibility for its attributes denoted.

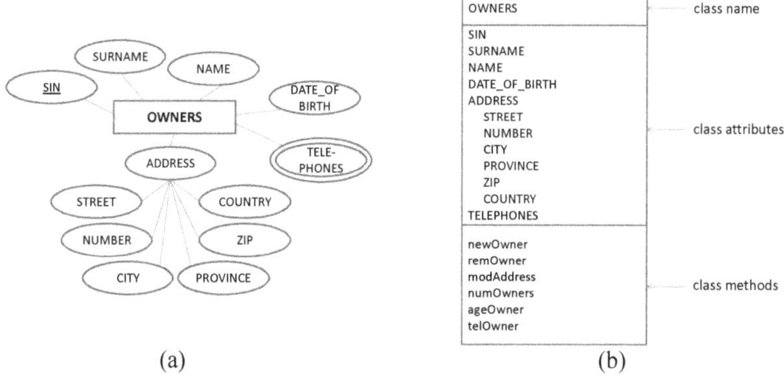



**Fig. 7.42** Representation of the class Owners in an EER-diagram (a) and a UML class diagram (b).

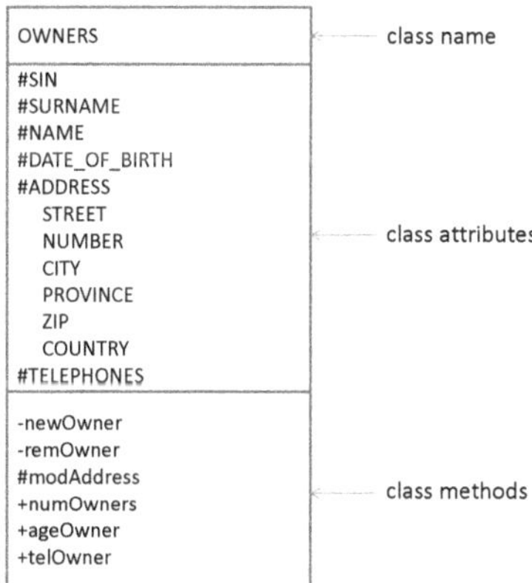



**Fig. 7.43** The class Owners with the levels of accessibility for its attributes and methods.

The *methods* of a class constitute the operations (functions) that can be applied on the class attributes. As described previously, these methods are part of the object definition. In a UML class diagram, the methods are also characterized as public, private or secured, depending on their accessibility level (Figure 7.43).

The *relationships* in a UML class diagram relate the classes to represent the application semantics. UML class diagram distinguish three types of relationships: (a) *associations*, (b) *aggregations*, and (c) *generalizations*.

An *association* represents the relationship between two distinct classes. An association can be of degree two or higher, depending on the number of classes that participate in it. The notation for associations of degree two in a UML class diagram is through a line segment connecting the two classes. Optionally, the line is assigned a name. Should the association have attributes, these are listed in a separate frame connected to the association line through a dashed line. The cardinality ratio of the classes participating in an association is denoted with the notation 'min…max' beside the class frame.

Figure 7.44 shows an example association (relationship type) between two classes (entity types) in the cadastral application as represented in an ER-diagram and in a UML class diagram. Figure 7.45 shows some example associations with different cardinality ratios and participation constraints in ER-diagrams and UML class diagrams. These examples were already discussed in Figure 6.5. Notice that the cardinality ratio is assigned a 0 and 1 in the min value to represent the partial and total participation, respectively.

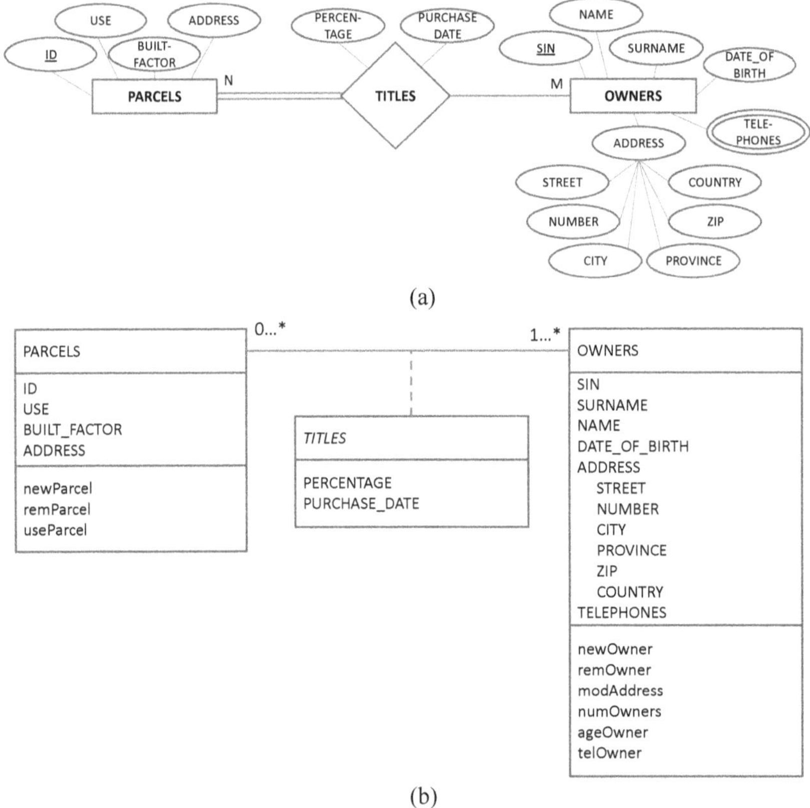


**Fig. 7.44** Example representation of an association in an ER-diagram (a) and a UML class diagram (b). To interpret the labels on the association line in (b) consider the following statements: "if I pick one parcel it is associated with 1 to many (1…*) owners" and "if I pick one owner s/he is associated with 0 to many (0…*) parcels".

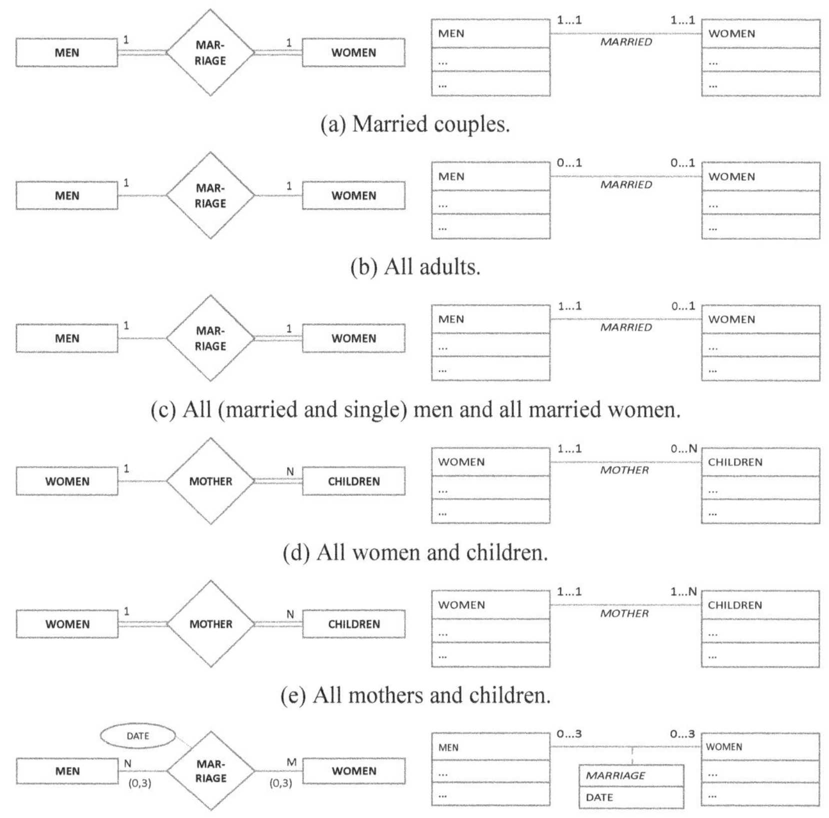

(a) Married couples.

(b) All adults.

(c) All (married and single) men and all married women.

(d) All women and children.

(e) All mothers and children.

(f) The history of all religious marriages for all adults.

**Fig. 7.45** Representation of the cardinality ratios and participation constraints in the ER-diagrams (left) and the UML class diagrams (right).

An *aggregation* represents the relationship between an object and its parts. In other words, an aggregation is the relationship type between the "whole" and its "components". The notation for an aggregation in a UML class diagram is through a line segment connecting the two classes with a diamond at the side of the class representing the "whole". An object instance in the "components" class can be part of either a single or multiple object instances of the "whole" class. Accordingly, the aggregation is characterized as *strong* or *weak*, and denoted by the black or white diamond, respectively.

Figure 7.46a presents an example of a *strong* aggregation, between the classes BUILDING_BLOCKS and PARCELS in the cadastral application. Each building block in an urban area consists of parcels (Figure 6.9) and a parcel falls into a *single* building block. On the other hand, Figure 7.46b shows a *weak* aggregation between the PARCELS and the LAND_USE_ZONES. In this example, the assumption that a parcel may be assigned to *multiple* land use zones is made. In other words, a parcel may be part of more than one zone (aggregations), e.g., both urban zone and industrial zone. The aggregation of parcels based on a single land use value lead to the formation of a homogeneous zone.

A *generalization* represents a relationship of superclass-subclass in a class hierarchy. The notation of a generalization in a UML class diagram is through line segments connecting the superclass with its subclasses and a triangle at the side of the superclass. Whether the generalization is *overlapping* or *disjoint* the triangle is black or white, respectively. Figure 7.47 shows the hierarchy of entity types in Figure 7.10 as represented in a UML class diagram. The class of owners is specialized into the subclasses persons and legal entities. This specialization is disjoint. The class of legal entities is specialized further into the classes of public and private legal entities. This is an overlapping specialization.

Figure 7.48 shows the UML class diagram describing a geometry schema adopted by the Open Geospatial Consortium (OGC).

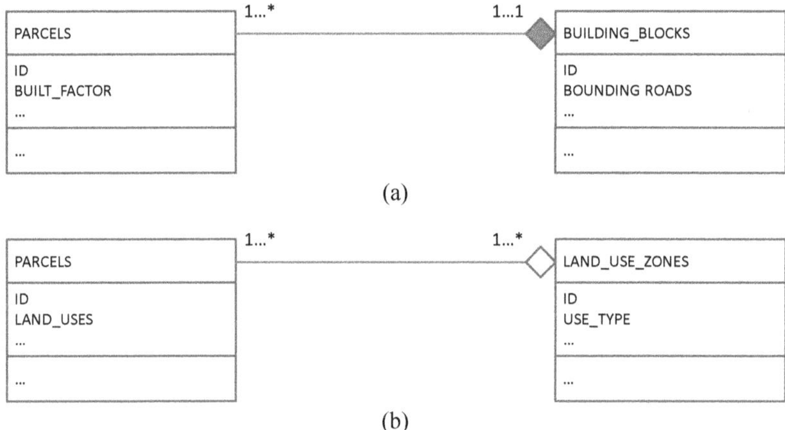


**Fig. 7.46** Example of a strong (a) and a weak (b) aggregation.

## *7.6.2 UML Class Diagrams versus EER-diagrams*

From the discussion in the previous Section it is obvious that UML class diagrams and EER-diagram have many similarities, but also many differences. Specifically, a class in a UML class diagram extends the concept of the entity type in an EER-diagram. A class encapsulates the attributes and methods in contrast to an entity type which encapsulates the attributes only. The definition of the methods is not part of an EER-diagram.

On the other hand, an entity type in an EER-diagram has always a key attribute defined, while a class in a UML class diagram has no reference on any key attribute. The reason is that in the object-oriented systems the key is system generated and attached automatically to all object instances in a database.

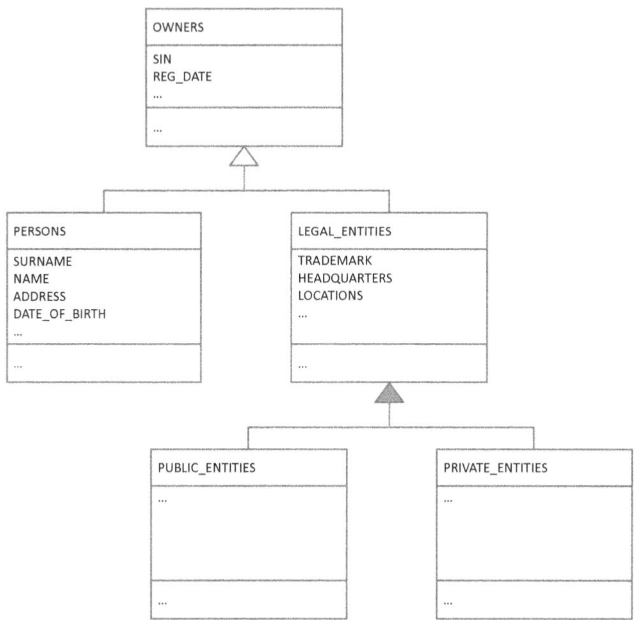

**Fig. 7.47** Example class hierarchy represented in a UML class diagram (see Figure 7.10).

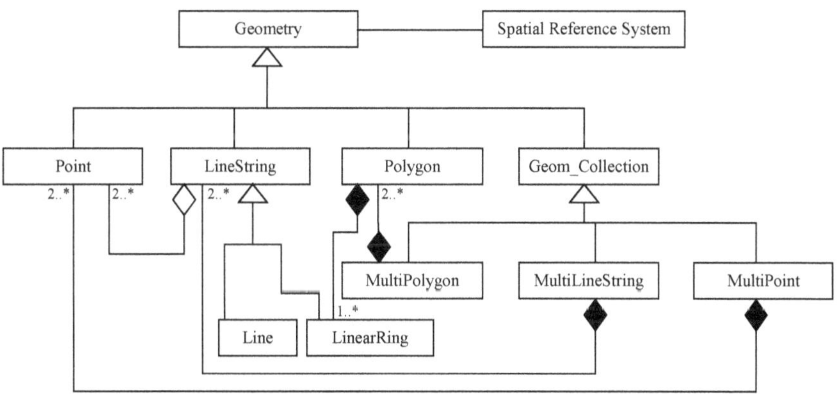

**Fig. 7.48** The UML class diagram of the basic geometric elements (Open Geospatial Consortium).

Both the entity types in an EER-diagram and classes in a UML class diagram can participate in relationships of type: association, aggregation, and generalization. The concepts of association and generalization are explicitly supported by the EER-model. However, the aggregation is supported implicitly by applying a typical relationship type and introducing some semantics through the name assigned to this type. An example is given in Figure 7.49 for the aggregations in Figure 7.46.

Finally, a UML class diagram does not handle the weak entity types differently as it occurs in an EER-diagram. Table 7.1 summarizes the concepts adopted by the two diagrams.

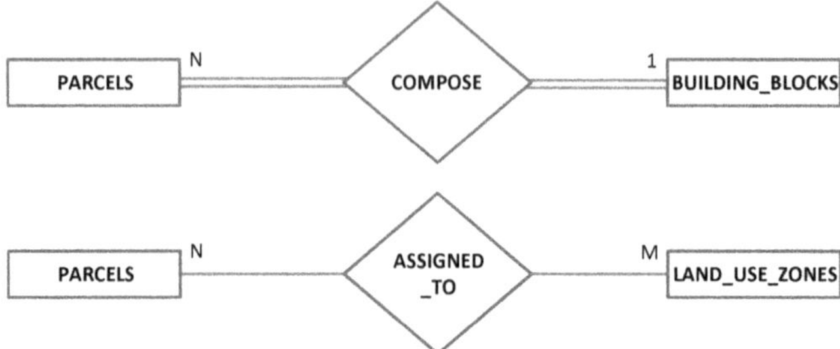


**Fig. 7.49** The aggregations of Figure 7.46 as represented in an EER-diagram.

**Table 7.1** Concept matching among UML class diagrams and EER-diagrams.

| UML Class Diagrams | | EER-diagrams | |
|---|---|---|---|
| Class | | Entity type | |
| | Attribute | | Attribute |
| | -- | | Key attribute |
| | Methods | | -- |
| Class | | Weak entity type | |
| Relationship | | Relationship | |
| | Association | | Relationship type |
| | Generalization | | Generalization |
| | Aggregation | | Relationship type |

# References and Further Reading


Booch, G., Rumbaugh, J., and Jacobson, I., 1999. *The Unified Modeling Language User Guide.* Addison-Wesley.

Cattell, R.G.G. (Ed.) 1994. *The Object Database Standard ODMG-93.* Morgan Kaufmann.

Cattell, R.G.G. (Ed.) 1997. *The Object Database Standard ODMG 2.0.* Morgan Kaufmann.

Date, C.J., 2003. *An Introduction to Database Systems.* Addison-Wesley.

Elmasri, R., and Navathe, S.B., 2000. *Fundamentals of Database Systems.* Addison-Wesley.

Garcia Molina, H., Ullman, J.D., and Widom, J., 2000. *Database Systems Implementation.* Prentice Hall.

Kim, W., 1995. *Modern Database Systems: The Object Model, Interoperability, and Beyond.* ACM Press, Addison-Wesley.

Lee, Y.C., 1992. *Geographic Information Systems.* Lecture Notes. Department of Geodesy and Geomatics Engineering, University of New Brunswick.

Molenaar, M., 1998. *An Introduction to the Theory of Spatial Object Modelling for GIS.* Taylor-Francis.

OGC, 2009. *Open Geospatial Consortium.* http://www.opengeospatial.org/

Ramakrishnan, R, and Gehrke, J., 2002. *Database Management Systems.* McGraw-Hill.

# 8 Data Structures and Algorithms

## 8.1 Introduction

*Algorithm* is an analytical description of a method to solve a problem. For example, the procedure to find the roots of a polynomial ($x^2$-3x-4) or to get cash from an ATM machine is an algorithm. An algorithm may correspond to either a human or a computer system.

In computer science, an algorithm is a step-by-step procedure to solve a problem that can be implemented in a computer program. An algorithm involves: (a) *data*, and (b) *computations*. In addition, most of the algorithms are accompanied with a description of how data is organized or structured at the physical level (storage media). This organization is called *data structure*. The procedure of solving a problem usually depends on the way data is organized at the physical level. Hence, algorithms and data structures are closely related subjects and need to be studied in combination.

Usually, there are several alternative ways to organize a dataset at the physical level. Notice that for the same data, alternative data structures may have different storage requirements. In addition, for the same computations, alternative data structures may offer a different efficiency, i.e., provide a faster or slower solution to the problem. Hence, it is important to choose the best combination of algorithm and data structure that leads to the lowest storage requirements with the fastest computational performance.

An example that shows the close relation between algorithms and data structures is the conventional phone book. The phone book is a database. The data in this book is the set of subscribers listed in alphabetical order by surname/name along with their telephone number. Hence, the data structure is a list of records ordered by the subscriber's surname/name. The phone book is there to assist users in solving the following problem (Figure 8.1): "What is the telephone number of subscriber X?" (John Smith's in the example). One algorithm to solve the problem would be to start from page 1/record 1 and compare the surname/name against X. If they match, report the number. If not proceed to the next record. Obviously, this process may take long to complete, especially if X's surname starts from a Z!

Because the records in the phone book are ordered, the user follows a more intelligent way to search in the book. S/he opens the book in the middle page. Based on the subscriber's surname and the surname found at the top of the middle page, the one half of the book can be ignored. This process is repeated recursively, until the subscriber is found (Figure 8.1). From the experience of a general user, this process is very fast, for whatever the subscriber's surname/name is. As it will be shown later in this Chapter, this algorithm can also be implemented in a computer program and it is called *binary search*.

Assume now that the user needs to solve the following problem, using the same phone book (Figure 8.2): "Who is the subscriber with telephone number Y?" Obviously, the only way to solve this problem is to search the book record-by-record, starting from page 1/record 1, and proceed until the telephone number in a record matches with Y. This algorithm can also be implemented in a computer program and it is called *sequential search*. Provided the organization of data in the book (data structure), there is no other option. Should similar queries be expected to be often posted by the customers, it would be wise for the telephone company to issue a second version of the phone book, where the records will be ordered by the telephone number. If so, a binary search algorithm could be applied to solve the second problem much faster.

The following Sections present the basic data structures and algorithms in handling alphanumeric datasets (thematic data). Chapters 12 and 13 focus on data structures and algorithms for handling the spatial dimension of geographic data (geometries).

**Fig. 8.1** Searching the telephone number of a subscriber in the phone book by applying the binary search.

**Fig. 8.2** Searching the subscriber's surname/name in the phone book based on the telephone number by applying the sequential search.

## 8.2 Algorithms

An algorithm has three sections: (a) *head*, (b) *variables*, and (c) *body*. The head includes the algorithm's name and the list of input/output parameters (arguments). The variables section defines the algorithm internal parameters (both their type and constraints). The body describes the actual operations (computations) of the algorithm.

An algorithm can be described in many different ways: *natural language, pseudo-code, programming languages,* and *flow charts*. No formalization applies in the natural language. The pseudo-code introduces some formalization, while the programming languages are well-formalized and do not allow any fuzziness. The flow charts provide a graphical representation of the algorithm using appropriate graphic notations to assist humans to better understand the processes involved in an algorithm.

Consider the problem of finding the average of the positive numbers in a set of N integers stored in a table structure P (Figure 8.3). An algorithm to solve this problem in *natural language* is as follows: Introduce a counter C for the positive integers in the table, an integer variable SUM for the sum of positive integers and a real variable AVG to store the result. Initialize all three variables to zero value. Scan the elements in table P starting from the first (I=1). Should the element I be a positive integer (i.e., P[I] > 0) add the element P[I] in the variable SUM (i.e., SUM := SUM+P[I]) and increase variable C by one (i.e., C:=C+1). Otherwise, if the element I is negative integer or 0 (i.e., P[I] ≤ 0) ignore it. Proceed to the next element (set I:=I+1). Repeat the process until all elements of table P have been scanned (i.e., I=N+1). Then, divide SUM by C and store the result to AVG (i.e., AVG:=SUM/C; if C=0 then AVG:=0). Report the value of AVG.

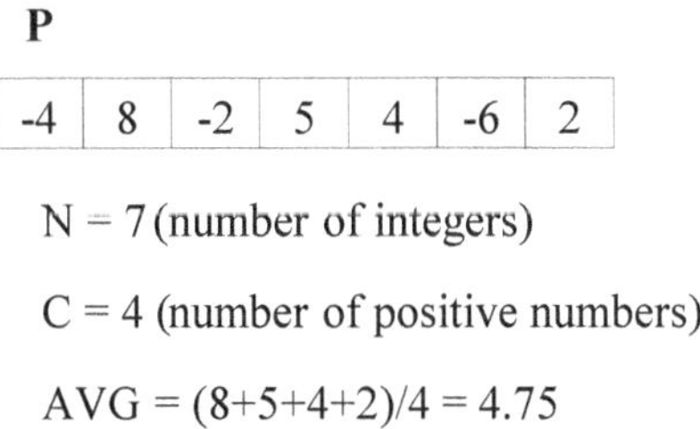

**P**

| -4 | 8 | -2 | 5 | 4 | -6 | 2 |
|----|---|----|---|---|----|---|

N = 7 (number of integers)

C = 4 (number of positive numbers)

AVG = (8+5+4+2)/4 = 4.75

**Fig. 8.3** Find the average of the positive numbers in a table of integers.

The previous paragraph is a description of the algorithm in natural language. Figure 8.4 shows the same algorithm in *pseudo-code*, while Figure 8.5 presents a version in *C programming language*. Figure 8.6 shows the *flow chart* of the algorithm. The graphical notation follows the ANSI specification. Specifically, the *ellipse* denotes the start and end of the algorithm; the parallelogram is used to represent the input data. The *rectangle* represents computations (e.g., I:=1) or processes (e.g., reporting). The *diamond* represents logical conditions and has two outputs:

yes is the condition is met and no otherwise. Finally, the *arrows* represent the flow of the processes. The iteration of operations (loops) is represented by arrows heading backwards.

```
ALGORITHM: FIND THE AVERAGE OF POSITIVE NUMBERS IN A TABLE OF N INTEGERS
  DATA
    P:TABLE[1,...,N] OF INTEGERS
    N,I,C,SUM INTEGERS
    AVG REAL
  BEGIN
    C:=0                      /* INITIALIZE THE VARIABLES */
    SUM:=0
    AVG:=0
    FOR I:=1 TO N REPEAT
       IF P[I]>0 THEN      /* CHECK IF POSITIVE */
          SUM:=SUM+P[I]
          C:=C+1
       END_IF
    END_FOR
    AVG:=SUM/C                 /* COMPUTE THE AVERAGE */
    PRINT(AVG)
  END
```

**Fig. 8.4** An algorithm in pseudo-code.

```c
#include <stdio.h>
  void find_mean(int P[N], int N)
  {
     int I,C,SUM;              /* DECLARATION OF VARIABLES */
     float AVG;
     SUM=0; C=0; AVG=0;        /* INITIALIZE THE VARIABLES */
     for (I=0; I<N; I++)
     {
        if (P[I]>0)            /* CHECK IF POSITIVE */
        {
           SUM=SUM+P[I];
           C=C+1;
        }
     }
     AVG=(float)SUM/C;         /* COMPUTE THE AVERAGE */
     printf("AVERAGE = %f\n", AVG);
  }
```

**Fig. 8.5** An algorithm in C programming language.

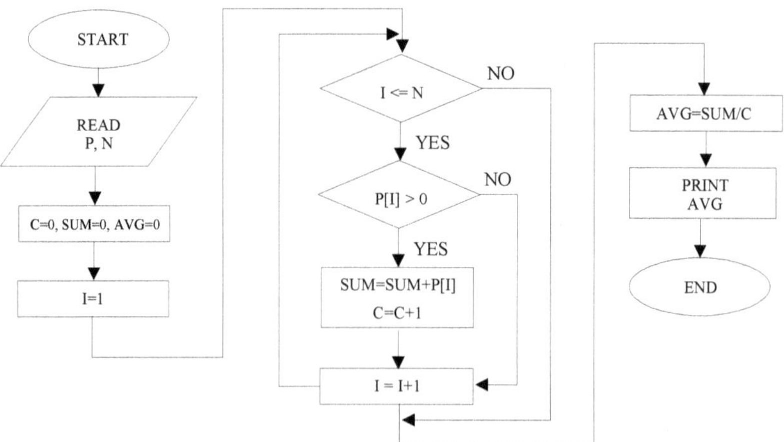


**Fig. 8.6** The flow chart of an algorithm.

# 8.3 Data Structures

A *data structure* is a way of storing and organizing data at the physical level (storage media) so that it can efficiently be used by computer programs. The most common data structures are: (a) *tables*, (b) *records*, (c) *linked-lists*, (d) *queues*, (e) *stacks*, and (f) *trees*. These structures are briefly presented next.

## 8.3.1 Tables

A *table* is a basic data structure. A table is described by its *dimensions*. A *one-dimensional* table (also called *vector*) consists of a specified number of elements of the same type (integers or characters, etc.), stored in a sequence in the computer storage (e.g., RAM). Each element can be accessed directly through a pointer. An example, one-dimensional table of five integers is shown in Figure 8.7.

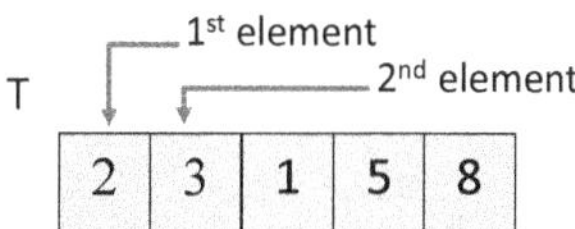


**Fig. 8.7** A one-dimensional table of five integers.

The retrieval of element i is achieved through the reference T[i], (e.g., T[4] = 5) and it is instantaneous (Figure 8.8). Provided that the table elements are stored in a sequence (one next to the other), given the memory address of the first element (e.g., memory address of T[1] is 3200) and the space occupied by each element (e.g., for an integer: 4 bytes), the address of each table element can be easily calculated. For example, the address of the fourth element equals the address of the first element (e.g., 3200) plus 3x4bytes = 12 bytes, where 3 the number of elements preceding the fourth element and 4 the space (number of bytes) occupied by each element. Hence, the address of the fourth element is equal to the address of the first shifted by 12 bytes (i.e., 3200+12=3212). Notice that for large tables, the elements might not all be stored in a sequence. In this case, the operating system is responsible to restore the sequence.

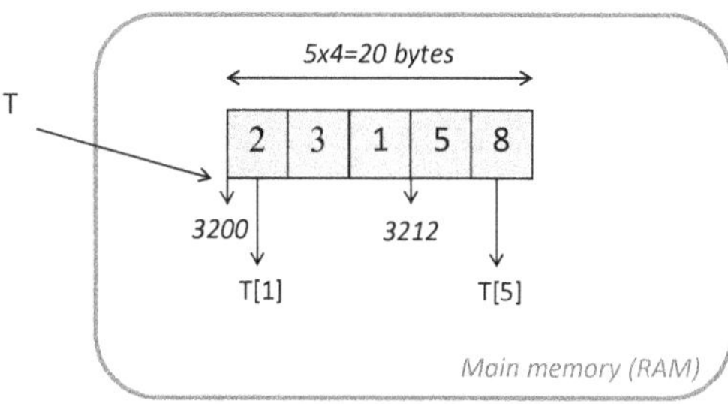


**Fig. 8.8** The table of Figure 8.7 as stored in the RAM.

A *two-dimensional table*, consists of a number of elements which are organized in a set of *rows* and *columns*. Each element is accessed through a pair of pointers denoting a row and a column in the table. Figure 8.9 shows an example two-dimensional table with three rows and five columns (i.e., table 3x5). Element i,j (i.e., row=i, column=j) is referred to as T[i][j] (e.g., T[2][3]=4).

Two or higher dimensional tables are linearly represented in the computer memory. The linearization of a two-dimensional table can be done either by column or by row as shown in Figure 8.10. Obviously, a two-dimensional table on N rows and M columns will result into a one-dimensional table of NxM elements.

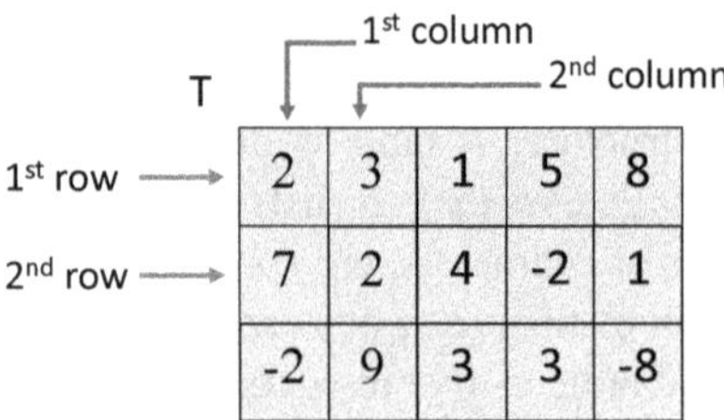


**Fig. 8.9** A two-dimensional table of 3 rows and 5 columns.

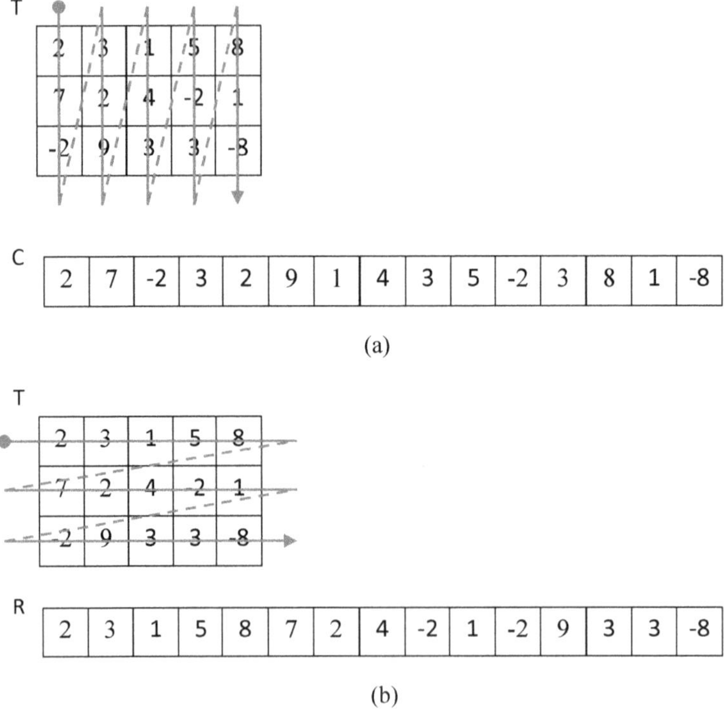


**Fig. 8.10** The table in Figure 8.9 linearized by column (a) and by row (b).

The element T[i][j] of a two-dimensional table of N rows and M columns (NxM) corresponds to the element R[(i-1)xM+j] of its linear representation for a linearization by row. For a linearization by column, the corresponding element is C[(j-1)xN+i]. For example, in Figure 8.10: (a) T[2][3]=4, C[(2-1)x5+3]=C[8]=4; and (b) T[2][3]=4, R[(3-1)x3+2]=R[8]=4.

Similarly to two-dimensional tables, tables of higher dimensionality (e.g., three-dimensional) can be defined. Their representation in the computer storage is done after they are linearized. Some applications involve tables with *special properties*, such as symmetric, diagonal, triangular, non-dense, etc. This property can be taken into consideration for minimizing the storage requirements. This discussion is out of the scope of this book.

## 8.3.2 Records and Tables of Records

All elements stored in a table are of the same type, e.g., integer in a table of integers, real in a table of real numbers, etc. A *record* is an alternative data structure, which comprises a finite number of related elements of the same or different type.

An example record is the collection of the attributes describing an owner in the cadastral application described in Chapter 5 (Figure 8.11). In such a record, data elements are of different type, e.g., variable length character string, date or integer. In addition, all these elements are related as they all describe a person (owner) in the cadastral database.

| SIN | SURNAME | NAME | DoB | STREET | No | CITY | PROV | ZIP | COUNTRY |
|---|---|---|---|---|---|---|---|---|---|
| 576129452 | COOPER | MARY | 11/23/1945 | QUEEN | 2581 | HALIFAX | NS | B3J2H8 | CANADA |

**Fig. 8.11** An example record.

A *table of records* is a combination of two structures: table and record. A table of records is a one-dimensional table, whose elements are of type record. An example table of records is the table of owners in the cadastral database (Figure 8.12). The following notation is used to access individual elements (atomic values) in the record i: Record[i].Element_name, e.g., Record[2].City='Halifax'; for the City name in the second record. Figure 8.12b shows the definition and storage requirements for each record and the definition of a table of 100 records.

## 8.3.3 Linked-lists

An alternative data structure for a sequence of data elements of the same type is the *linked-list*. In a linked-list, the data elements are not stored in a sequence at the physical level (as it occurs in tables) but in randomly selected spots. On the other hand, the elements are ordered in the list. This is implemented as follows. Each *el-*

*ement* is accommodated in a *node* structure. The node also hosts a *pointer* which points to the next node in the list. The first node in the list is pointed through a special pointer which is called *head* pointer. The list is terminated to a node with no-successor node (its pointer points to null). Figure 8.13a presents the node structure and Figure 8.13b shows an example of a linked-list of four elements.

| SIN | SURNAME | NAME | DoB | STREET | No | CITY | PROV | ZIP | COUNTRY |
|---|---|---|---|---|---|---|---|---|---|
| 464197351 | SMITH | JOHN | 08/15/1952 | GEORGE | 191 | FREDERICTON | NB | E3A5A3 | CANADA |
| 576129452 | COOPER | MARY | 11/23/1945 | QUEEN | 2581 | HALIFAX | NS | B3J2H8 | CANADA |
| ... | ... | ... | ... | ... | ... | | | ... | ... |

(a)

```
struct owner {            size in bytes          struct date {        size in bytes
    int ID;                   4                       short day;       2
    char SURNAME[20];        20                       short month;     2
    char NAME[12];           12                       int year;        4
    date BIRTH_DATE;          8                   }
    char STREET[30];         30
    int NUMBER;               4                   owner O[100];        definition of a
    char P_CODE[5];           5                                        table of records
    char CITY[12];           12                   sizeof(owner) = 95 bytes (each record)
}
```

(b)

**Fig. 8.12** An example table of records (a); its definition in C and storage requirements (b).

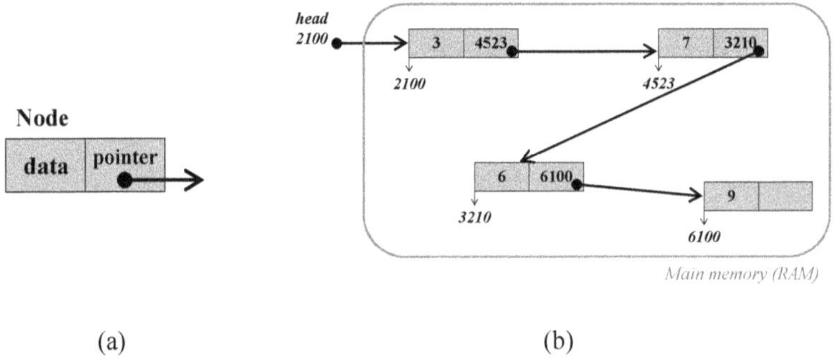


**Fig. 8.13** The structure of a linked-list node (a). An example linked-list with four nodes as stored in the main memory (b).

The structure of a linked-list has several advantages and disadvantages over the structure of a table. Starting from the advantages, the definition of a linked-list (unlike that of a table) does not determine the maximum number of elements in the list. Hence, the list has no limits on the number of elements that it is able to accommodate. Additionally, the storage space occupied is always relevant to the current number of elements in the list. The linked-list may grow or shrink, while appropriate defensive programming may prevent any issues as regard to memory overflow. Another significant advantage of the structure is that it can easily be updated. Specifically, the reordering, insertion or deletion of elements only involves a rearrangement of node pointers, unlike tables where the elements must be relocated in the memory. Figure 8.14 presents three examples of restructuring the nodes in the list of Figure 8.13b.

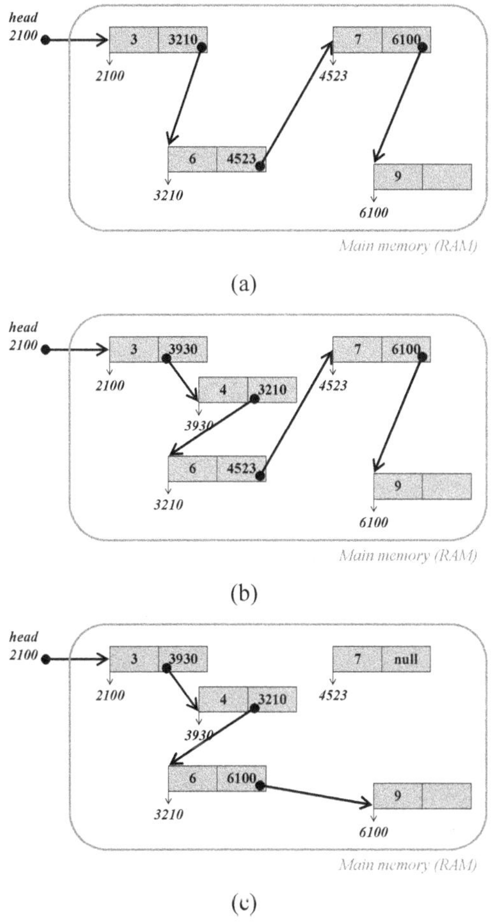

Fig. 8.14 Operations in linked-lists: (a) reordering the nodes in the list of Figure 8.13 (3,7,6,9) → (3,6,7,9); (b) insert value 4 between 3 and 6; and (c) delete value 7 from the list.

A drawback of the linked-list structure is that it does not support fast access to its elements. The only way to access an element in the list is through the head node (the node pointed by the head pointer) and a sequential scanning of the next nodes through the pointers until the reference node is reached. For example, the fourth node in Figure 8.15a (the node hosting data element D) can only be accessed through the node pointed by the head pointer (node with data element A). Following the pointer of that node, the next node is reached (node with element B) and so on until the fourth node in the list is visited. This happens because the nodes are stored randomly in the memory and their location is only recorded in the previous node of the list. On the other hand, the elements in a table structure can be accesses directly (e.g, P[4] in Figure 8.15b).

Various alternative implementations of the linked-list structure can be found in the literature to serve specific application requirements. The most common is the *double-linked-list* (Figure 8.16a) and the *circular-linked-list* (Figure 8.16b). In the former, each node maintains two pointers, one to the next and one to the previous node, while in the second, the end node of the list points to the head node of the list.

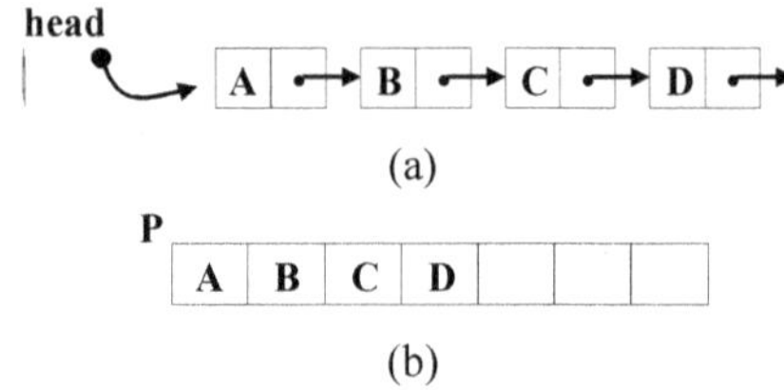

Fig. 8.15 Representation of the same elements in a linked-list (a) and in a table structure (b).

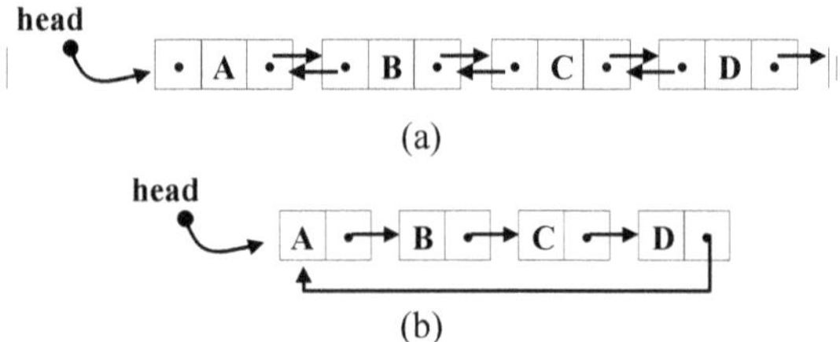

Fig. 8.16 Alternative linked-list structures: double-linked (a) and circular (b).

### 8.3.4 Stacks

The *stack* data structure is a specific type of list or table which allows for insertions and deletions of nodes at one end only: the top or free end. Hence, an element may be either *inserted* (push) or *deleted* (pop) from the top of the stack. This structure is also known as *LIFO* (Last-In-First-Out), and its implementation is accomplished with either a linked list or table.

Figure 8.17 shows a representation of stack using a table. The stack has a pointer called *top* which points to the position of the last entry. This pointer is re-arranged after any insertion or deletion. Figure 8.17 shows an empty stack on the left side. Then three elements are inserted into the stack (operation: push): A, B, C. Finally, an element is deleted (operation: pop) from the stack. The top pointer always points to the top element in the stack, which is the most recently inserted element still in the stack.

The stack is an ideal structure to support certain functions in computer programming. Two typical examples are: (a) recording of a series of subroutine invocations when a program runs, and (b) solving of a mathematical expression with parentheses. Focusing on the second example, consider the mathematical expression:

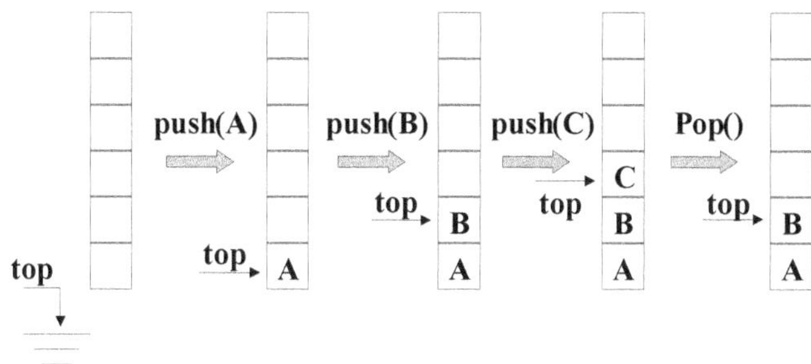


**Fig. 8.17** Structure of the stack and the operations push and pop.

```
5 x ( ( ( 7 + 3 ) x ( 6 - 4 ) ) + 12 ) = ?
```

A stack can support the execution of operations in order based on the brackets. Specifically, the processing starts from left to right and successively traverses the expression data. Each number is inserted into the stack (operation: push). When a right bracket is reached, the last two numbers are retrieved from the stack (operation: pop) and the corresponding operator is applied between them. The result is imported in the stack (operation: push) push. When the stack gets empty, the result is reported. Figure 8.18 shows schematically the processing of the above expression.

---

```
                    5 x ( ( ( 7 + 3 ) x ( 6 - 4 ) ) + 12 ) = ?
Push(5)
Push(7)
Push(3)
Push(Pop() + Pop())
Push(6)
Push(4)
Push(Pop() - Pop())
Push(Pop() x Pop())
Push(12)
Push(Pop() + Pop())
Push(Pop() x Pop())
Print(Pop())
```

---

**Fig. 8.18** Solving a mathematical expression with parentheses using a stack.

## *8.3.5 Queues*

The *queue* structure is similar to the stack. It also supports two basic operations: insertion and deletion of nodes. But in contrast to the stack, new elements can be *inserted* (push) at one end (the tail) of the queue only, while existing elements can be deleted (pop) from the other end (head) only. This structure is also known as *FIFO* (First-In-First-Out), and its implementation is achieved by either a linked list or table.

The queue structure simulates the customer service in a bank. The customer who arrives first in the queue is served first (first come first served). Figure 8.19 shows an implementation of a queue using a table. The queue has two pointers the top and tail. Figure 8.19 shows an empty queue on the left side. Then three elements are inserted into the queue (operation: push): A, B, C. Finally, one element is deleted (operation: pop). In each operation, the pointers are reallocated accordingly.

The queue data structure finds many applications in computer programming. Specifically, the control of sequential execution of functions in a complex process is often implemented by a queue structure.

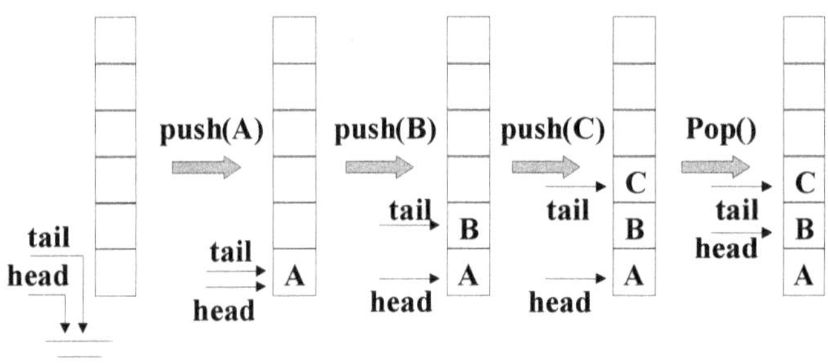


**Fig. 8.19** The structure of the queue and the operations push and pop.

## *8.3.6 Trees*

All the structures presented so far are linear (or one-dimensional). In other words, all the elements are placed sequentially one after the other. Even in the case of multidimensional tables, their representation in computer memory is linear.

On the other hand, a tree is a non-linear (or multidimensional) data structure. It is similar to a linked list in which each node can point to more than one node. Schematically, a tree consists of *nodes* and *edges* (Figure 8.20). Each node has a name and hosts data. Each edge is a directed link between two nodes: from a *parent* (ancestor) node and to a *child* (descendant) node. Edges are implemented through pointers (see next).

In a tree structure *paths* are defined. Each path is a list of distinct nodes of the tree connected by edges. Each tree has a node, called the *root*. From the root, which is unique in a tree, there are outgoing edges only. Moreover, there is a unique path from the root to each node of the tree. The access to the nodes of the tree is done through the root only. Note that each node of the tree is the root of a subtree, which hosts all its descendants.

The nodes of a tree, except the root, are either (a) *leaf*, or (b) *intermediate* (or non-leaf) nodes. Leaf nodes have incoming edges only. Intermediate nodes have both incoming and outcoming edges.

Each node of the tree is characterized by a *degree*, which represents the number of its children. *Degree of a tree* is called the maximum degree of its nodes. *Height* of the tree is called the largest (in number of nodes) path from the root to a leaf of the tree. Figure 8.20 shows a simple tree and illustrates these properties.

*Binary tree* is a tree with degree equal to two (2) for each node. The children of each node in a binary tree are named as *left child* and *right child*. Correspondingly, the subtrees of each node are called *left* and *right subtree*. Figure 8.21 shows a binary tree, and the structure of nodes in that tree.

*Full tree* is a tree whose all levels (except maybe the last one) have the maximum possible number of nodes. *Balanced tree* is a tree whose leaf nodes lie all at the same level. Figure 8.22 shows a full and balanced binary tree. The height (h) of a full binary tree with k nodes (elements) is equal to: $h=\log_2(K+1)-1$ . Table 8.1 shows the number of nodes per tree level. The total number of nodes in the tree is equal to: $2^h+(2^h-1)$. Hence:

$$2^h+(2^h-1) \ = \ K \ \Rightarrow \ 2^{h+1} \ = \ K+1 \ \Rightarrow \ h \ = \ \log_2(K+1)-1$$

A variant of the binary tree is the *binary search tree (BST)*, which has the following properties: (a) it is a binary tree, (b) the data value of the right child (subtree) is greater than that of the parent node, and (c) the data value of the left child (subtree) is less than or equal to that of the parent node. This condition is shown in Figure 8.23.

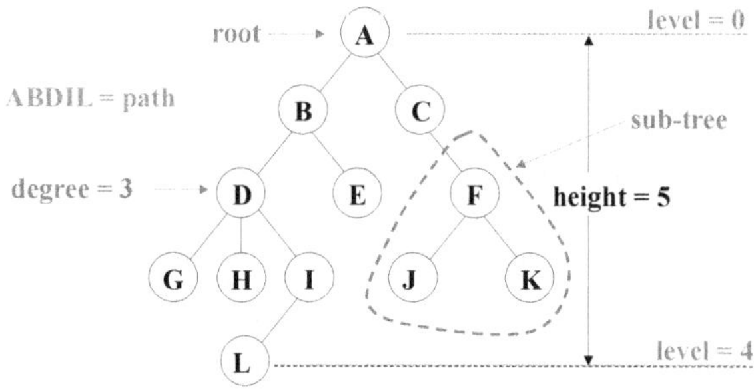


**Fig. 8.20** The tree data structure.

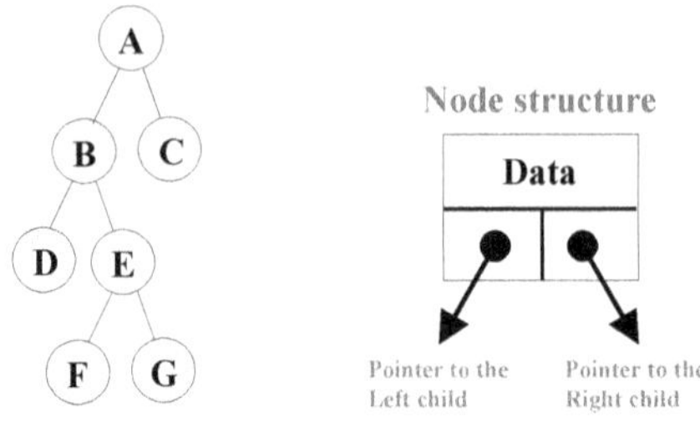


**Fig. 8.21** An example binary tree (a); and the structure of a node in the binary tree (b).

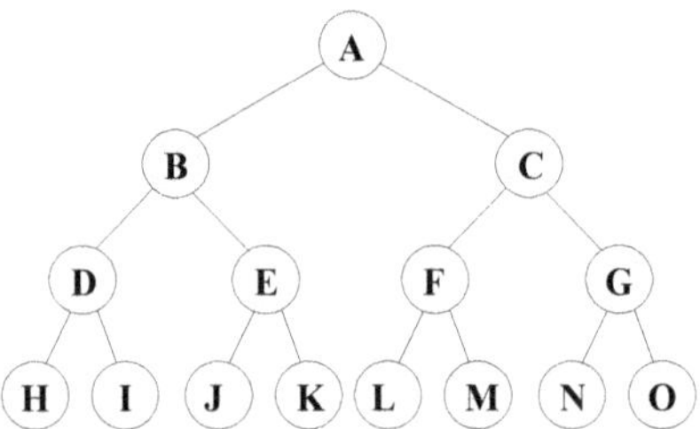


**Fig. 8.22** A full and balanced binary tree.

**Table 8.1** Number of nodes and height of a full and balanced binary tree.

| Tree level | Nodes in this level | Sum of nodes from the roo to this level | Height |
|---|---|---|---|
| 0 (root) | $2^0 = 1$ | $2^0$ | 0 |
| 1 | $2^1$ | $2^1 + (2^1 - 1)$ | 1 |
| 2 | $2^2$ | $2^2 + (2^2 - 1)$ | 2 |
| ... | ... | | ... |
| h | $2^h$ | $2^h + (2^h - 1)$ | h |

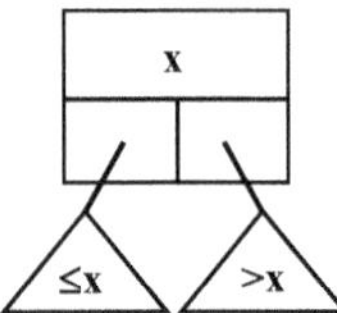


**Fig. 8.23** The data values of a node and its subtrees in a binary search tree.

To *build* a binary search tree the following algorithm is applied. For a sequence of data values (e.g., integers), place the first value as the root of the BST. Compare the second value against the root value. If it is greater than the value of the root, insert the new value as right child of the root. Otherwise, insert the new value as a left child of the root. Compare each next value against the root. If it is greater (or less), and there is no right (left) child, put the value as a right (left) child of the root. If there is a right (left) child follow the path to the right (left) subtree of the root and recursively repeat the process. Figure 8.24 shows the process for the sequence of integers {15, 18, 6, 3, 4, 7, 13, 2, 20, 9, 17}.

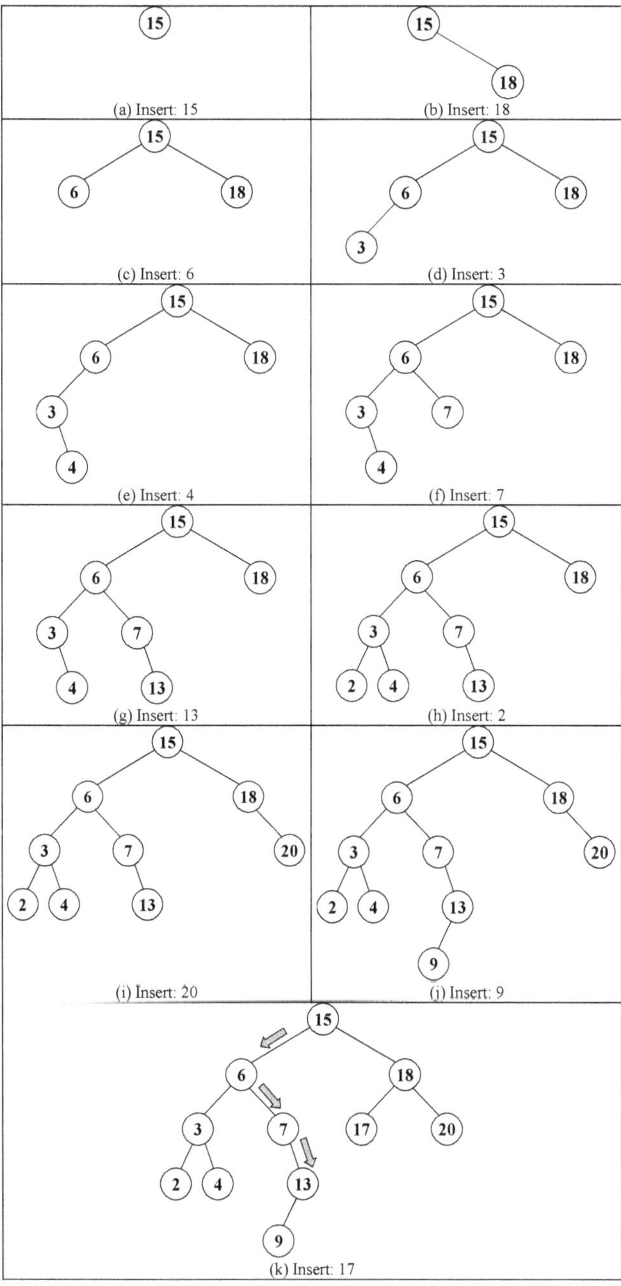

**Fig. 8.24** Steps to build a binary search tree for the sequence of integers: {15, 18, 6, 3, 4, 7, 13, 2, 20, 9, 17}. Arrows in (k) represent the steps to locate number 13 (search algorithm).

The *search* algorithm in a BST exploits the properties of the tree. Following, the search algorithm is demonstrated through an example. Specifically, to find the integer 13 in the tree of Figure 8.24k, the following steps should be carried out. Read the data value (15) of the root node. As 13 is less than 15, visit the left subtree of the root (ignore the right one). Compare 13 against 6 (data value of the root's left child). As 13 is greater than 6, head to the right subtree. Compare 7 against 13. Because 13 is greater head to the right subtree, where 13 resides.

## 8.4 Recursion

*Recursion* is a basic concept in both mathematics and computer science. Focusing in computing, a program (algorithm or function) is called recursive when it references itself. In a recursive program, the presence of a termination condition is necessary, which – when satisfied – the next call is cancelled and the recursion terminates.

An example of a recursive function is the factorial of an integer. For an integer N, this function is defined as follows:

```
N ! = N x (N-1) !    ,     for N ≥ 1  with  0! = 1
```

The calculation of the factorial of 5 is achieved by a recursive reference to the function as follows (bold numbers represent a new reference):

```
5!   = 5 x 4! =
     = 5 x 4 x 3! =
     = 5 x 4 x 3 x 2! =
     = 5 x 4 x 3 x 2 x 1! =
     = 5 x 4 x 3 x 2 x 1 x 0! =
     = 5 x 4 x 3 x 2 x 1 x 1  = 120
```

Figure 8.25a shows a recursive algorithm to compute the factorial of an integer N. Notice that it is possible to implement the same function without recursion, and through an iterative process, as shown in Figure 8.25b.

```
int factorial(int N)              /* Recursive function */
  {
     if ( N == 0) return 1;       /* Termination condition */
     return N * factorial(N-1);
  }
```

(a)

```
int factorial(int N)              /* Non-recursive function */
  {
     int I, K = 1;
     for (I = 2; I <= BST; I++) { K = K * I; }
     return K;
  }
```

(b)

**Fig. 8.25** Implementation of factorial integer in C programming language with a recursive (a), and repetitive (b) function.

## 8.5 Complexity of Algorithms

Most problems can be solved using multiple different algorithms. The challenge in computer science is to recognize the most efficient algorithm to solve a problem. To achieve this, some measures are applied that help assess the effectiveness of an algorithm. The measures that are applied most often are: (a) the execution time of an algorithm, and (b) the total number of transactions (e.g., comparisons or multiplications) executed to reach the solution.

The first measure is dependent on the performance of the computer processor. This is not the case in the second one, which offers a more objective factor of effectiveness. The discussion that follows focuses on this measure.

Solving a specific problem does not "cost" always the same. For example, the sequential search for a number in a list of N numbers will be achieved: (a) in a single comparison (if the number is located at the beginning of the list), (b) after N comparisons (if number is located at the end of the list), and (c) after N/2 comparisons in average. In practice, the worst case is considered, and it is the one that characterizes the efficiency of an algorithm.

The efficiency of an algorithm is often expressed as a function of a key parameter N, which is typically either of the following: (a) the degree of a polynomial to be solved, (b) the number of records in a file to sort or search, (c) the number of nodes of a tree in search, (d) the number of nodes in a graph to find the optimum path, etc.

Algorithms running time (worst case) are described with one of the following functions *f*, also called *complexity functions*:

| | |
|---|---|
| **K** | fixed time, independent of the data |
| **logN** | logarithmic complexity |
| **N** | linear (proportional to the data) complexity |
| **NlogN** | N times logarithmic complexity |
| $N^2$ | quadratic complexity |
| $N^3$ | cubic complexity |
| $2^N$ | exponential complexity |

Algorithms are classified according to the complexity function. Table 8.2 shows the values of these functions for different values of N. Obviously, for large values of N, algorithms with quadratic, cubic or exponential complexity become too costly. However, those with logarithmic complexity are still very effective.

**Table 8.2** Complexity function values for different orders of magnitude of the number of data items (N).

| log(2)N | [log(2)N]^2 | N | N*log(2)N | [N*log(2)N]^2 | N^2 | N^3 | 2^N |
|---|---|---|---|---|---|---|---|
| 3 | 11 | 10 | 33 | 110 | 100 | 1000 | 1024 |
| 7 | 44 | 100 | 664 | 4414 | 10000 | 1000000 | 1,26765E+30 |
| 10 | 99 | 1000 | 9966 | 99317 | 1000000 | 1E+09 | 1,0715E+301 |
| 13 | 177 | 10000 | 132877 | 1765633 | 100000000 | 1E+12 | >> |
| 17 | 276 | 100000 | 1660964 | 27588016 | 1E+10 | 1E+15 | >> |
| 20 | 397 | 1000000 | 19931569 | 397267426 | 1E+12 | 1E+18 | >> |

According to the mathematical definition, the complexity of an algorithm is the growth of runtime, depending on the number of data items N. Given an algorithm (function) with running time T(N), it is referred to as T(N)=O($f$(N)), if there are two non-negative constants c, $N_o$, such that T(N) $\leq$ c $\cdot f$(N) for N $\geq$ $N_o$. The values of T(N) and $f$(N) are positive.

Hence, the classification of an algorithm as of its complexity depends on the magnitude of the operations performed for large values of N. This means that if the number of operations is N/2 or 2N, the complexity is in both cases O(N).

Next, two categories of algorithms that are closely related to data structures are presented: (a) *sorting* algorithms, and (b) *searching* algorithms.

## 8.6 Sorting Algorithms

*Sorting algorithms* solve the following problem: given a number of N data items (e.g., integers) stored in an array in random order, reorder these data items according to some rule (e.g., in ascending or descending order).

To solve this problem various algorithms can be applied. The complexity is usually O($N^2$), where N is the number of items to be sorted. The optimum complexity that can be achieved is O(NlogN).

### *8.6.1 Sort by Counting*

*Sort by counting* algorithm is the simplest sorting algorithm. Given N items in a linear array, each item i is compared with all others. The k items that precede i (e.g., that are less than i, for an ascending ordering) are counted. The position of the element i in the sorted array will be at spot k +1.

Figure 8.26 shows schematically the algorithm for a set of seven integers. The item 5 is compared with the other six integers. There are two integers with value less than 5 (i.e., items: 1 and 2). This means that 5 will occupy the third position in the sorted array. As the process is repeated for all N items of the initial table and N-1 comparisons are applied for each item, the total number of transactions is equal to Nx(N-1). Therefore, the complexity of the algorithm is O($N^2$).

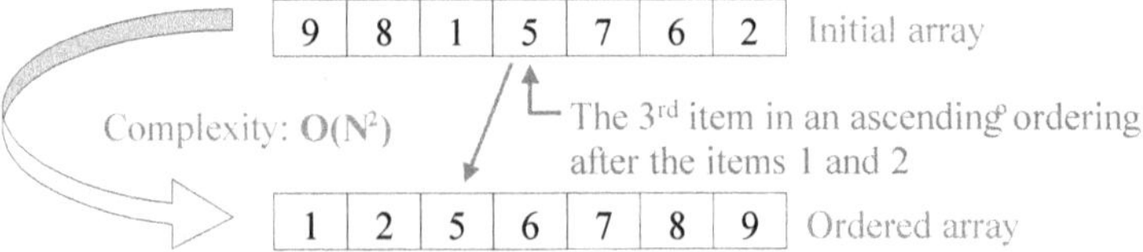


**Fig. 8.26** Example of sorting by counting.

## 8.6.2 Select Sort

In the *select sort* algorithm, the N items are first stored in an array A and then repeatedly scanned from left to right. In the $k^{th}$ scan, the item with the minimum (maximum) value among the items A[K], ..., A[N] is selected. This item is swapped with item A[K]. After the $k^{th}$ scan is completed, the first K items of the array A[1], A[2], ..., A[K] are sorted in ascending (descending) order.

The steps of the algorithm are presented through an example in Figure 8.27. In this case, the complexity equals to $O(N^2)$, as (N-1) scans are carried out. In each scan, the minimum item from (N-1) to 1 items (N/2 items on average) is sought. Therefore, (N-1)x(N/2) operations are executed in total.

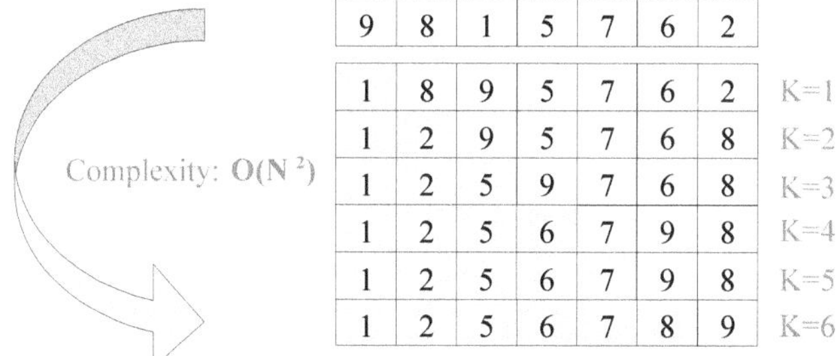


**Fig. 8.27** Example of select sort.

## 8.6.3 Quick Sort

As mentioned above, the optimal complexity of sorting N items is $O(N\log_2 N)$. Next, an effective algorithm of this complexity is presented.

The *quick sort* algorithm is based on the tactic of "divide and conquer", under which the problem (here the sorting of N items) is split into smaller problems, each of which is further split into even smaller ones, etc. The solution to the original problem results from an appropriate combination of the solutions of all smaller problems.

The algorithm selects an item, called *pivot*, and groups the remaining items into two sets. All items, with a value less that the pivot, reside into one group (left of the pivot), and all others into the second group (right of the pivot). The process is repeated recursively for the two new subsets.

One challenge is the selection of the pivot for a given set. The selection can be done either randomly or by applying some rule. If the content of the set is known in advance, an appropriate selection of the pivot may speed up the process. Specifically, a value close to the medium value of the set would be ideal. If the content is unknown, the pivot value is just picked up randomly, e.g., the first element of the set.

After the pivot value is chosen, the initial set is divided into two subsets. This is implemented by the simultaneous scanning of the set from the two ends towards the center by means of two pointers. The left (right) pointer stops when it encounters an element which is greater (less) than the pivot. When both pointers stop, their values are swapped and the process continues until the two pointers meet.

Once the pointers meet, two subsets are created: the one left of the pivot and the one right of the pivot. The algorithm is repeated recursively to these two sets. At the end, the union of all subsets gives a sorted array. Figure 8.28 shows the division of the original set into two subsets after a simultaneous scanning of the set with the two pointers and using as pivot the 22.

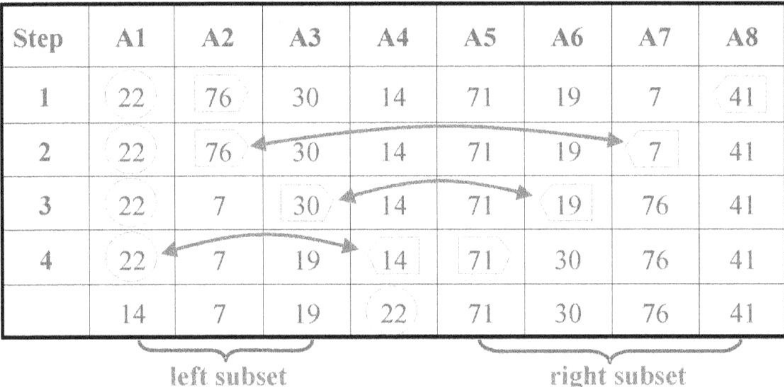

| Step | A1 | A2 | A3 | A4 | A5 | A6 | A7 | A8 |
|---|---|---|---|---|---|---|---|---|
| 1 | 22 | 76 | 30 | 14 | 71 | 19 | 7 | 41 |
| 2 | 22 | 76 | 30 | 14 | 71 | 19 | 7 | 41 |
| 3 | 22 | 7 | 30 | 14 | 71 | 19 | 76 | 41 |
| 4 | 22 | 7 | 19 | 14 | 71 | 30 | 76 | 41 |
|  | 14 | 7 | 19 | 22 | 71 | 30 | 76 | 41 |



**Fig. 8.28** Example of quick sort algorithm. Division of the initial set into two subsets. The circle represents the pivot. The pointers are represented with ▷ (left pointer) and ◁ (right pointer). Arrows represent the swaps.

## 8.7 Searching Algorithms

*Searching algorithms* are the second major category of algorithms in computer science. As discussed in the introduction of this Chapter, the approach followed by a searching algorithm is directly dependent on the organization (structure) of the data. Next, some common searching algorithms and methods are presented.

### 8.7.1 Sequential Search

*Sequential search* is the simplest search algorithm yet most time consuming. It is applied when data items are not organized, but simply stacked serially one after the other (raw data).

Given a set of N elements (e.g., integers) randomly organized in a list, to seek the presence of an element K in the list, a sequential search examines successively the elements starting from the first and until either the element K or the last element of the list is reached. Obviously, the complexity of the sequential search is O(N).

## 8.7.2 Binary Search

If the elements of the set are ordered, then a more efficient searching approach can be applied to find element K in the list. This approach is similar to the one humans follow to find a subscriber's telephone number in a phone book (Figure 8.1). This approach applies the tactic of "divide and conquer".

Specifically, for a set of N data items (e.g., integers) arranged in ascending order, the algorithm of *binary search* works as follows: the middle element (M) of the set is retrieved and compared with K. If K is equal to M, the element is found. Otherwise, if K is smaller than M (K <M) the right subset of M can be ignored and the process is repeated to the left subset. Else, if K is larger than M (K> M) the left subset of M is ignored and the process is repeated to the right subset.

An example of the binary search over an ordered list of 11 integers is given in Figure 8.29. To find element K=9 the following steps are performed. First, K is compared against the middle element of the array, i.e., M = 8 (11/2 $\rightarrow$ the sixth element). As K>M, all elements on the left side of M can be ignored. The search focuses on the elements lying at the right of M. This process is repeated recursively until K is found.

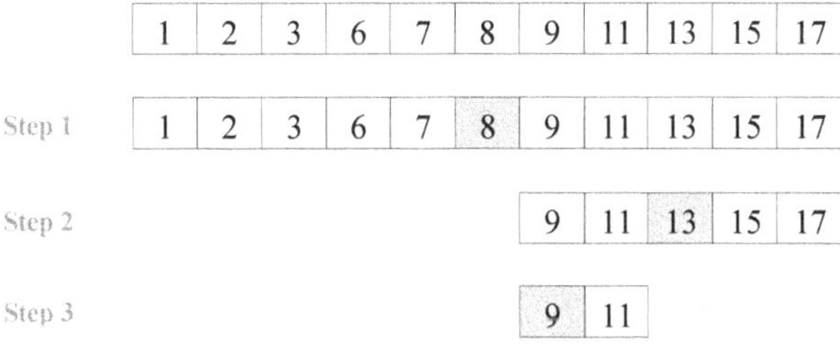

**Fig. 8.29** Binary search over an array of 11 integers; searching for element K=9.

The complexity of the binary search for an element K in an array of N elements is equal to $O(\log_2 N)$ and this is what best can be achieved. Note that this complexity results into only 20 comparisons (steps) for a total of one million elements (Table 8.2). The proof of the above complexity is simple. Specifically, at

| | | | |
|---|---|---|---|
| at step 1 | there are | $N = N/2^0$ | elements |
| at step 2 | there are | $N/2 = N/2^1$ | elements |
| at step 3 | there are | $N/4 = N/2^2$ | elements |
| ... | | | |
| at step Z | there is | $1 = N/2^{Z-1}$ | element |

Solving the last expression: $1 = N/2^{Z-1} \Rightarrow 2^{Z-1} = N \Rightarrow Z = \log_2 N + 1$. Note that Z steps will be executed in the worst case. For example, in Figure 8.29 three steps are executed for the retrieval of 9. However, 13 is reached in two steps and 8 in one. So, the complexity of the binary search algorithm is $O(\log_2 N)$.

The above complexity relates to items that are ordered and structured in a table structure and stored in the main memory (RAM) of a computer. As described in a previous section, the access to the middle element of a table or sub-table (throughout the binary search) is immediate. This complexity is obviously not feasible for ordered data structured in a linked list, as the access to any middle element is done sequentially (after one half of the list is scanned). In other words, the binary search only applies to data elements organized into a table structure.

### 8.7.3 Tree Structures for Quick Search

As shown in the previous Section, the *binary search tree (BST)* is a structure that can support a quick search. On the other hand, the shape of this tree is directly dependent on the order the data elements were inserted into the tree. Figure 8.30 presents three BSTs for the same set of numbers. These trees are resulted from a different order of inserting the same numbers into them.

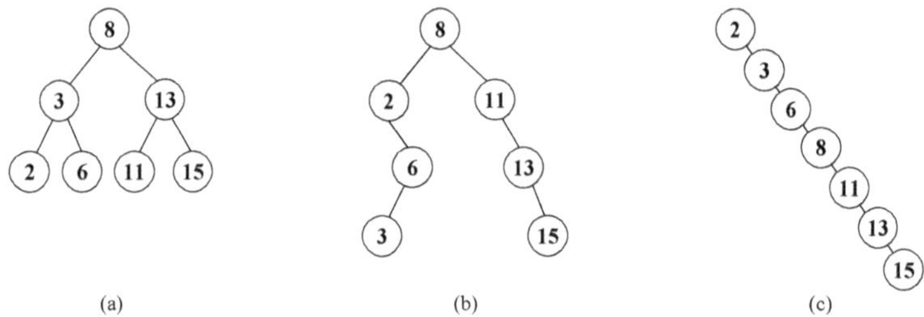

(a)                                                 (b)                                                 (c)

**Fig. 8.30** Three binary search trees for the same set of numbers.

Figure 8.30a,c show the two extreme cases. The tree in Figure 8.30a is a full tree, while the tree in Figure 8.30c was degenerated into a simple linked list (this occurs when the data is entered in ascending or descending order into the tree). On the other hand, Figure 8.30b presents a medium case.

The tree in Figure 8.30a can support the binary search, as its height (maximum path from root to leaf) equals to $h = \log_2(N + 1) - 1$, where N is the number of elements (see the proof in a previous Section). Although a full BST can be built for a static set of data (by inserting elements with an appropriate order into the tree), this does not apply in the general case, where a BST accommodates dynamic elements (i.e., data items inserted and deleted often). In the latter case, the resulting BST is not full and has a height usually larger than $\log_2 N$ (e.g., Figure 8.30b). Hence, the performance of the search is much slower that the desired logarithmic function.

### 8.7.3.1 2-3-Trees

A *2-3-tree* is the basic tree structure that can support a search with logarithmic complexity, even in dynamic datasets. Such a tree has the following properties: (a) leaf nodes host data in ascending order from left to right, (b) all leaf nodes lie at the same level and therefore each path from the root to a leaf has the same length (which equals the height of the tree), (c) any intermediate node has 2 or 3 child nodes, (d) intermediate nodes of the tree are ancillary in the sense that they do not host the actual data, but pointers to the data, (e) intermediate nodes host the smallest element of the second subtree, and (if applicable) the smallest element of the third subtree, and (e) the number of leaf nodes varies from $2^{h-1}$ to $3^{h-1}$, where h is the height of the tree.

Figure 8.31 shows a 2-3-tree hosting the following integers {2,5,7,8,12,16,19}. Notice that all integers are accommodated in the leaf nodes, which all lie at the same level (the tree is balanced). Intermediate nodes and the root have either 2 or 3 children and can accommodate two values: the smallest element of the second subtree and (if applicable) the smallest element of the third subtree. In particular, the root has three children (subtrees) and hosts the values 7 and 16. This means that integers less than 7 lie in the left (first) subtree, integers greater than or equal to 7 and less than 16 lie in the middle (second) subtree, and, finally, integers greater than or equal to 16 lie in the right (third) subtree.

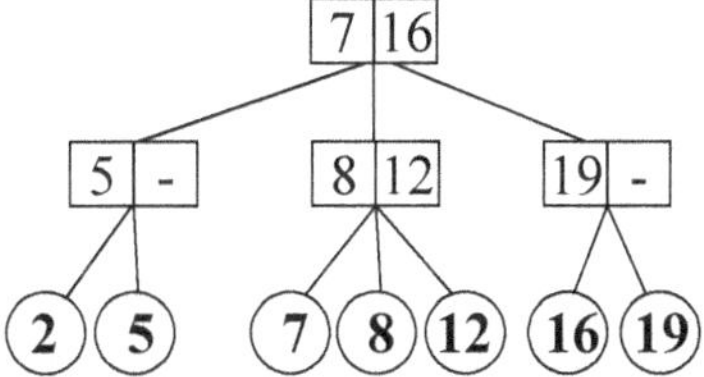


**Fig. 8.31** An example 2-3-tree.

The searching for an element K in the 2-3-tree is as follows: Starting from ancillary data hosted at the root node, if K is less than the smallest descendant of the second subtree, the search proceeds to the leftmost subtree. Else, if K is less than the smallest descendant of the third subtree, the search proceeds to the second (middle) subtree. Else, if none of the above occurs, the search proceeds to the right most subtree. The process is repeated recursively until a leaf node is reached and element K is found.

The logarithmic search complexity in a 2-3-tree is ensured by the fact that the tree is always balanced, i.e., all leaf nodes are lying on the same level, while the height of the tree is maintained short. This is achieved by the application of specific algorithms to insert and delete data elements to and from the tree. The basic difference of the 2-3-tree from the trees considered previously is that it grows and shrinks from the root, rather than from the leaf nodes.

Figure 8.32 shows by way of example (quote of algorithms is avoided in this book) the steps of inserting two new integer numbers (6 and 10) in the 2-3-tree of

Figure 8.31. The insertion of the first integer (6) is simple. Initially, a search is applied to locate the node to host it. Here the position is in the left subtree. A check follows to see if there is space in the tree to accommodate the new number. Initially, there are only two integers grouped together: 2 and 5. As there is space for 6, it is simply grouped with them (Figure 8.32b), while the ancillary data of the ascendant nodes are updated. The insertion of the second integer (10) follows by searching first its position in the tree (Figure 8.32c). In this case, there is no space available to accommodate the new element, as there are already three integers grouped under a non-leaf node. Hence, the immediate ancestor is split into two new nodes (Figure 8.32d), each of which hosts two of the four integers. The two new ancestors along with others must be connected to the tree through their own ancestor node (here the root). But as there is no space (Figure 8.32d), the root will also be split into two nodes to accommodate the four offspring (two in each new node) and a new root will be created for the tree. Obviously, the height of the tree will be increased by one level (Figure 8.32). This is the updated 2-3-tree structure, after the insertion of the two integers.

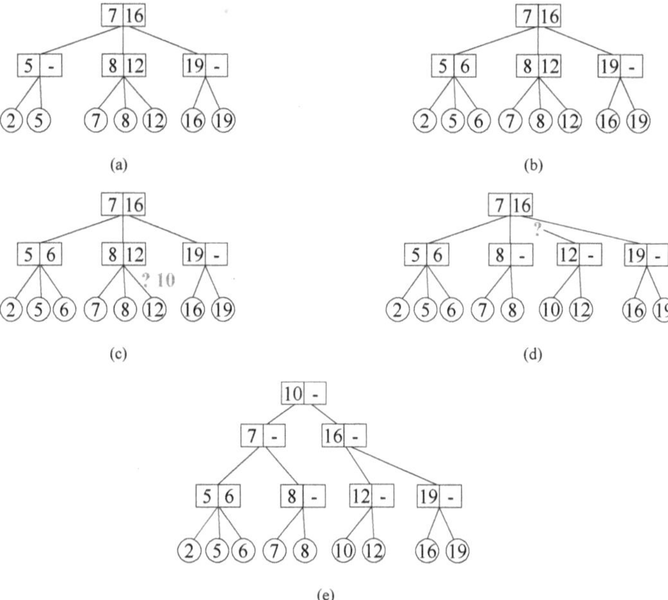

**Fig. 8.32** Insertion of integers 6 and 10 in the 2-3-tree of Figure 8.26: (a) the original tree; (b) insertion of 6; (c) identification of the node to host 10; (d) splitting the node into two new nodes; and (e) splitting the root and increasing the height of the tree.

Note that a similar algorithm can be applied for deleting data elements from a 2-3-tree. Any deletion of elements leads to a new 2-3-tree with a possible shrinkage of the tree (reduction of tree height) from the root.

The cost of searching an element in a 2-3-tree equals to $O(\log_2 N)$, where N is the number of data elements in the tree. This is proven as follows. Each node of the tree has 2 or 3 children. The tree will have its maximum height (which corresponds to the worst search performance), when all its nodes have 2 children. Then the following applies:

| | |
|---|---|
| at level 0 (leaf) | there are $N/2^0 = N$   nodes |
| at level 1 | there are $N/2^1 = N/2$ nodes |
| at level 2 | there are $N/2^2 = N/4$ nodes |
| ... | |
| at level h (root) | there is  $N/2^h = 1$   node |

Solving the last equation: $N/2^h = 1 \Rightarrow h = \log_2 N$, where h+1 the number of nodes to visit when searching (it is equal to the root-leaf path length). So the complexity of search is $O(\log_2 N)$, where N is the number of data elements in the tree.

Note that 2-3-tree and its extensions thereof combine the advantages of both basic data structures: table and linked list. Hence, they provide a rapid (logarithmic) access to data (as in the structure of a table), while they can effectively manage the updates in a dynamic environment (as in the structure of a linked list).

### 8.7.3.2 B+tree

The $B^+tree$ is the most popular structure in traditional (alphanumeric) databases. It is an extension of the 2-3-tree with the following properties: (a) all data values are accommodated in the leaf nodes of the tree, which lie at the same level, (b) the root of the tree has at least two children, (c) each internal node has between $\lceil (M/2)$ and M children, and (d) each leaf node has a pointer to the next leaf node on its right.

With the exception of (d), which is discussed next, the 2-3-trees are $B^+$trees, with M=3 ($\lceil 3/2=2$). Figure 8.33 shows an example $B^+$tree with M=4. This means that the intermediate nodes of the tree have between 2 and 4 children. Notice that in the example of Figure 8.33 each intermediate node and the root accommodate 4 ancillary numbers and four pointers. Each number equals to the maximum value in the subtree of the corresponding pointer.

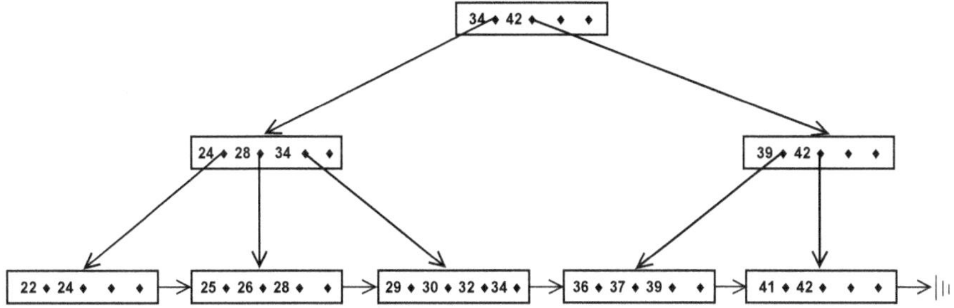


**Fig. 8.33** An example $B^+$tree with M=4.

The algorithms to insert and delete data elements in B$^+$tree are similar to those of the 2-3-tree, described in the previous Section. The search algorithm is also similar and of logarithmic complexity. Specifically, the complexity of searching an element in B$^+$tree is O $(\log_{\lceil(M/2)\rceil}N)$. The proof is similar to the one described for 2-3-trees in the previous Section.

As mentioned above, the leaf nodes of a B$^+$tree are connected to form a linked list. This helps in performing searches of data elements in a *range*. For example, the search of the integers in the interval [26,37] (Figure 8.33) is achieved by locating 26 and then scanning the linked list until 37 is reached. So the complexity of a range search is O($\log_{\lceil(M/2)\rceil}N$ + K), where K the number of data elements in the interval (range).

The leaf nodes of a B$^+$tree form an *index set*, as these nodes can in turn be indexing the records in a database table to support fast (logarithmic) search on these entries. An example is given in Figure 8.34. The field ID (identity) is used for indexing the table records through a B$^+$tree. This way, given the identity of a person, other information (e.g., name, address, etc.) related to this person can be retrieved from a table of records in logarithmic time. This technique is commonly applied in relational databases.

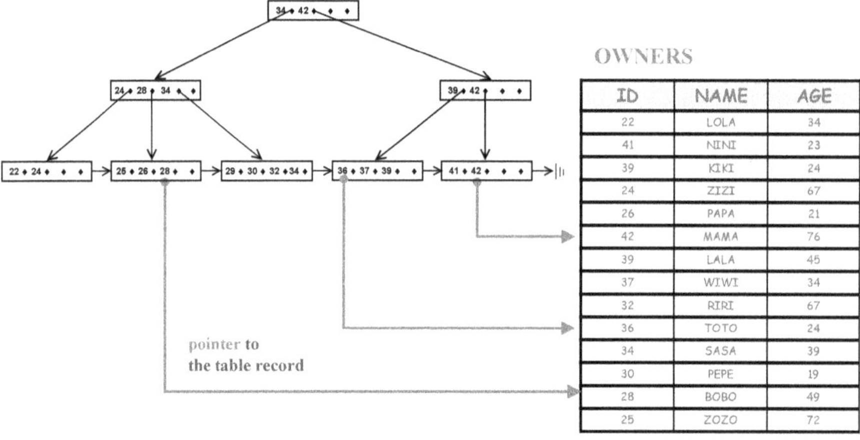


OWNERS

| ID | NAME | AGE |
|----|------|-----|
| 22 | LOLA | 34 |
| 41 | NINI | 23 |
| 39 | KIKI | 24 |
| 24 | ZIZI | 67 |
| 26 | PAPA | 21 |
| 42 | MAMA | 76 |
| 39 | LALA | 45 |
| 37 | WIWI | 34 |
| 32 | RIRI | 67 |
| 36 | TOTO | 24 |
| 34 | SASA | 39 |
| 30 | PEPE | 19 |
| 28 | BOBO | 49 |
| 25 | ZOZO | 72 |

**Fig. 8.34** Indexing the records of a table using a B$^+$tree. Index based on the owners' ID values.

A B$^+$tree can be implemented in either the *main* or *secondary* memory of a computer. In the latter case, which occurs in practice, each node of the tree occupies a *page* in the storage media (disk). To minimize both the pages reserved for storing the tree and the number of input/output (I/O) pages from the secondary to the main memory (note that is a time-consuming task), an appropriate value for M is chosen so that the maximum number of data elements can fit into a disk page. For example, in a computer system with a page size equal to 1024 bytes, a B$^+$tree built on top of 4-byte integers using 4-byte pointers would choose an M equal to 1024:(4+4)=128. Actually, M should be chosen slightly smaller, to leave room for some ancillary data stored per node; such as the current number of entries in the tree, the address of node in the disk, the address of the next node for leaf nodes, etc. The complexity in this case is (assuming M=128) $O(\log_{64}N)$ and relates to the number of pages of the secondary memory being retrieved (read) in the main memory.

## 8.8 Graph Theory

The *theory of graphs* is derived from mathematics and provides algorithmic solutions to many geographic problems, such as finding optimal routes, extracting topological relations, etc. In the following Sections, an introduction to the theory is provided and some representative algorithms are presented. Graphs will be revisited in Chapter 12.

### *8.8.1 Definition of a Graph*

A *graph* G(N,E) consists of a set of vertices or *nodes* (N) and a set of arcs or *edges* (E). An edge is defined by two nodes located at the ends of the edge. The set of nodes in a graph is finite and non-empty. The set of edges may be empty, as it occurs in a point database.

An example graph is given in Figure 8.35. The graph consists of six nodes and eight edges. A graph can simulate the road network of a city or the topological relations (e.g., adjacency) between the countries in a continent. In the first case, each node represents an intersection of the network and each edge a road segment connecting two intersections. In the second case, each country is represented by a node in the graph. Countries (nodes) sharing a boundary are connected with edges.

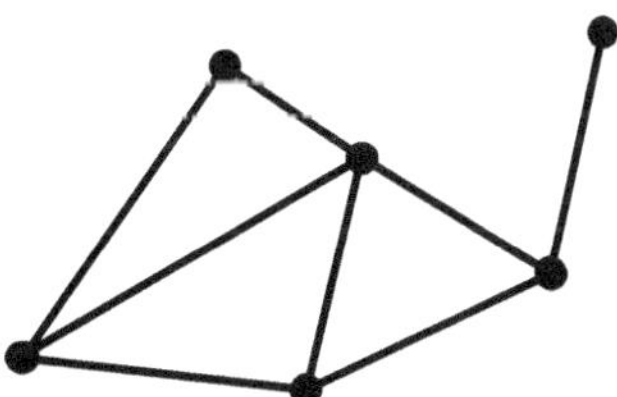

**Fig. 8.35** Example of a graph with six nodes and eight edges.

Figure 8.36 brings together some simple graphs. Each edge starts and ends at a node of the graph. An edge terminating to a node with no connection to another edge is called *blind edge*. Two edges are called *adjacent edges*, if they share a node. On the other hand, two nodes are called *adjacent nodes*, if there is an edge that connects them directly.

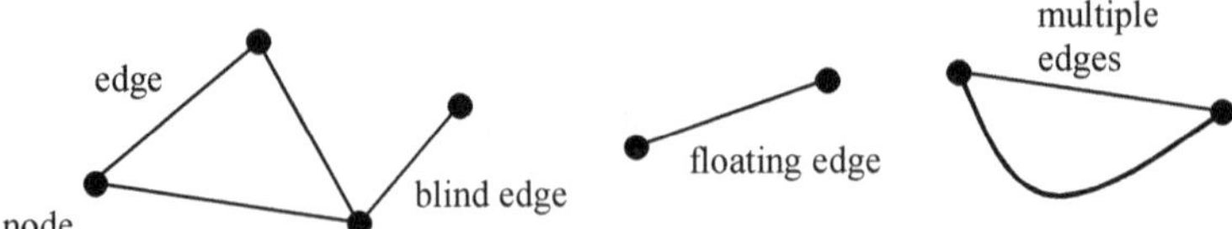



**Fig. 8.36** Types of edges in graphs.

*Degree* of a node is called the number of edges sharing that node. For example, the degree of the node to the floating end of a blind edge is equal to one. An edge connecting two nodes of degree one is called *floating edge*, and it is not connected to any other edge of the graph. The edges connecting the same pair of nodes are called *multiple edges*.

A graph is called *connected*, if there is a continuous *path* defined by a sequence of edges for any pair of nodes (Figure 8.35). Otherwise, the graph is called *non-connected* and it consists of parts called *components* (graph in Figure 8.36 as a whole; a graph with three components). A graph which includes directions along the edges is called *directed* graph (Figure 8.37b). Otherwise, the graph is called *non-directed*.

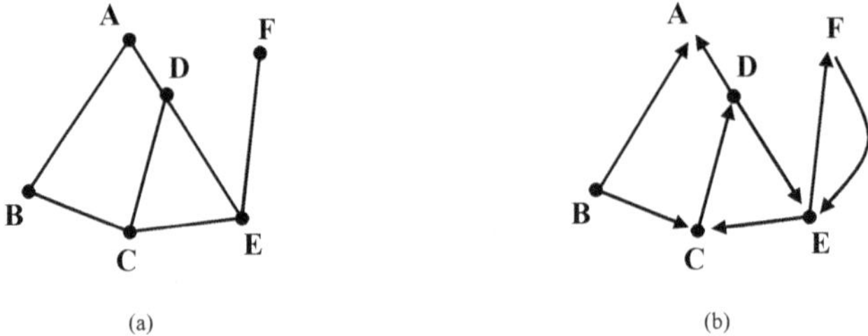



**Fig. 8.37** Example of a non-directed (a), and a directed (b) graph.

The mathematical definition of the graph does not include its graphical representation (figure). In fact, a graph G(N,E) can be drawn in different ways, as shown in Figure 8.38.

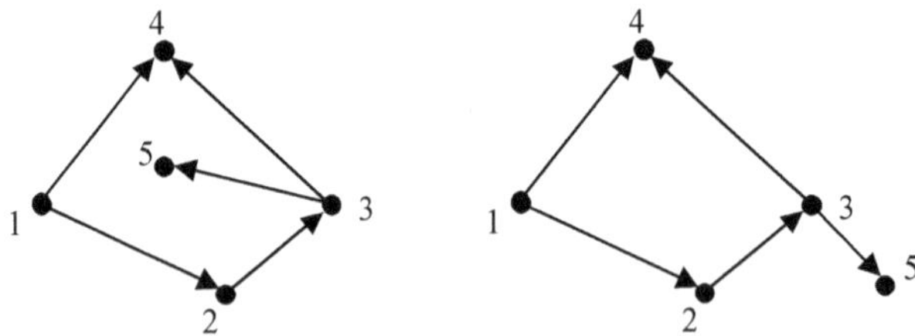



**Fig. 8.38** Two graphical representations of the same directed graph G(N,E). Set of nodes N: {1, 2, 3, 4, 5}; set of edges E: {(1,2), (2,3), (3,4), (1,4), (3,5)}

A graph, in which weights are assigned to its edges, is called *weighted graph*. These weights could represent the length of the edge, cost of traversal, and so on. Figure 8.39 shows an example of a graph with weights.

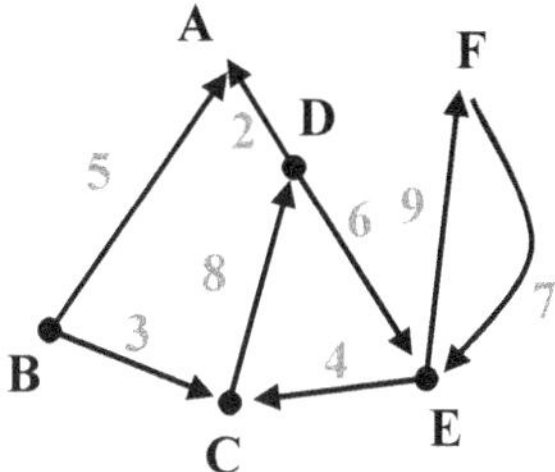

**Fig. 8.39** Example of a graph with weights. The cost of transition from B to E (via C and D) equals to: 3+8+6=17.

## 8.8.2 The Concept of Polygon

A closed chain of edges forms a *polygon* (Figure 8.40). Note that the strict definition of a graph G(N,E) does not include the concept of the polygon. But, as the polygon is a basic geometric element in geographic applications, it is under consideration. Each edge of a closed ring forms a boundary of exactly two polygons. The polygon lying on the left side of the edge (according to its direction) is called *left polygon* and the one lying on the right is called *right polygon*.

A polygon may contain *islets* (Figure 8.40), which in turn can contain other islets. The islets can be *simple* or *composite*. In the first case, the islet consists of an edge, whose start and end nodes are identical (circular islet). In the second case, the islet is delineated by two or more edges.

*Bridge* (Figure 8.40) is an edge whose deletion would lead to the creation of an islet. A bridge has the same polygon on both sides. When an islet is degenerated into a node (a floating node), the blind and floating edges are bridges.

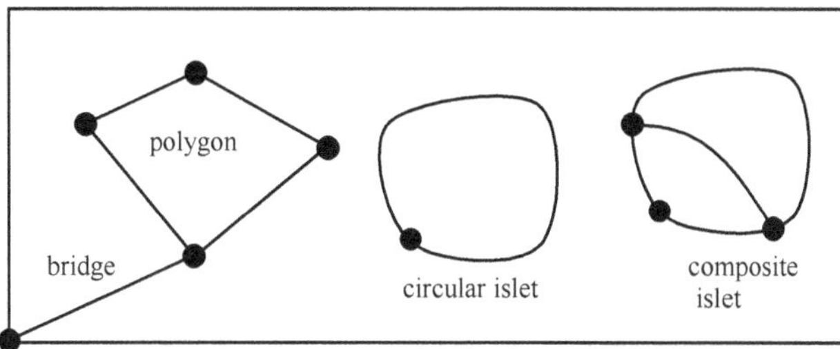


**Fig. 8.40** Polygons in graphs.

## *8.8.3 Planar and Non-Planar Graphs*

Graphs are divided into two categories, depending on whether the nodes and edges are co-planar or not. If they are, they form a *planar* graph; else a *non-planar* graph.

Figure 8.41a shows a non-planar graph. Note that there no nodes at the intersections of some edges. This is not the case in a planar graph (Figure 8.41b).

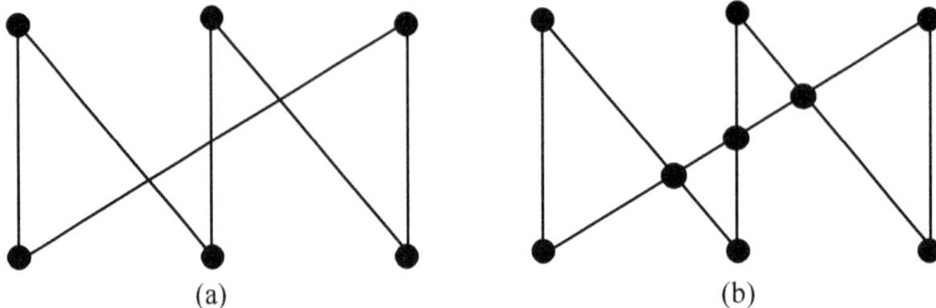

(a)           (b)

**Fig. 8.41** A non-planar graph (a) and a planar graph (b).

### 8.8.3.1 Euler's Condition

It turns out that for a planar graph, all possible representations consist of the same number of polygons. Specifically, if a planar graph has N nodes, E edges, and P polygons, the following equation applies:

```
P - E + N = 2
```

For example, a triangular graph has three nodes, three edges and two polygons (interior and exterior) (Figure 8.42a). According to the equation above: 2–3+3=2. This equation is known as *Euler's condition* and applies to any planar graph (Figure 8.37b).

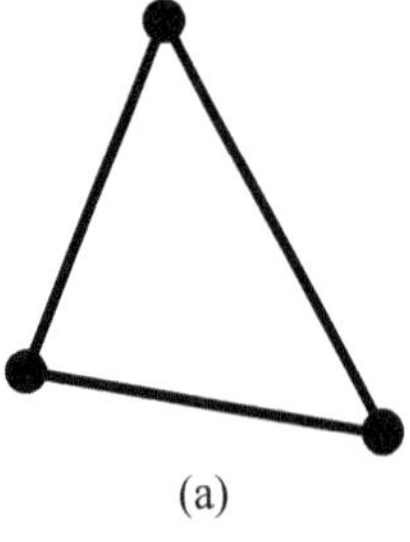
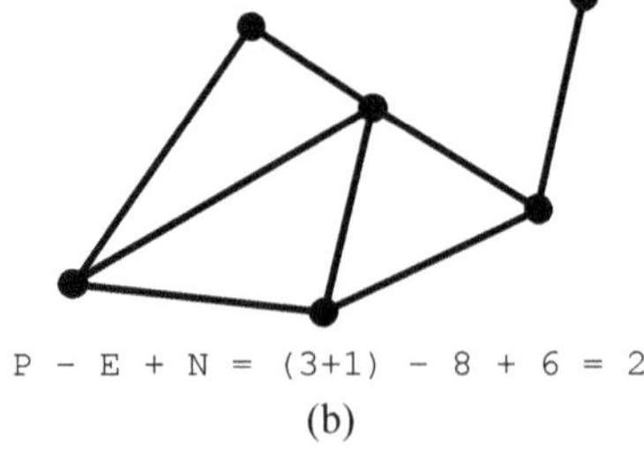



(a)           (b)

**Fig. 8.42** Euler's condition: an example.

Euler's condition can be generalized to include non-connected graphs. In particular, if C is the number of components (associated sub-graphs) of a non-connected graph, including the islets, the Euler's condition takes the following general form:

```
P - E + N - C = 1
```

This equation for the example graph in Figure 8.40 gives:

```
6 - 11 + 9 - 3 = 1
```

Note that Euler's condition provides a check for the *non-planarity* of a graph. However, it is a sufficient but not a necessary condition for planarity. This means that a graph which does not satisfy the condition is non-planar. On the other hand, a graph that satisfies the condition is not necessarily planar either. Euler's condition is commonly applied in GIS (in combination with appropriate algorithms) to detect digitizing errors (such as non-closed polygons, etc.).

## *8.8.4 Representation of a Graph*

To represent a graph in a computer program, appropriate data structures need to be used. The simplest representation is a two-dimensional *table*. Figure 8.43 shows an example representation of a non-directed graph without weights in a table structure. One row and one column of the table are reserved for each node in the graph. Cells in the table are binary, and host an 1 if there is an edge connecting the node in row with the node in column, and a 0 otherwise. Note that the table is symmetric. The space requirements for a graph with N nodes in a table structure are $N^2$, and it is independent on the number of edges in the graph.

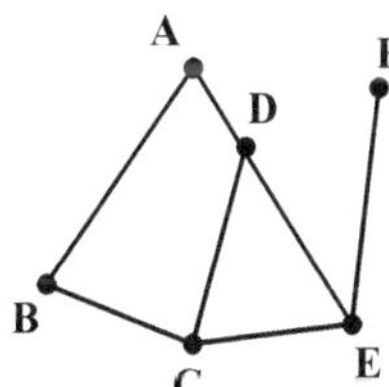

|   | A | B | C | D | E | F |
|---|---|---|---|---|---|---|
| A | 1 | 1 | 0 | 1 | 0 | 0 |
| B | 1 | 1 | 1 | 0 | 0 | 0 |
| C | 0 | 1 | 1 | 1 | 1 | 0 |
| D | 1 | 0 | 1 | 1 | 1 | 0 |
| E | 0 | 0 | 1 | 1 | 1 | 1 |
| F | 0 | 0 | 0 | 0 | 1 | 1 |

**Fig. 8.43** Example representation of a non-directed graph without weights in a table structure.

If the graph is directed, the representation is similar, except that nodes in the rows of the table are assigned the role "from", while those in the columns are assigned the role "to". An example representation of a directed graph without weights is given in Figure 8.44.

In case weights are assigned to the edges of the graph, the representation takes the form shown by example in Figure 8.45. Notice that the cell values of the table are no longer binary, but accommodate the weights of the corresponding edges. The absence of a directed edge connecting two nodes is symbolized by an infinite ($\infty$) weight.

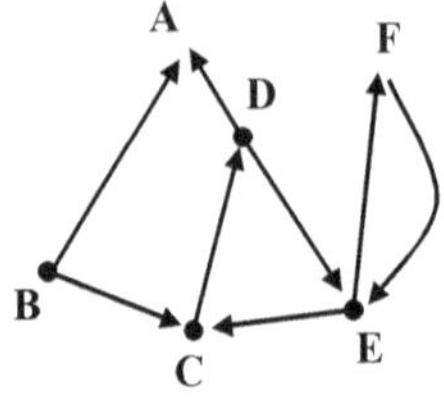

to

| from | A | B | C | D | E | F |
|---|---|---|---|---|---|---|
| A | 1 | 0 | 0 | 0 | 0 | 0 |
| B | 1 | 1 | 1 | 0 | 0 | 0 |
| C | 0 | 0 | 1 | 1 | 0 | 0 |
| D | 1 | 0 | 0 | 1 | 1 | 0 |
| E | 0 | 0 | 1 | 0 | 1 | 1 |
| F | 0 | 0 | 0 | 0 | 1 | 1 |

**Fig. 8.44** Example representation of a directed graph without weights in a table structure.

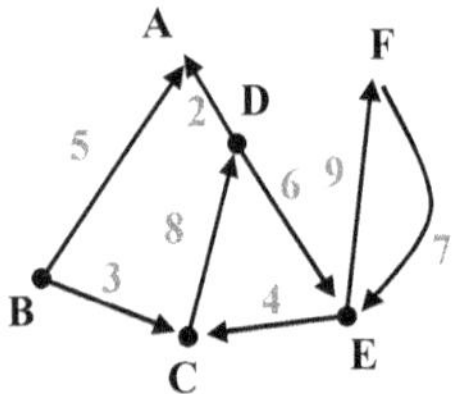

to

| from | A | B | C | D | E | F |
|---|---|---|---|---|---|---|
| A | 0 | $\infty$ | $\infty$ | $\infty$ | $\infty$ | $\infty$ |
| B | 5 | 0 | 3 | $\infty$ | $\infty$ | $\infty$ |
| C | $\infty$ | $\infty$ | 0 | 8 | $\infty$ | $\infty$ |
| D | 2 | $\infty$ | $\infty$ | 0 | 6 | $\infty$ |
| E | $\infty$ | $\infty$ | 4 | $\infty$ | 0 | 9 |
| F | $\infty$ | $\infty$ | $\infty$ | $\infty$ | 7 | 0 |

**Fig. 8.45** Example representation of a directed graph with weights in table structure.

An alternative structure for the representation of a graph in a computer program is the *linked list*. In this case, a linked list is assigned to each node of the graph, hosting all nodes with a direct access through an edge. In case of non-directed graph the list hosts all its neighbor nodes.

Figure 8.46 shows an example representation of a directed graph in linked lists. The storage requirements depend on the number of both the nodes and edges of the graph; and equals to (N+E) linked list nodes. Figure 8.47 shows an example representation of a directed graph with weights in linked lists. In this case, each linked list node also accommodates the weight of the corresponding edge.

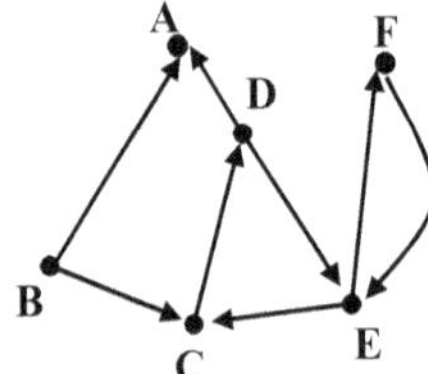

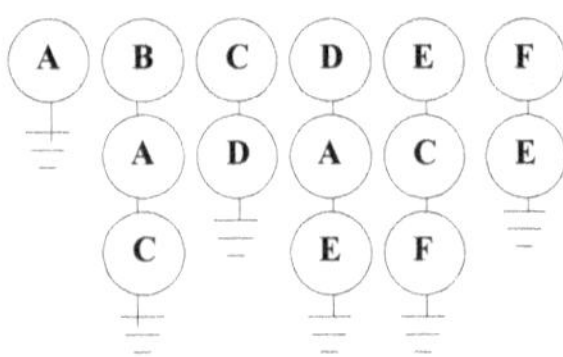

**Fig. 8.46** Representation of a directed graph without weights in linked lists.

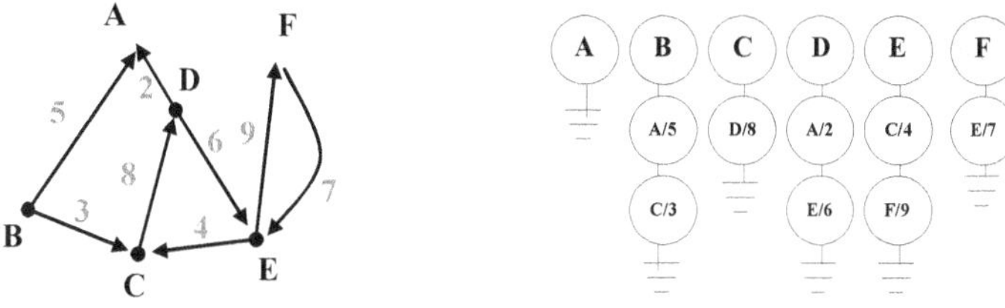

**Fig. 8.47** Representation of a directed graph with weights in linked lists.

## 8.8.5 Graph Algorithms

Graphs have been adopted for the representation of many geographic phenomena and effective algorithms have been developed on top of graphs to solve complex geographic problems. A popular class of algorithms includes those that find the optimal path and the cost of moving from a node of the graph to other nodes in the graph. One of them is presented next.

### 8.8.5.1 Dijkstra's Algorithm

*Dijkstra's algorithm* solves the following problem. Consider a directed graph $G(N,E)$, with weight $W_{i,j}$ to the edge $(i, j)$, where $W_{i,j} \geq 0$, for all $i, j$. Given a reference node $N_o$, the minimum cost for the transition to each node $N_i$ of the graph from the node $N_o$ is sought. For example, given the directed graph with weights in Figure 8.48, and the origin node A, what is the minimum cost of transition from A to the other nodes of the graph?

Dijkstra's algorithm performs the following steps to solve this problem. Initially, it assigns an infinite (very big) cost to all nodes of the graph but the origin node A which is assigned a zero cost. Then, it creates two ancillary sets: (a) the set VN for the nodes that have already been visited during the execution of the algorithm (the set is initially empty), and (b) the set NVN concerning the neighbor nodes of all visited nodes not in VN. A temporal variable CN (current node) hosts the NVN node with minimum accumulated cost throughout the execution. Table 8.3 presents the steps of the algorithm, the successive states of the sets VN, NVN, CN, and the intermediate and final cost of transition from the origin node to the other nodes in the graph of Figure 8.48.

At each step, the algorithm selects the node CN and calculates/updates the cost of the nodes in NVN. For example, in step 2 the cost of the node C is updated to 6 (from 10 in step 1), which is equal to the cost of the path ABC (in replacement of cost of the path AC). Similarly, the cost of the node E is updated. The process is repeated and ends when the set VN equals to the set of nodes N, i.e., all nodes of the graph are visited.

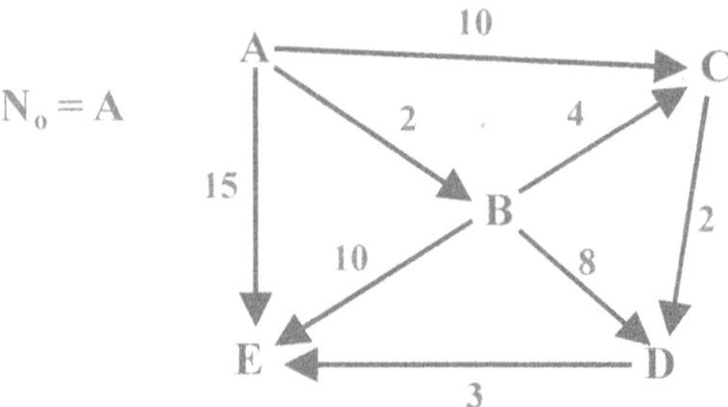



**Fig. 8.48** Example directed graph with weights. Dijkstra's algorithm with origin node: node A.

**Table 8.3** Steps to resolve the Dijkstra's algorithm.

| Step | Visited VN | Next NVN | Curr CN | A | B | C | D | E |
|---|---|---|---|---|---|---|---|---|
| 0 | ∅ | A | | 0 | ∞ | ∞ | ∞ | ∞ |
| 1 | A | BCE | A | 0 | 2 | 10 | ∞ | 15 |
| 2 | AB | CED | B | 0 | 2 | 6 | 10 | 12 |
| 3 | ABC | ED | C | 0 | 2 | 6 | 8 | 12 |
| 4 | ABCD | E | D | 0 | 2 | 6 | 8 | 11 |
| 5 | ABCDE | ∅ | E | 0 | 2 | 6 | 8 | 11 |

The implementation of Dijkstra's algorithm, as described above, gives as output the cost of transition from the starting (origin) node $N_o$ to each node $N_i$ of a linear network. A method of finding the optimal path from the node $N_o$ to a particular (destination) node $N_d$ of the graph, based on the accumulated costs is described next.

From the accumulated cost (C) of the destination node $N_d$, subtract the weight $w_i$ (i=1, ..., k) of the k incoming edges to the node and select as previous node the node $N_i$ with accumulated cost $C_i=C-w_i$. Repeat this process until the origin node $N_o$ is reached.

Consider the graph in Figure 8.48 with $N_o=A$ and $N_d=E$. Based on the results given in Table 8.3 (last step), the optimal route is as follows (in the reverse order of nodes): E (11-3 = 8) → D (8-2 = 6) → C (6-4 = 2) → B (2-2 = 0) → A. Hence, the path is ABCDE.

Dijkstra's algorithm does not apply if the edges are assigned *negative weights*. In this case the graph theory offers alternative algorithms, such as the Ford's algorithm. Obviously, this algorithm prevents the presence of negative rings, i.e., circular paths with a negative overall cost (Figure 8.49).

The complexity of Dijkstra's algorithm is $O(N^2)$. This complexity is prohibitive when the graph is composed of a large number of nodes. To find the optimal paths in big graphs alternative algorithms, which adopts methods from Artificial Intelligence are available. The most popular one is $A^*$ *algorithm* which applies the following basic idea.

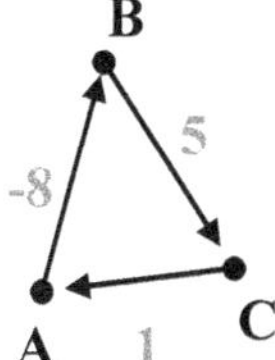


**Fig. 8.49** Example of a negative ring in a graph with weights. The cost of a circular path ABC (after the first passing) is equal to -2.

Initially, an estimate of the transition from the origin node $N_0$ to the destination node $N^d$ is made. This estimate is indicative and not necessarily accurate. Then, in correspondence with Dijkstra's algorithm commuting costs to neighboring nodes of the origin node are calculated recursively. $A^*$ introduces an early exclusion of paths that have an accumulated cost greater than that initially estimated for the entire travel. For a detailed description of the $A^*$ algorithm, the reader is referred to artificial intelligence or expert systems textbooks.

### 8.8.5.2 Travel Salesman Problem

The *travel salesman problem (TSP)* is about finding the path to be followed by a salesman distributing products to a network of cities (or stations), so that s/he visits all the cities and traces a path with minimal cost. The cities and the roads connecting them are represented by a graph, and graph algorithms are applied to solve the problem. The complexity of solving the TSP is non-polynomial as all possible paths must exhaustively be considered (Figure 8.50). Hence, the solution to the problem is highly time consuming and prohibitive for large graphs.

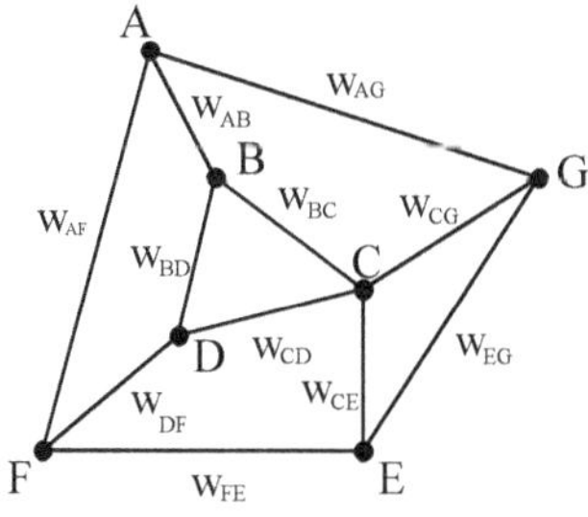


(a)

| path 1 | A | F | D | B | C | E | G | A |   |
|--------|---|---|---|---|---|---|---|---|---|
| path 2 | A | B | D | F | E | C | G | A |   |
| path 3 | A | B | C | E | F | D | C | G | A |
| path 4 | A | G | E | C | D | F | A | B | A |
| . . . | . . . | . . . | . . . | . . . | . . . | . . . | . . . | . . . | . . . |

(b)

**Fig. 8.50** Example of the travel salesman problem (a) and some of the possible routes with A as starting/ending node (b).

For this reason, several algorithms have been proposed to solve the problem based on *heuristics*. These algorithms can offer a good, but not the best solution. A heuristic method to solve the travel salesman problem could be the following.

Initially, a node is selected as origin node and it is denoted as v. Then, the non-visited outgoing edge from v with the minimum weight is chosen in priority, e.g., the edge (v,u). After chosen, the edge is labeled as visited and takes a lower priority in a later re-crossing. The neighboring node u is renamed to v and the process repeats until all nodes are visited and the starting node is reached again.

### 8.8.5.3 Coloring Polygons in Thematic Maps

Another problem solved with graphs and appropriate algorithms is that of *coloring the polygons* in a thematic map with the minimum number of colors. For example, the problem could be the following: "how many colors are needed to color the countries of Europe?" Obviously, there is a limitation as two countries sharing a boundary cannot be assigned the same color.

This problem can be modeled as a graph problem, in which nodes correspond to the polygons (countries) and edges represent the topological relationship of adjacency (touch) between them. Figure 8.51 shows a representation of a map of polygons with a graph. The algorithms to choose a color for them are not included in this textbook.

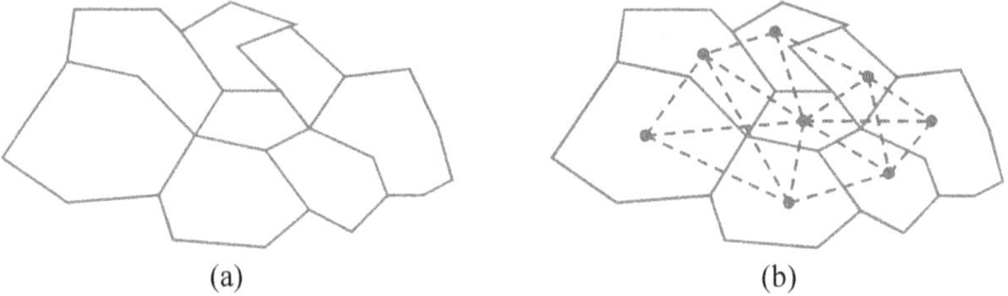

(a)                             (b)

**Fig. 8.51** Representation of a map of polygons (a) with a graph (b).

## References and Further Reading


Esakov, J., and Weiss, T., 1989. *Data Structures: An Advanced Approach Using C.* Prentice Hall.

Horowitz, E., and Sahni, S., 1999. *Fundamentals of Data Structures.* Freeman & Co.

Knuth, D.E., 1973. *The Art of Computer Programming.* Addison-Wesley.

Sedgewick, R., 1990. *Algorithms.* Addison-Wesley.

Stefanakis, E., and Kavouras, M., 2002. *Navigating in Space Under Constraints. International Journal of Pure and Applied Mathematics (IJPAM).* Vol. 1, pp. 71-93. Academic Publications.

Van Wyk, J.C., 1990. *Data Structures and C Programs.* Addison-Wesley.

# PART III

# GEOGRAPHIC DATABASE SYSTEMS TECHNOLOGY

# 9 Geographic Representation Models

## 9.1 Introduction

Assume a conversation by phone with a friend of ours, who asks us to describe the image that we see from our window. How would we describe what we see? An approach that we usually follow in our discourse is the segmentation of the landscape into units, such as buildings, roads, hills, valleys, rivers, sea, etc. The description of each of these units is enriched with the characteristics of the features, e.g., the building is tall, the road has two lanes, the hill is green, etc., and the use of geo-reference terms, such as in, right, left, in front of, at the top of the hill, etc.

This way we transmit through words and statements (in English) a picture of reality to our partner. This is accomplished using appropriate *constructs*, which make up a *representation model* of the reality.

Obviously, the *constructs* (i.e., the building blocks of the model), which are used to describe the reality depend on its content, or more accurately the image we want to convey to our friend (i.e., selected elements of reality); but are also influenced by the experience and intellectual culture, both ours and that of our partner.

If our partner is replaced by a computer or a particular software program, we must follow the rules and restrictions imposed and the underlying software technology, in modeling the reality.

Generally, the term *model* denotes a set of appropriately selected concepts or constructs for mapping parts of a field, the *source field*, in another field, the *target* (or *image*) *field* (Figure 9.1).

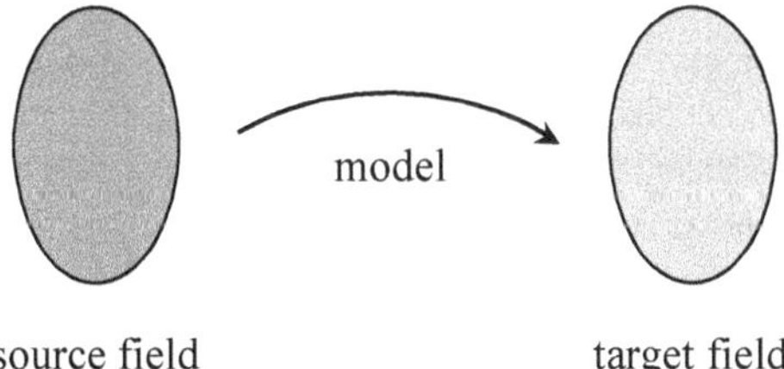


**Fig. 9.1** Mapping a source field to a target (image) field.

The representation of reality in a computer system involves two models (Figure 9.2): (a) *representation model*, and (b) *data model*. The representation model results in a picture of reality as a human perceives it. The data model reflects a human's perception in a computer representation.

The models are divided into *lossy* and *lossless* models. A lossy model provides a target field which cannot reproduce the source field as data is lost or ignored in the mapping process (Figure 9.1). In contrary, a lossless model maps all properties of the source field to the target field and an inverse mapping can reproduce the source field. Typical examples of lossless models are the monotone mathematical functions in $\Re$, e.g., $y = x\text{-}5 \Leftrightarrow x = y+5$.

The task of representing the reality or portion thereof, e.g., in the telephone conversation mentioned above, undoubtedly involves a lossy model. This means that the target field (representation of reality) cannot be used to reconstruct the source field (reality) with completeness. Therefore, the target field differs from reality itself. There are several reasons for this lossy representation. These include: (a) the subjectivity of the observer, (b) the deficiencies of data collection techniques, (c) the deliberate omission of details of reality that are irrelevant for the application (the caller), and (d) the weak concepts and constructs of the representation model fail to accommodate all aspects of reality.

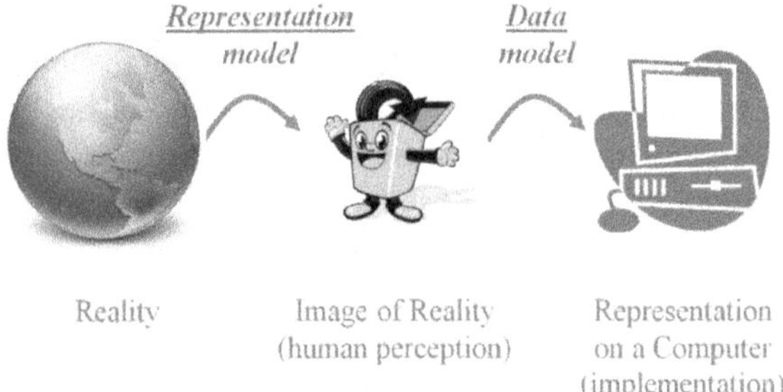


**Fig. 9.2** Models for the representation of reality in a computer system.

This Chapter focuses on alternative ways of describing the reality, i.e., the representation models of reality. As the perception of reality varies according to users and applications, there is no single representation model that meets all needs. The advantages and disadvantages of alternative models are highlighted in the following Sections.

Note the clear distinction between the representation models of reality from the data models (Figure 9.2) discussed in the previous Chapter. The first relate to the human perception of reality (abstraction, representation, and classification of geographic phenomena in the human mind), while the second is inherent with building this perception into a database by applying standardized constructs depending on the level of design (conceptual, logical, physical; see Chapter 6).

## 9.2 Representation Models

The reality is composed of geographic data, which can be perceived at a glance out the window. As discussed in Chapter 1, the reality (or miniworld) comprises a set of entities, which have three dimensions: identity, spatial, and thematic. These dimensions may vary over time and be assigned quality measures.

In the following Sections, the discussion focuses initially on a *static space*, which comprises entities with static spatial and thematic dimension. The discussion evolves to the representation of *space-time*, i.e., entities with dynamic spatial and thematic dimension. Finally, representation models which consider the *quality* of geographic data are examined.

## *9.2.1 Representation Models of Space*

The representation models of space describe a static world in which the temporal variation of geographic entities is ignored. There are two basic questions that these models must answer: (a) *what is it*; and (b) *where is it* located.

When the description of reality (e.g., a room, city, or country) is sought, different approaches may be applied to generate the target field. The two extreme approaches are (Figure 9.3): (a) the perception of space as a set of *discrete entities*, and (b) the perception of space as a *continuous field*.

These two approaches are complementary. This means that neither of them can substitute the other and provide an efficient representation for all geographic data across geographic applications.

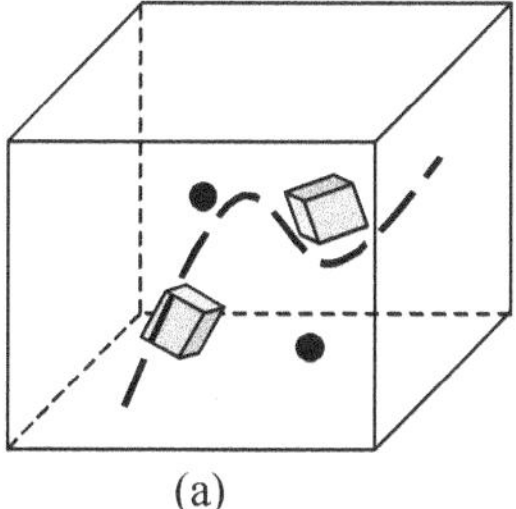
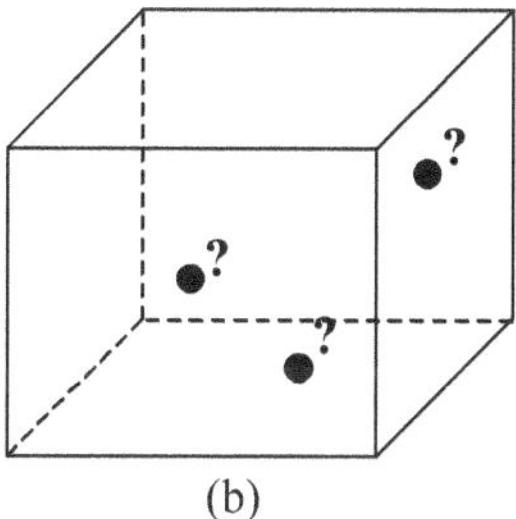

(a)          (b)

**Fig. 9.3** The two basic approaches for the perception of space.

### 9.2.1.1 Space as a Set of Discrete Entities

The first model describes the space as a set of *discrete entities*. Specifically, the space is considered empty, except for the area occupied by discrete entities. For example, a village consists of roads, houses, cars, and other discrete entities. A key feature of this approach is that the entities are distinct from each other and can be counted; for example, the village has 150 houses. Generally, entities in this model are required to be: (a) *identifiable*, (b) *relevant* to the application, (c) with *sharp boundaries*, and (d) amenable of *description*, i.e., to have attributes relevant to the application.

The space has three dimensions (or four in a dynamic environment). Often, for practical reasons or simplicity, the space is depicted on a sheet of paper in two dimensions as it is shown to an observer from above (Figure 9.4). In this view, the discrete entities are represented by points, lines, and polygons, or a combination of these. Correspondingly, the discrete entities are of *geometry type* point (zero-dimensional, e.g., a well), line (one-dimensional, e.g., a road), or polygon (two-dimensional, e.g., a forest). Obviously, the geometry type depends on the entity form and the scale of the representation (*resolution*). For example, an airport in a high, low, or an even lower scale can be represented with a polygonal, linear, or point geometry, respectively (Figure 9.5).

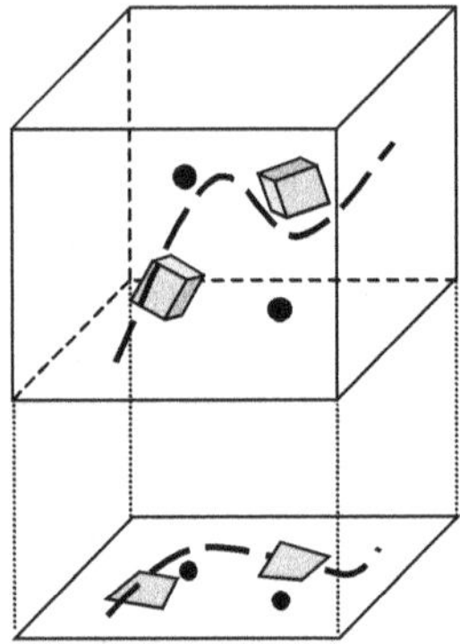

**Fig. 9.4** Mapping of three-dimensional space in two dimensions.

**Fig. 9.5** An airport represented with different geometry types in different scales.

In many cases the perception of space as a set of discrete entities contains ambiguities. It is not always easy or possible to define an entity. Although it is simple to define artificial objects (e.g., a car, a bridge) or living beings (e.g., a bird, a mammal), this does not apply to other entities, such as a mountain, a forest, or a swamp. What is a mountain? Where is its boundary? What does differentiate it from a hill? Moreover, it is not always easy or possible to assign values to attributes that describe an entity. What type of vegetation would be assigned to the entity mountain? Obviously, the vegetation varies from point to point. Hence, the description of the world with discrete entities is not always straightforward or even feasible.

### 9.2.1.2 Space as a Continuous Field

The alternative model considers the space as a continuous surface or a *continuous field*. Each position of this field can be assigned values for one or more attributes (variables). For simplicity, the three-dimensional space is restricted to a plane or a surface on which the behavior of a phenomenon (recorded as attribute values) is examined (Figure 9.6). For example, a continuous field can describe the vegetation (variable: vegetation type), soil type (variable: ground cover), land use (variable: land use), or elevation (variable: altitude).

As discrete entities are categorized based on their geometry type into point, line, or polygon entities, continuous fields can be classified by (a) *what varies*, and (b) *how smoothly it varies*. The values in a continuous field can vary smoothly between neighboring points, as in the case of field elevation (excluding cliffs); or abruptly (discontinuously), as in the case of the field of land use (across the boundary separating two different uses).

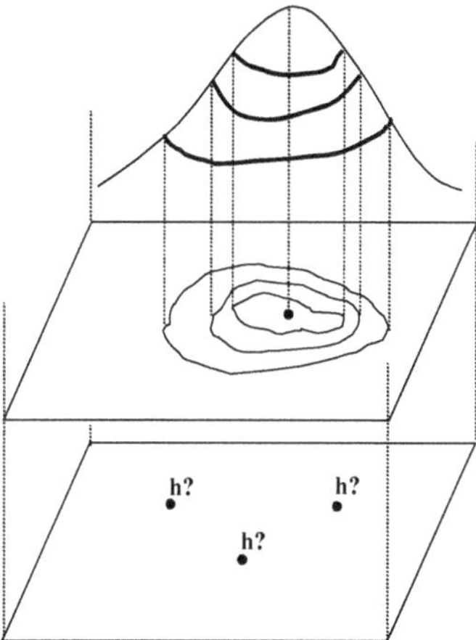

**Fig. 9.6** The space as a continuous field.

## 9.2.2 Thematic Layers

Often, it is convenient to consider the study area as a set of *sub-spaces* (Figure 9.7), each of which hosts a category of data and differs from others by *theme* (thematic dimension). For example, a settlement may be regarded as a composition of sub-spaces: roads, houses, people, etc. By doing so, the distinction of detail (i.e., a single theme) from the total is achieved.

Each sub-space is called *thematic layer* and is represented as a continuous field or a collection of discrete entities (of this theme). The individual thematic layers can overlay each other, in other words be combined location by location or entity to entity, to yield findings related to the correlation of geographic data.

## 9.3 Spatial Model

A *representation model*, introduced by *Tomlin*, handles in an elegant manner both perceptions of geographic space (Figure 9.3): (a) collection of discrete entities; and (b) continuous field.

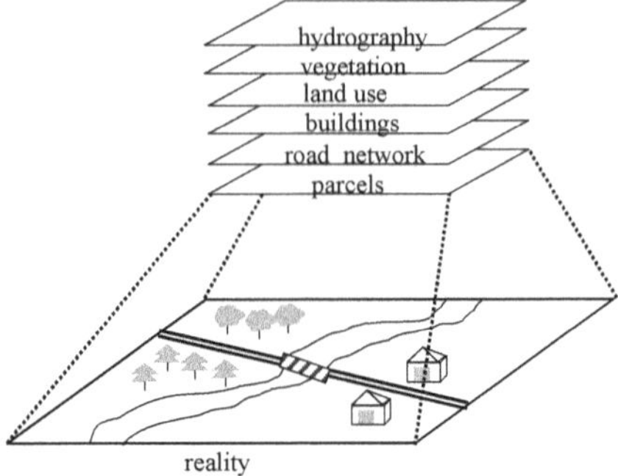



**Fig. 9.7** Example thematic layers.

In this general representation model, geographic information can be viewed as a hierarchy of data (Figure 9.8). At the highest level, there is a *map* which is a collection of *thematic layers* (more commonly referred to as *layers*), all of which are in registration (i.e., they have a common coordinate system) (Figure 9.9a). Each layer is partitioned into *zones* (regions), where a zone is a set of *individual locations* with the same attribute value (Figure 9.9b).

Examples of layers are the land-use layer, which is divided into land-use zones (e.g., wetland, river, urban, desert, city, park and agricultural zones), and the road network layer, which contains the roads that intersect the portion of space that is covered by the layer.

Tomlin's model is very effective as it applies regardless of the approach adopted for the representation of reality (Figure 9.3). Specifically, this is accomplished by introducing the concepts of zone and location in the hierarchy (Figure 9.8). Thus, each discrete entity, e.g., a road or a parcel, can be represented as a zone in this model. The zone is defined as the set of locations that make up the road or parcel. On the other hand, a continuous field, e.g., field of vegetation, is represented in the model as a thematic layer which comprises all locations in space.

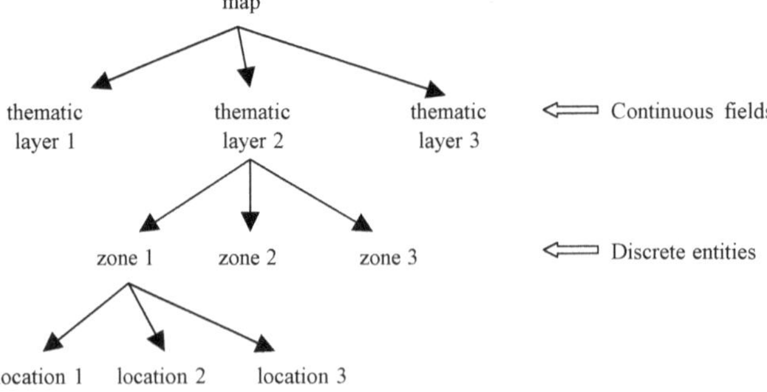



**Fig. 9.8** The hierarchy data model by Tomlin.

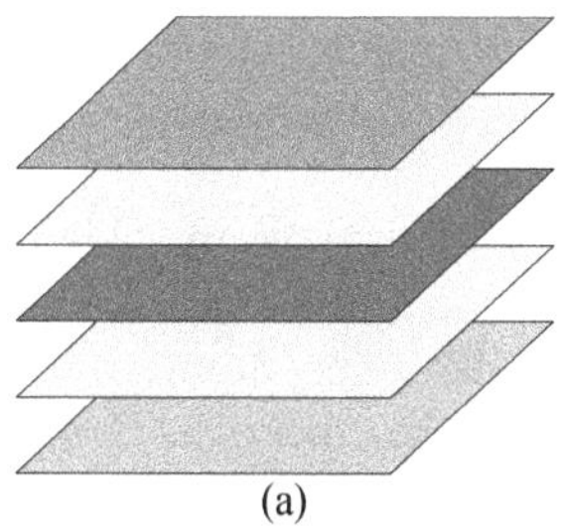 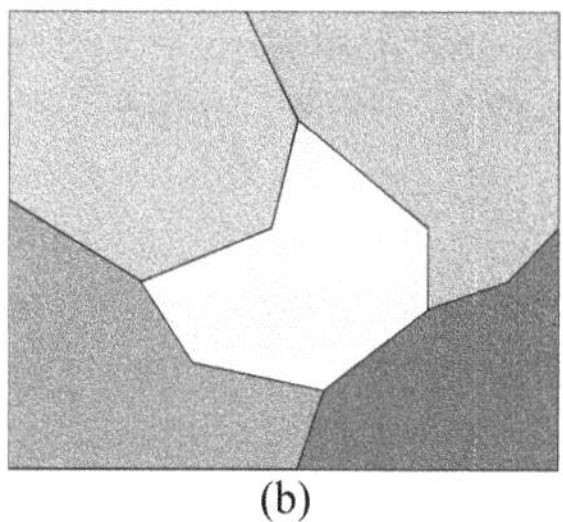

(a)          (b)

**Fig. 9.9** Tomlin's model: the map layers (a), and the zones of a layer (b).

## 9.4 Spatial Relations

The *spatial relations* among geographic entities originate from their relative positions in space (e.g., north of, adjacent to, within, near, etc.). The most important spatial relations are as follows:

- *Topological relations:* all relations that remain invariant in topological transformations, such as rotation, translation, and scale change. Example topological relationships are the proximity and overlap (see next section).
- *Direction relations:* all relations that are derived after the definition of a basic direction and the extraction of relevant direction patterns (e.g., front of, left of, etc.). Representative group of direction relations are these that classify entities based on the direction of North, e.g., North of, West of, etc. Each direction relation that applies between two entities is also described by an equivalent inverse relationship that relates them. For example, the relation A in front of B (A in North of B), is the inverse of the relationship B back of A (B in South of A).
- *Metric relations:* all relations that provide distance estimations between the entities in space. For example, the relations close by or far apart relate geographic entities and return a distance estimate between them. Note that these relations are inherently vague concepts, e.g., near or far are both relevant and subjective terms.

The spatial relations are generally numerous and complex. In the following Section, topological relations are examined closely.

### *9.4.1 Topological Relations*

*Topology* refers to the geometric relations among objects, which remain unaffected by the imposition of topological *transformations*: *shift, rotation,* and *scaling.* Example topological relations in two dimensional space are shown in Figure 9.10.

Consider the geometric objects a, b, c, d, ..., as represented on a rubber sheet. Topology relates to all those properties of geometric relations thereof kept unchanged when exerting stresses in the elastic sheet causing creasing, twisting, stretching, or shrinking, etc., except the *partition*.

Observing the objects in Figure 9.10, before and after the two alternative deformations, it is evident certain spatial relations among the objects have substantially been modified. For example, the lengths of the edges, the areas of polygons as well as their direction relations were affected. On the other hand, some relations remain unchanged. Specifically, object c still lies inside b. Also, objects a and b continue to be adjacent and still share an edge. These are examples of topological relations.

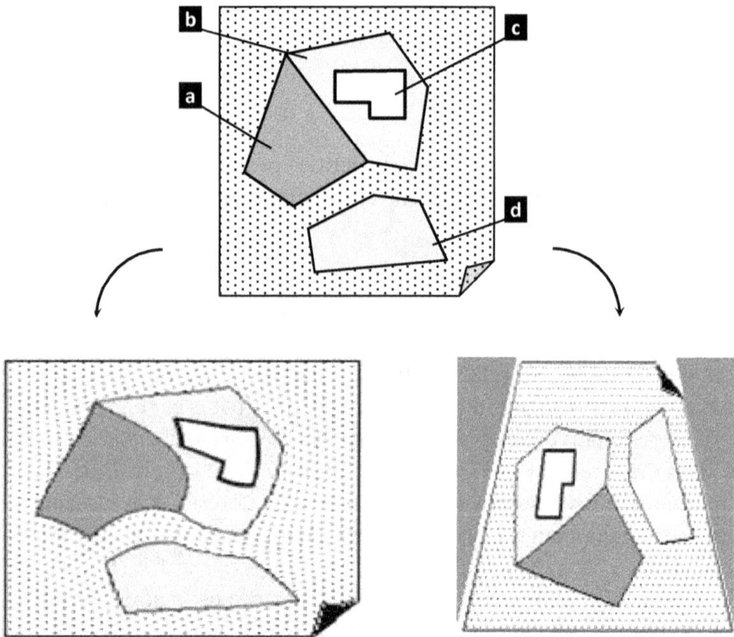


**Fig. 9.10** Deformations and topological relations.

In the past, several models have been proposed for the representation of topological relations. An efficient model is described next. This model takes into account the outline and the inner area of each entity, as it is represented in two-dimensional space. In comparing two entities, to identify their topological relations, the intersections between outlines and inner areas of the two entities are checked.

Specifically, if $O_A$, $O_B$ are the outlines (border lines) of entities A and B, and $I_A$, $I_B$ are their inner areas, respectively, the following four intersections are checked: (a) intersection of outlines: $O_A \cap O_B$; (b) intersection of inner areas $I_A \cap I_B$; (c) intersection of the outline of the first entity with the inner area of the second $O_A \cap I_B$; (d) the intersection of the inner area of the first entity with the outline of the second $I_A \cap O_B$. Depending on whether these intersections give or not an empty set as a result, they derive $2^4=16$ possible combinations. These correspond to eight basic topological relations (known as *Egenhofer's relations*) between two polygonal entities A and B listed in Table 9.1.

Figure 9.11 shows the topological relations among two entities in two-dimensional space. Note that these relations are extended accordingly in the space of three or more dimensions. Also, each polygonal entity (A or B) can be degenerated into a line or point entity to describe the topological relations between entities of different geometric types, e.g., line versus polygon, point versus line, point versus polygon, etc.

**Table 9.1** The eight basic topological relations between polygons in the space of two-dimensions.

| | $O_A \cap O_B$ | $I_A \cap I_B$ | $O_A \cap I_B$ | $I_A \cap O_B$ |
|---|---|---|---|---|
| A disjoint B | $\varnothing$ | $\varnothing$ | $\varnothing$ | $\varnothing$ |
| A meets B | $\neq\varnothing$ | $\varnothing$ | $\varnothing$ | $\varnothing$ |
| A overlaps B | $\neq\varnothing$ | $\neq\varnothing$ | $\neq\varnothing$ | $\neq\varnothing$ |
| A inside B | $\varnothing$ | $\neq\varnothing$ | $\neq\varnothing$ | $\varnothing$ |
| A contains B | $\varnothing$ | $\neq\varnothing$ | $\varnothing$ | $\neq\varnothing$ |
| A covers B | $\neq\varnothing$ | $\neq\varnothing$ | $\neq\varnothing$ | $\varnothing$ |
| A covered by B | $\neq\varnothing$ | $\neq\varnothing$ | $\varnothing$ | $\neq\varnothing$ |
| A equals B | $\neq\varnothing$ | $\neq\varnothing$ | $\varnothing$ | $\varnothing$ |

# 9.5 Spatio-Temporal Representation Models

In previous Sections the assumption that the world is static, i.e., does not change over time, was made. This assumption although compatible with several geographic applications, it is not compliant with the representation of *dynamic phenomena*. In the latter case, the space (set of discrete entities or continuous field) must be accompanied by *temporal information*.

Next, two simplified spatio-temporal applications are presented. Then, the concepts of time-stamp and temporal relations are introduced. Finally, an extended spatial model to also include the temporal dimension and describe the dynamic space is discussed.

## *9.5.1 Spatio-temporal Applications*

Two simplified applications involving geographic entities that vary over time are briefly described next to highlight the requirements for the modeling and analysis of spatio-temporal entities. The first application refers to moving point objects, while the second to dynamic polygon features.

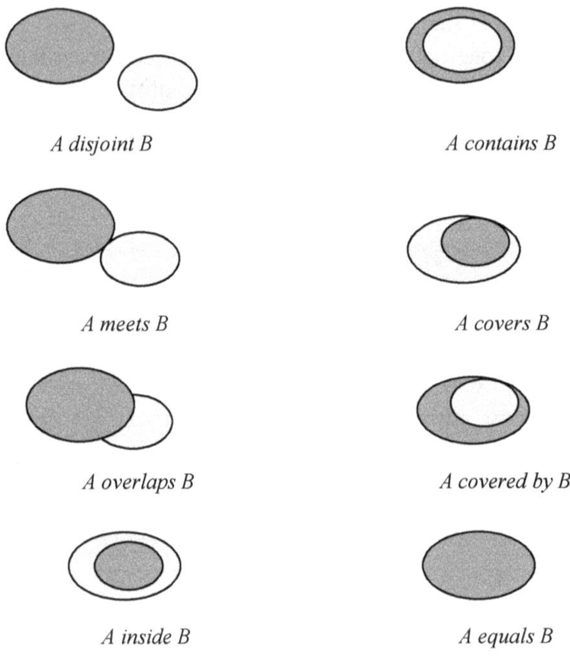



**Fig. 9.11** Topological relations among two polygons (A and B) in the two dimensional space (A refers to the dark polygon).

## *Case 1: Sea Vessel Trajectories*

In maritime world there is a need for modeling and analysis of vessel trajectories. This problem can be viewed as the representation and reasoning on point objects whose position changes over time, i.e., moving point objects. Figure 9.12a illustrates a simplified example of two vessel trajectories (A, B) in the territorial waters (TW) of a country. Assuming a navigation on a plane surface, each trajectory consists of a set of triplets $(X_i, Y_i, T_i)$, where $(X_i, Y_i)$ indicates the location of the vessel at time $T_i$. Some typical queries regarding this situation are:

1.   Does the vessel pass through point (X,Y)?
2.   When does the vessel pass through point (X,Y)?
3.   Where is the vessel at time $T_k$?
4.   Does the vessel cross the national boundary (NB) on July the 3rd?
5.   Which is the shortest distance of the vessel from the sea shore (SSh)?
6.   Do two trajectories (A, B) intersect?
7.   Do the two vessels collide? (a distance threshold is involved in this query)

8. What is the shortest distance between two trajectories (A, B)?
9. What is the shortest distance between the two vessels?
10. What is the vessel velocity at time $T_k$?
11. What is the length of the vessel trajectory between two locations?
12. What is the duration of the trip between two locations?

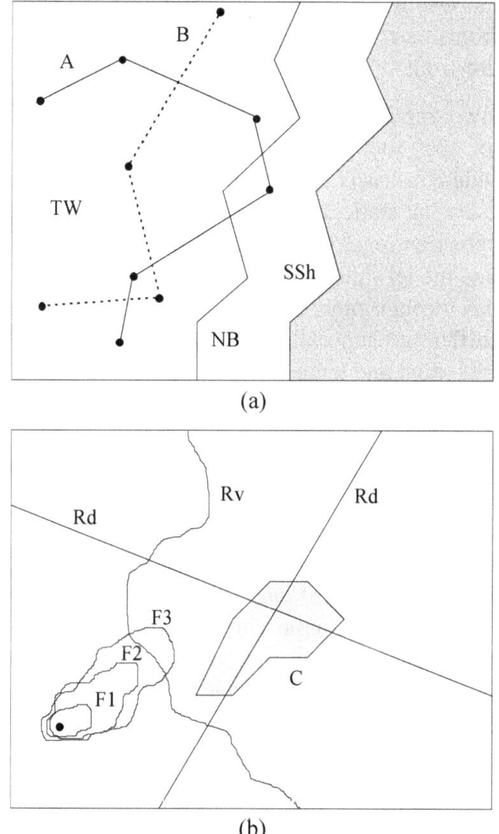


**Fig. 9.12** Example applications with spatio-temporal entities.

## Case 2: Forest Fire Spread

In environmental applications, there is a need to represent objects whose position, shape and size change over time. Focusing on forest management, the representation and analysis of areal units that change over time, such as the burning area in case of fire, is a common example. Figure 9.12b illustrates three snapshots ($F_1$, $F_2$, $F_3$) of the burned area in a forest. Assuming a fire on a plane surface (a projection of the earth surface), the burned area consists of a set of pairs ($GEO_i$, $T_i$), where $GEO_i$ describes the geometry of the area at time $T_i$. In general, $GEO_i$ is a composite polygon (areal unit), possibly with holes (islands), or multiple components (compound object). Some typical queries regarding this situation are:

1.  Does the fire pass through point (X,Y)?
2.  When does the fire pass through point (X,Y)?
3.  Where is the fire front at time $T_k$?
4.  What is the burned area at time $T_k$?
5.  Does the fire cross the road (Rd) or the river (Rv) at time $T_k$?
6.  What is the shortest distance of the fire from the city (C)?
7.  Do two fire-fronts meet (in case of more than one fires)?
8.  What is the fire-front velocity at time $T_k$?

Examining the two applications above, several of the requirements for modeling and analysis of spatio-temporal data are highlighted. Specifically, the data model should provide constructs for the representation of change over time. Most geographic entities are not static, as considered in conventional GIS, but dynamic. That is, both their position (e.g., vessel) and shape (e.g., burned area) change over time. Hence, the data model should be able to represent: (a) the location and shape of individual entities through time; (b) the state of a geographic area (i.e., collection of entities) at different temporal instants; and (c) the distribution of geographic entities at various spatial and temporal scales (resolutions).

## 9.5.2 Time-stamps

Temporal information is represented through *time-stamps*. There are three basic types of time-stamps: (a) *instant*, (b) *interval*, and (c) *spans*.

An *instant* time-stamp is a simple chronon identifier (e.g., on 2014.06.05 at 10:32am UTC). An *interval* time-stamp consists of two instant time-stamps, one for the starting instant of the interval and one for the terminating instant of the interval (e.g., on 2014.06.05 from 10:32am to 10:42am UTC). The temporal distance between these two specifies the *duration* of the interval (e.g., 10 minutes). If the two instances correspond to the same time, the interval is degenerated to an instant time stamp (zero duration). A *span* is a directed duration of time with known temporal length but no specific starting and ending instances. For example, the duration of "one week" is known to have a length of seven days, but can refer to any block of seven consecutive days in time. Time-stamps may be associated either with discrete entities (their attributes) or continuous fields.

## 9.5.3 Temporal Relations

Time is a one-dimensional (linear) space and the time-stamps are either point or linear elements in this space, depending on their type. Among the time-stamps *temporal topological relations* apply to characterize either distinct entities and their attributes or continuous fields.

Specifically, seven topological relations (Allen's temporal relations) take place between two interval time-stamps A and B (Figure 9.13). Note that any of the intervals A, B can be degenerated into an instant time-stamp (point in time with zero duration).

| Temporal relationship | Definition |
|---|---|
| *A* before *B* | 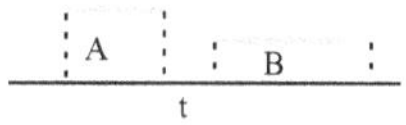 |
| *A* equal *B* | 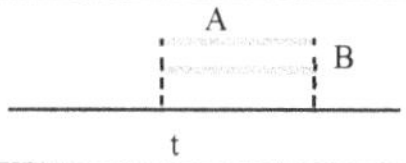 |
| *A* meets *B* | 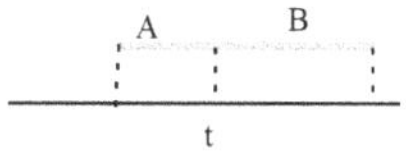 |
| *A* during *B* | 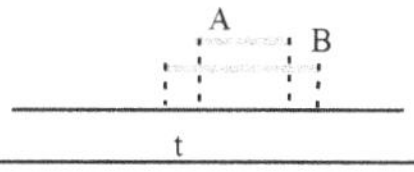 |
| *A* overlaps *B* | 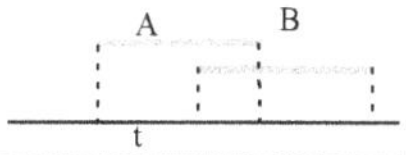 |
| *A* ends *B* | 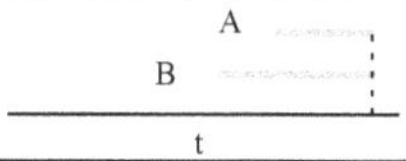 |
| *A* starts *B* | 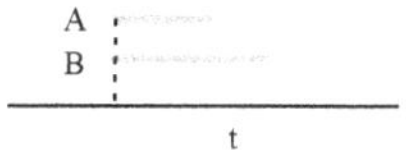 |

**Fig. 9.13** Topological relationships between two interval time-stamps A and B.

## 9.5.4 A Spatio-temporal Model

This Section shows how the temporal dimension of geographic features may be incorporated into the general spatial data model introduced by Tomlin and briefly presented in a previous Section. In that model *individual locations* constitute the primitive units and *attribute values* regarding various *layers* (themes) are assigned to them.

The assignment of an attribute value to an individual location is assumed to last forever and no reference to its appearance or duration is made. This assumption is not compliant with most geographic applications, where the attributes characterizing individual locations change over time (e.g., change in crop type of a land).

In order to represent temporal changes there are two options: (a) assignment of *time-stamps* to the layers (themes) associated with the study area; and (b) assignment of *composite attributes with temporal descriptions* (i.e., list of temporal signatures) to individual locations of each layer.

The first option is the most trivial one. It refers to the maintenance of *snapshots* for the themes associated with the study area. Figures 9.14a and 9.15a illustrate two examples of this representation for the applications presented previously (Figure 9.12). Notice that this model provides images of the region of interest at different points in time.

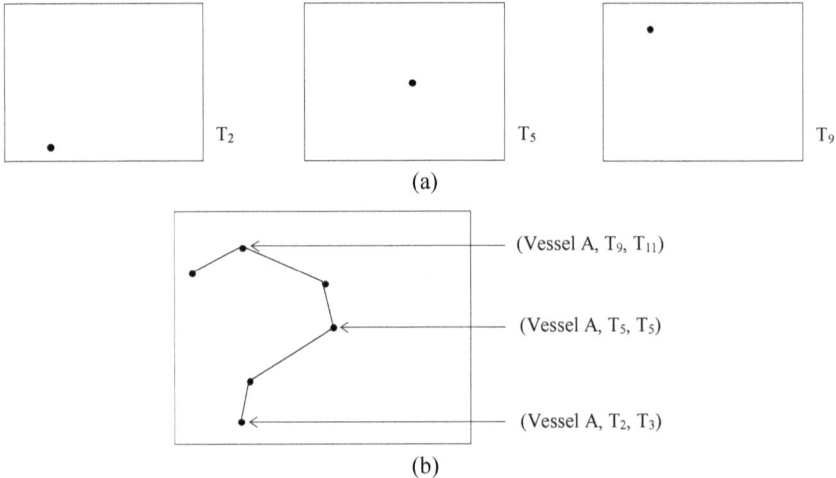



**Fig. 9.14** Two alternative representations for a moving point object.

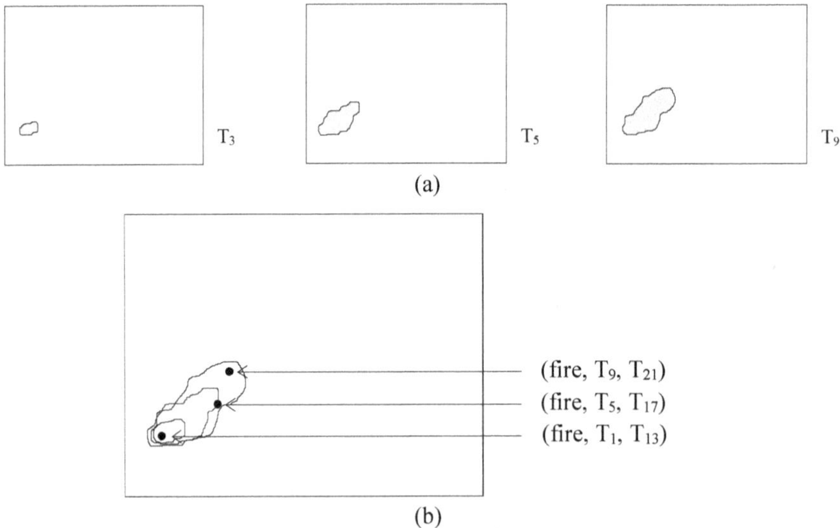



**Fig. 9.15** Two alternative representations for an areal unit which changes over time.

The second option explicitly incorporates the temporal dimension of geographic features into the data model. The single attribute value, which is assigned to each individual location of a layer, is replaced with composite attribute values with time-stamps. Specifically, each individual location on a layer is assigned a list of *temporal attributes*, that is a set of ordered triplets: $(AV, T_{from}, T_{to})$, where AV denotes the attribute value characterizing that location in the corresponding layer during the interval time-stamp $[T_{from}, T_{to}]$. Figures 9.14b and 9.15b illustrate two examples of this representation for the applications presented previously (Figure 9.12).

With the latter representation any change to the attribute value (e.g., crop type) of an individual location is recorded as a new triplet in the list of triplets (Figure 9.16) associated with that location in the corresponding layer (e.g., layer of crop). Hence, minor changes in the study area can easily be recorded, in contrast to the first solution (i.e., the maintenance of snapshot images), in which major changes can only be recorded in order to avoid the generation of a massive set of layers.

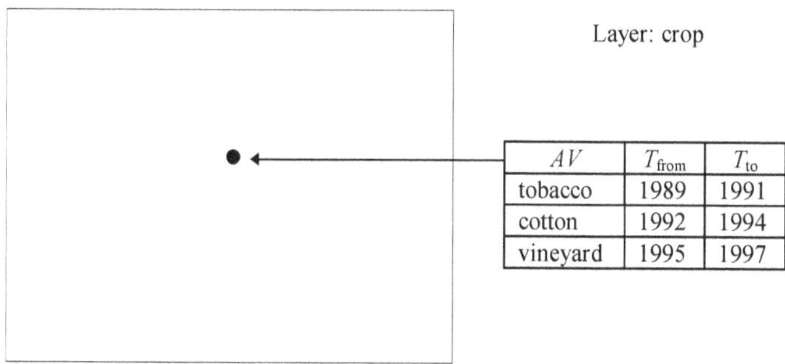


| $AV$ | $T_{from}$ | $T_{to}$ |
|---|---|---|
| tobacco | 1989 | 1991 |
| cotton | 1992 | 1994 |
| vineyard | 1995 | 1997 |

**Fig. 9.16** Representation of change using a list of temporal signatures (triplets).

The concept of zone also changes in the spatio-temporal model. Specifically, a *temporal zone* is defined as the set of individual locations on a layer with a common value in the dynamic attribute during an interval time-stamp. Notice that this time-stamp must concur but is not necessarily identical with the time-stamps of the individual locations composing the zone and may be degenerated into an instant time-stamp (i.e., for an instant zone).

In summary, the spatio-temporal model consists of a library of *layers*, all of which are in registration. Each layer corresponds to a specific theme of interest and consists of a set of individual *locations*. Each individual location in a layer is assigned a *list of temporal attributes*, where a temporal attribute is defined as an attribute value with an interval time-stamp. A temporal attribute is a set of ordered triplets: $(AV, T_{from}, T_{to})$, where AV denotes the attribute value characterizing a location in the corresponding layer during the interval time-stamp $[T_{from}, T_{to}]$. All individual locations on a layer with common attribute values assigned to them during a specified interval time-stamp constitute a *temporal zone*.

Figure 9.17 shows the hierarchy by Tomlin (Figure 9.8) as extended in space-time. Each individual location is extended to a *spatio-temporal location* (i.e., a location in space-time with two coordinates for space and one for time); and each thematic layer is extended to a *spatio-temporal cube* (with two coordinates for space and one for time). A schematic example of a space-time cube, which comprises a set of spatio-temporal locations, is shown in Figure 9.18. Each space-time location is assigned a value for each attribute of the cube. All spatio-temporal locations with common attribute values during an interval time-stamp define the *spatio-temporal zones*.

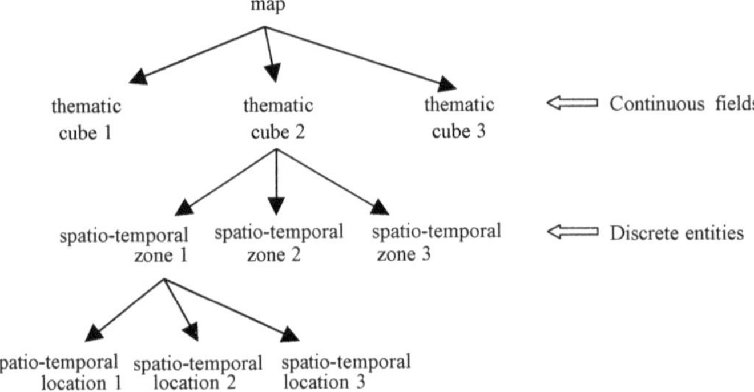


**Fig. 9.17** The spatio-temporal data hierarchy (exntends Tomlin's spatial model in Figure 9.8).

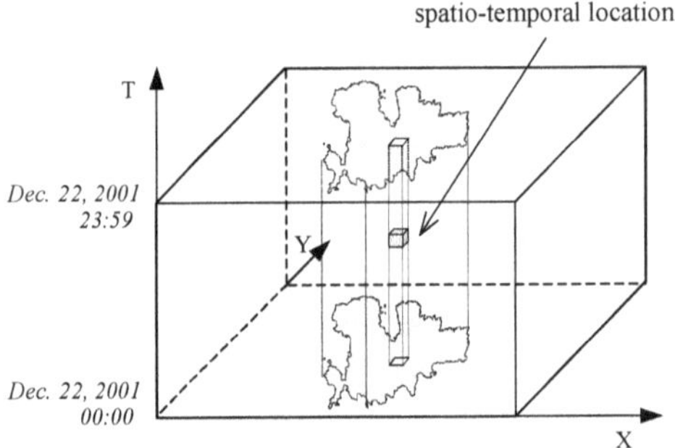


**Fig. 9.18** Example spatio-temporal layer (cube): recording the insolation during a day.

## 9.6 Modeling the Uncertainty of Geographic Data

*Uncertainty* (imprecision or vagueness) refers to the imperfect and inexact knowledge concerning some domain of interest. The uncertainty is an inherent feature of geographic data. It may arise from (a) incomplete information, (b) highly variable attribute values, (c) complex relationships, and/or (c) qualitative descriptions, associated with geographic entities.

Traditional methods for the representation and analysis of geographic information do not tolerate uncertainty. This is largely due to the underlying membership concept of the *classical set theory* (i.e., Boolean logic), according to which a set has sharp boundaries and an element has either full or no membership in a given set.

In the GIS literature it is argued that Boolean logic (the all-or-nothing system) is not an appropriate foundation for the representation of geographic data. This logic imposes an artificial precision on geographic information and processes. Hence, it fails to determine and communicate the extent of imprecision and error. In addition, Boolean logic is not adequate to model neither human cognition nor natural languages, which are imprecise in nature.

Fuzzy logic has been proposed as an alternative foundation to overcome these impediments. The associated *fuzzy set theory* seems to be more appropriate in modeling geographic data as it provides the constructs to represent the uncertainty and the methodologies to support reasoning on data with uncertainty.

Next, a framework of incorporating uncertainty in the spatial representation model is examined. After a short introduction to the concept of uncertainty, the negative impact of classical set theory in the representation and analysis of geographic data is highlighted. Then, the alternative rationale offered by the theory of fuzzy sets is presented. Finally, the extension of Tomlin's model to accommodate the uncertainty of geographic data is discussed.

## *9.6.1 Uncertainty and Classical Set Theory*

The classical set theory an entity either has an attribute entirely or does not have it at all. No third situation is allowed. Hence, all geographic entities are assigned a single attribute values regarding a theme (thematic layer).

Consider the values "forest" and "bare land" describing the theme "land cover". A piece of land with sparse trees has to be classified exclusively into either forest or bare land. Obviously, such a classification fails to describe the actual land cover accurately and imposes sharp boundaries in the representation of the land cover.

As it will further be discussed in Chapter 11, the representation of geographic data based on the classical set theory has a restrictive effect on reasoning and analysis processes. Specifically, the selective search is intended to provide as a result the set of entities (i.e., individual locations) whose attribute values satisfy absolutely a constraint posed by decision-makers. In addition, decision-makers are obliged to express their constraints through arithmetical terms and mathematical symbols in crisp relationships (e.g., slope < 10%), as they are not allowed to use natural language lexical terms (e.g., level land). Finally, there is no qualitative factor available for ordering the qualified locations derived from a sequence of analytical operations.

These problems are commonly faced by decision-makers in practical situations. The employment of a sequence of basic GIS operations to support a real world problem, such as that of a residential site selection (see Chapter 11), is accompanied with all problems of an *"early and sharp classification"*. First, the overall de-

cision is made in steps which drastically and sharply reduce the intermediate results. Any constraint is accompanied by an absolute threshold value and no exception is allowed. For instance, if the threshold for a level land is slope=10%, a location with slope equal to 9.9% is characterized as level, while a second location
with slope equal to 10.1% is characterized as non-level (steep). Moreover, for decisions based on multiple criteria, it is usually the case that an entity (i.e., an individual location), which satisfies quite well the majority of constraints and is mar-
ginally rejected in one of them, to be selected as valid by decision-makers.
However, based on Boolean logic, a location with slope 10.1% will be rejected (as
non-level), even if it satisfies all other constraints posed by decision-makers. Finally, another effect of classical set theory is that the selection result is flat, in the
sense that there is no overall ordering of the valid entities as regard to the degree
they fulfill the set of constraints. For instance, dry-level layer highlights all locations which satisfy the constraints: dry land (threshold 20%) and level ground
(threshold 10%). However there is no clear distinction between a location with
moisture=10% and slope=3% and another with moisture=15% and slope=7%.


## *9.6.2 Fuzzy Set Theory*


In *classical set theory* (Boolean logic) an individual either is or is not a member of
any given set. Specifically, the degree to which an individual observation z is a
member of a set A is expressed by the *membership function* (Membership Function in Boolean logic, $MF^B$), which can take the value 0 or 1, i.e.,

$$MF^B(z) = 1 \quad , \quad \text{if} \quad b_1 \le z \le b_2$$
$$MF^B(z) = 0 \quad , \quad \text{if} \quad z < b_1 \text{ or } z > b_2$$

where $b_1$ and $b_2$ define the exact boundaries of set A. For instance, if the
boundaries between "gentle", "moderate", and "steep" land were set at $b_1$=30%
and $b_2$=60%, then the membership function defines all "moderate slope lands"
(Figure 9.19a). Notice that classical sets allow only binary membership functions
(i.e., TRUE or FALSE).

*Fuzzy set theory*, introduced by Zadeh in 1965, is an extension of the classical
set theory. A fuzzy set A is defined mathematically as follows: If Z={z} denotes a
space of objects, then the fuzzy set A in Z is the set of ordered pairs:

$$A = \{z, MF^F_A(z)\} \quad , \quad z \in Z$$

where the *membership function* (in Fuzzy logic, $MF^F$) $MF^F_A(z)$ is known as the
*"degree of membership (d.o.m.) of z in A"*. Usually, $MF^F_A(z)$ is a real number in
the range [0,1], where 0 indicates no-membership and 1 indicates full membership. Hence, $MF^F_A(z)$ of z in A specifies the extent to which z can be regarded as
belonging to set A.

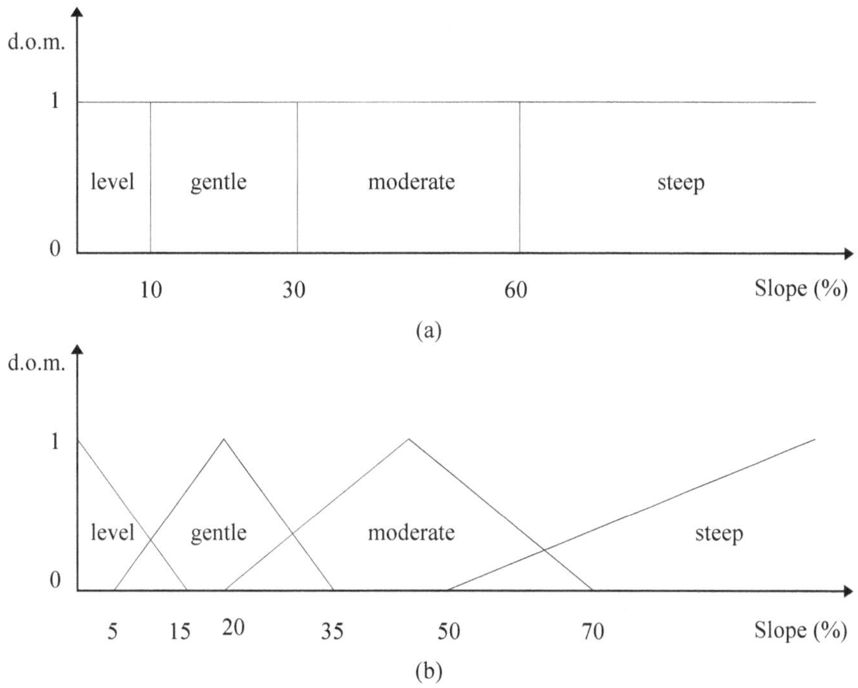

**Fig. 9.19** An example of the classic (a) and fuzzy (b) classification for ground slope.

The choice of the membership function, i.e., its shape and form, is crucial and significantly affects the results of the analysis. Two options are available for choosing the membership functions for fuzzy sets: (a) by applying an *"expert's model"*; and (b) by a *data-driven method*. These options are briefly described next.

### 9.6.2.1 Membership Function using an Expert's Model

The membership function can be chosen based on prior knowledge. Using this function, individual entities (locations) will be assigned a d.o.m. regarding a lexical value characterizing a theme. Several mathematical functions are available for transforming physical values to d.o.m. values. Each of these functions has its own characteristics and can simulate better or worse various physical phenomena. In fact, the choice of the best mathematical function to simulate a physical phenomenon combines both mathematical background and expertise regarding that phenomenon. For simplicity, the linear transformation function is examined next. This function consists of three components:

1. *Linear increasing:* It is used in cases where a straight-forward mapping of physical values to the fuzzy domain is needed. The fuzzy set A is defined as the ordered pair:

$$A = \{\, z\,, \mathrm{LI}(z)\,\}\ ,\quad z \in [c_0\,, c_1]$$

where
$$\mathrm{LI}(z) = \frac{z - c_0}{c_1 - c_0}$$

2. *Linear decreasing:* The fuzzy set A is defined as the ordered pair:

$$A = \{\, z\,, \mathrm{LD}(z)\,\}\ ,\quad z \in [c_0\,, c_1]$$

where
$$\mathrm{LD}(z) = \frac{z - c_0}{c_0 - c_1} + 1$$

3. *Triangular:* The set of physical values is divided into k parts: $[c_0,c_1],[c_1,c_2]$, ...,$[c_{k-1},c_k]$. The fuzzy set A is defined as the ordered pair:

$$A = \begin{cases}
\{z, \mathrm{TR}_1\}, & z \in & [c_0, c_1] \\[2mm]
\{z, \mathrm{TR}_2\}, & z \in & \left[c_i, \dfrac{c_i + c_{i+1}}{2}\right] \\[2mm]
\{z, \mathrm{TR}_3\}, & z \in & \left[\dfrac{c_i + c_{i+1}}{2}, c_{i+1}\right] \\[2mm]
\{z, \mathrm{TR}_4\}, & z \in & [c_{k-1}, c_k]
\end{cases}$$

where
$$\mathrm{TR}_1(z) = \frac{z - c_0}{c_0 - c_1} + 1$$

$$\mathrm{TR}_2(z) = \frac{2(z - c_i)}{c_{i+1} - c_i}$$

$$\mathrm{TR}_3(z) = \frac{2(c_i - z)}{c_{i+1} - c_i} + 1$$

$$\mathrm{TR}_4(z) = \frac{z - c_{k-1}}{c_k - c_{k-1}}$$

Next the classification of individual locations of a layer based on the slope values of the ground (physical values) is examined. Four *lexical values* are used to describe the slope: *[level,gentle,moderate,steep]*. The transformation functions are linear decreasing and increasing for the first and last lexical values respectively, and triangular for the rest of them. Figure 9.19 illustrates the conventional (Figure 9.19a) and fuzzy classification (Figure 9.19b) of slope values. Notice that the conventional way to classify slope involves discrete classes with disjoint ranges, while fuzzy classification captures the gradual transition between classes (lexical values), providing a better way to categorize imprecise concepts, such as gentle and steep land. Based on the fuzzy classification, a location with slope 6% is assigned a d.o.m. of 0.6 for level, 0.1 for gentle, 0 for moderate, and 0 for steep.

### 9.6.2.2 Membership Function using a Data-driven Method

The choice of the membership function can alternatively be based on a data-driven approach in the sense that it is locally optimized to match a given data set. This method is analogous to cluster analysis examined in Chapter 12. One method frequently used is the *fuzzy k-means*. An algorithm for computing the *fuzzy k-means* was introduced by Bezdek. This algorithm computes the membership $\mu$ of the i[th] object to the c[th] cluster as follows:

$$\mu_{ic} = \frac{\left[(d_{ic})^2\right]^{-1/(q-1)}}{\sum_{c'=1}^{k}\left[(d_{ic'})^2\right]^{-1/(q-1)}}$$

with d the distance measure used for similarity, and q the fuzzy exponent determining the amount of fuzziness. The computation of fuzzy k-means proceeds by iteration (similarly to K-Means algorithm; see Chapter 12).

## *9.6.3 Fuzzy Set Algebra*

Fuzzy set operations provide the counterpart operations to those of classical set theory. Logical operations with fuzzy sets are generalizations of usual Boolean algebra and applied to observations that have partial membership in more than one set. The standard operations of *union, intersection,* and *complement* of fuzzy sets A and B, defined over some domain C, create a new fuzzy set whose membership function is defined as:

- Union: $\qquad MF^F_{A\cup B}(z) = \max\{ MF^F_A(z), MF^F_B(z) \}, \forall\, z \in C$
- Intersection: $\quad MF^F_{A\cap B}(z) = \min\{ MF^F_A(z), MF^F_B(z) \}, \forall\, z \in C$
- Complement: $MF^F_{A'}(z) = 1 - MF^F_A(z), \forall\, z \in C$

Consider the previous classification of individual locations on a layer based on the slope values *[level,gentle,moderate,steep]*; and a second classification based on the land moisture with the following lexical values: *[dry,moderate,wet,water]*. For each individual location l (e.g., d.o.m. for level = 0.8 and d.o.m. for dry = 0.4) the d.o.m. value which provides an overall measure regarding:

- level ground and dry land is derived by:
  $\min\{ MF^F_{level}(l), MF^F_{dry}(l) \}$, (e.g., $\min\{0.8, 0.4\} = 0.4$)
- level ground or dry land is derived by:
  $\max\{ MF^F_{level}(l), MF^F_{dry}(l) \}$, (e.g., $\max\{0.8, 0.4\} = 0.8$)
- non-level ground is derived by:
  $1 - MF^F_{level}(l)$, (e.g., $1 - 0.8 = 0.2$)

A problem that arises in this case is that only one of the participating d.o.m. values dominates by assigning its value to the whole decision criterion. This way the contribution of the other d.o.m. values is eliminated. Several other functions can be found in the literature for handling the logical operations on observations that have partial membership on more than one set. These functions are more complex to compute and interpret, however they provide more expressive and accurate results. Next, the discussion is confined to the presentation of the *energy metric* (introduced by Gupta), in which k fuzzy sets ($A_1$, $A_2$, ..., $A_k$) defined over some domain C, create a new fuzzy set E with membership function given by the formula:

$$MF^F{}_E(z) = \sum_{i=1}^{k} \left[ MF^F{}_{A_i}(z) \right]^q$$

where q is a positive integer. By applying this equation (e.g., for q=2; quadratic measure) the big weight values (d.o.m.) are amplified, while the small values are nearly eliminated.

Assuming the previous example, the overall measure characterizing each individual location ($l$) of a region, regarding *level ground* and *dry land* using the energy function, is given by the following formula:

$$MF^F{}_{level-dry}(l) = \left[ MF^F{}_{level}(l) \right]^2 + \left[ MF^F{}_{dry}(l) \right]^2$$

Notice that the energy measure derived by the previous formula should be normalized in the fuzzy domain [0,1].

The last equation is more flexible and expressive than the intersection one, because its overall measure is explicitly affected by the d.o.m. values of individual locations on both fuzzy sets A and B. Specifically, assume two individual locations $L_1$ and $L_2$ with the following d.o.m. values: $L_1$ {d.o.m. for level=0.8 and d.o.m. for dry=0.4} and $L_2$ {d.o.m. for level=0.6 and d.o.m. for dry=0.4}. The overall measure given by the classical intersection equation is 0.4 for both locations, while that derived by the latter equation is 0.80 for $L_1$ and 0.56 for $L_2$. Clearly the energy metric provides an ordering of the two locations. It says that $L_1$ satisfies better the two criteria posed by decision-makers (i.e., level and dry land).

This feature of energy metric is very beneficial for decision criteria which combine multiple sets and lexical values, while ordering of the qualified entities (i.e., individual locations) is required (e.g., find the five most level and dry locations of a region).

Obviously, in fuzzy logic methodologies, contrary to traditional logic, reasoning is based on a *"late and flexible classification"*, and consequently several of the problems presented in the previous Section are overcome. The fuzzy operations are discussed in Chapter 11.

## *9.6.4 Representation of Fuzziness and the Extended Spatial Model*

The purpose of this Section is to show how fuzziness, which models the uncertainty related to geographic data, may be incorporated into the general spatial model introduced by Tomlin and presented in earlier in this Chapter. In that model individual *locations* constitute the primitive units and attribute values regarding various *layers* (themes) are assigned to them. Each layer is described through a set of attribute values and each individual location on it is assigned only one of these values. The assignment of an attribute value to an individual location indicates its *full membership* regarding this feature in the corresponding layer.

In fuzzy set theory the concept of full membership is replaced by that of *partial membership* and consequently the representation of individual locations should be changed. Specifically, the data model must be extended in order to represent the partial memberships of each individual location in the fuzzy sets of attribute values characterizing a theme.

Figure 9.20 illustrates the classical and fuzzy representation of individual locations on a layer. Four attribute values (i.e., soil, grass, fruit-trees, forest) are adopted to characterize these locations based on a specific theme (i.e., vegetation type). In the classical representation each individual location (e.g., B) is assigned one of the four attribute values (i.e., grass). This assignment is unique for that location regarding a theme (i.e., vegetation type) and indicates its full membership in the corresponding set (i.e., grass land). On the other hand, in the fuzzy representation each individual location (e.g., B) is assigned one d.o.m. value for each attribute value (i.e., 0.1 for soil, 0.6 for grass, 0.3 for fruit-trees, and 0 for forest) characterizing the theme (i.e., vegetation type). These d.o.m. values may be derived from sampling techniques, such as the percentage of participation of each attribute value (i.e., vegetation type) in the area covered by the individual location (by introducing the factor of resolution).

The incorporation of fuzziness into the spatial data model forces the redefinition of the components forming the hierarchical data model (Figure 9.8). Specifically, while in conventional set theory the individual locations in a layer are assigned the attribute values (e.g., soil, grass, fruit-trees, forest) characterizing a theme (e.g., vegetation), in fuzzy set theory they are assigned d.o.m. values for each attribute value (e.g., 0.3 for soil) characterizing a theme (e.g., vegetation). Hence the number of layers increases, since each theme is represented by as many layers as the number of attribute values associated to it. These attribute values are referred to as *lexical values*. Figure 9.21 illustrates the extended Tomlin's model to incorporate the fuzziness of geographic data. At the top of the hierarchy is the *map* (library) of *thematic layers*. Each thematic layer is composed of a set of *lexical layers* (one layer for each lexical value describing the theme). Each lexical layer comprises a set of *individual locations*, which in turn are grouped to form the *fuzzy zones*.

The concept of zone has changed in the fuzzy representation. Specifically, a *fuzzy zone* is defined as the sets of individual locations in a lexical layer with a d.o.m. value greater than zero (alternative definitions may apply different values). For instance, a forest area (road) consists of all individual locations that are assigned a non-zero d.o.m. value in the layer "forest" ("road") characterizing the theme "vegetation" ("development"). Hence, all individual locations in a layer with non-zero d.o.m. values constitute a fuzzy zone.

| ind. locations | attribute values | | | |
|---|---|---|---|---|
| | soil | grass | fruit-trees | forest |
| A | 1 | 0 | 0 | 0 |
| B | 0 | 1 | 0 | 0 |
| C | 0 | 0 | 1 | 0 |
| D | 0 | 0 | 0 | 1 |

( a )

Layer:   *vegetation*

Attribute Values: *{*
   *soil,*
   *grass,*
   *fruit-trees,*
   *forest }*

| ind. locations | attribute values | | | |
|---|---|---|---|---|
| | soil | grass | fruit-trees | forest |
| A | 0.8 | 0.2 | 0 | 0 |
| B | 0.1 | 0.6 | 0.3 | 0 |
| C | 0.3 | 0 | 0.7 | 0 |
| D | 0 | 0.2 | 0 | 0.8 |

( b )

**Fig. 9.20** Assignment of attribute values to individual locations on a layer according to conventional (a) and fuzzy (b) classification.

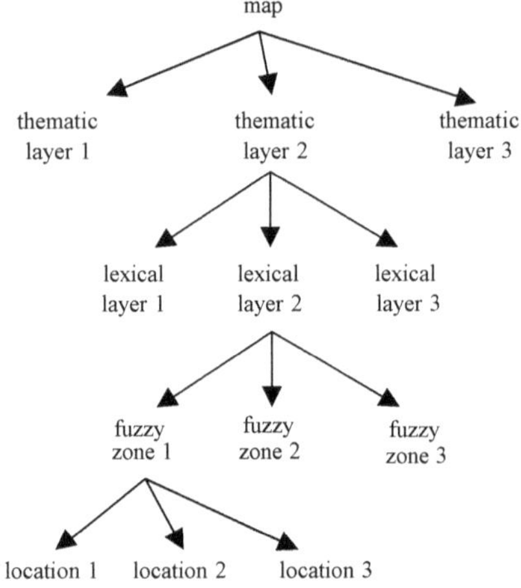


**Fig. 9.21** Extended Tomlin's model to incorporate fuzziness.

Figure 9.22b illustrates the procedure of generating the lexical layer(s) characterizing a theme (thematic layer) in the extended model. Based on both the appropriate membership functions (as chosen by decision-makers) and the knowledge provided by the experts, earth measurements and results derived from sampling techniques are processed and transformed into d.o.m. values for the predefined attribute (lexical) values characterizing a theme. Notice that in the classical set theory the product is one layer (Figure 9.22a), while in fuzzy set theory it is a set of layers (i.e., one for each lexical value).

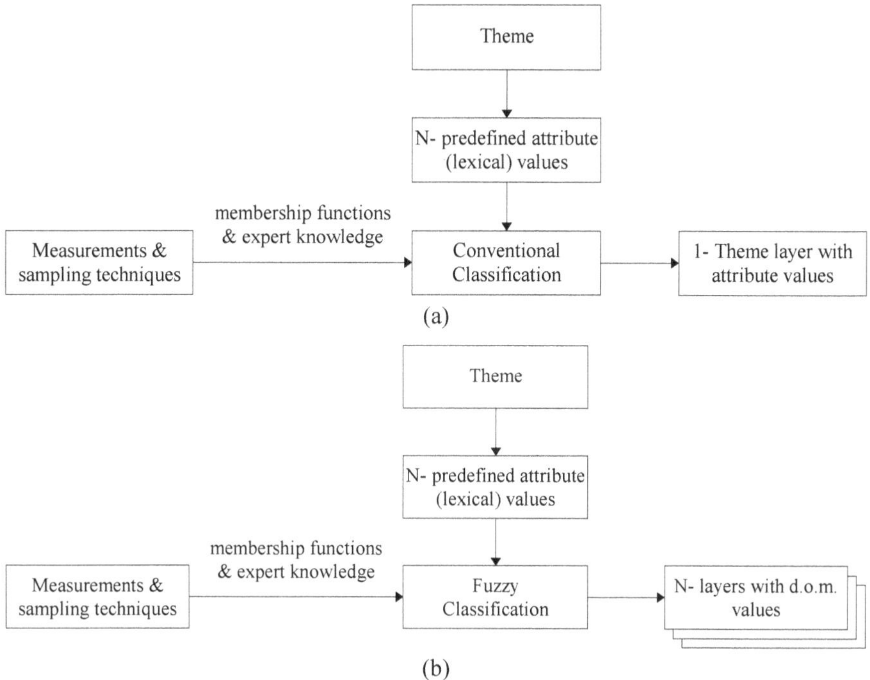

Fig. 9.22 Procedure of generating the layer(s) characterizing a theme in conventional (a) and fuzzy (b) set theory.

## 9.7 Representation of Dynamic Data with Fuzziness

Closing this Chapter a schematic representation of an extended spatial model, to host both dynamic and fuzzy geographic data is given. Figure 9.23 illustrates the hierarchy of data in that model.

At the top of the hierarchy there is a library of *thematic cubes* (map). Each thematic cube corresponds to a theme and represents the space-time (e.g., two co-ordinates for space and one for time). Each thematic cube is composed of as many *lexical cubes* as the number of lexical values in the corresponding theme. Any lexical cube comprises *spatio-temporal locations* (e.g., two coordinates for space and one for time). These locations are grouped into *fuzzy spatio-temporal zones*. A fuzzy spatio-temporal zone is defined as the set of locations in a lexical cube, with a d.o.m. greater than zero during an interval time-stamp, which in turn characterizes the zone (alternative definitions are also possible).

Figure 9.24 shows an example of thematic cube representing the vegetation, with the corresponding lexical cubes {forest, orchards, vineyards, uncultivated area}. Also, the d.o.m. values for these lexical values of a spatio-temporal location of Crete in the 20th Century are shown.

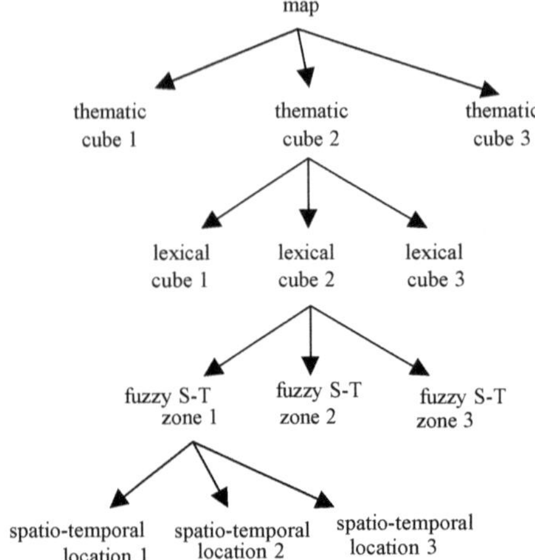



**Fig. 9.23** The hierarchy of dynamic and fuzzy geographical data.

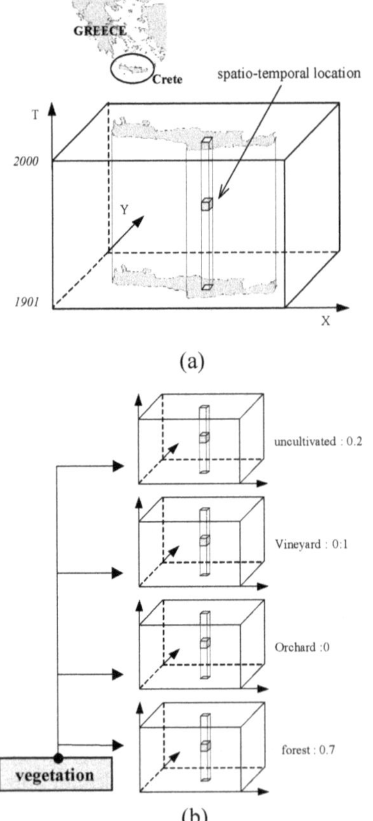



**Fig. 9.24** Example of fuzzy thematic cube of vegetation (a) and the corresponding lexical cubes (b).

# References and Further Reading


Allen, J.F., 1983. Maintaining knowledge about temporal intervals. *Communications of the ACM*. Vol. 26, pp. 832-843.

Burrough, P.A., and Frank, A.U. (Eds), 1996. *Geographic Objects with Indeterminate Boundaries*. Taylor & Francis.

Burrough, P.A., and McDonnell, R.A., 1998. *Principles of Geographic Information Systems*. Oxford University Press.

Chorochronos, 2000. *Research Network on Spatio-Temporal Database Systems*. http://www.dbnet.ece.ntua.gr/~choros/

Egenhofer, M., 1989. A formal definition of binary topological relationships. In the *Proceedings of the International Conference on Foundations of Data Organization and Algorithms*. pp. 457-472.

Leung, Y., and Leung, K.S., 1993. An intelligent expert system shell for knowledge-based GIS: 1. The tools, 2. Some applications. *International Journal of Geographical Information Systems*. Vol.7, pp. 189-213. Taylor-Francis.

Longley, P.A., Goodchild, M.F., Maguire, D.J., and Rhind, D.W., 2001. *Geographic Information Systems and Science*. Wiley.

Stefanakis, E., 2001. A unified framework for fuzzy spatio-temporal representation and reasoning. In the *Proceedings of the 20th International Cartographic Conference*. Beijing, China, pp. 2678-2687.

Stefanakis, E., and Sellis, T., 2000. *Towards the design of a DBMS repository for temporal GIS*. In ESF- GISDATA Series Book: Life & Motion of Socio-economic Units (Eds A.U. Frank, J. Raper and J.-P. Cheylan). Taylor & Francis.

Stefanakis, E., Vazirgiannis, M., and Sellis, T., 1999. Incorporating fuzzy set methodologies in a DBMS repository for the application domain of GIS. *International Journal of Geographical Information Science*. Vol.13, pp.657-675. Taylor-Francis.

Tomlin, C.D., 1990. *Geographic Information Systems and Cartographic Modeling*. Prentice Hall.

Zadeh, L.A. (1965). Fuzzy sets. *Information and Control*. Vol. 8, pp. 338-353.

# 10 Geographic Data Models

## 10.1 Introduction

In the previous Chapter the representation models of reality were examined. Two basic models for representing the geographic space were presented: (a) collection of discrete entities, and (b) continuous field. Both models reflect the human perception of reality and cannot be mapped into a digital environment (e.g., a computer system) as they are.

To achieve the representation of human perception into a software system a number of models can be applied. These models are called *data models* (Figure 9.2) and are presented in this Chapter.

## 10.2 Spatial Data Models

According to the definition of continuous fields (Chapter 9) every and each location in space is assigned a distinct value for a variable theme (attribute) in the field. However, the study area occupies an infinite number of locations which cannot be hosted in a computer system. The same problem occurs in the model of discrete entities, where the full description of a phenomenon requires an infinite number of location data. For example, the full representation of coastline geometry requires an infinite number of vertices.

In computer systems there is a requirement for *discretization* (generalization) of continuous phenomena, so as to enable their representation in digital form. Two basic models of spatial data have been proposed in the past and adopted by commercial and open source software systems (Figure 10.1):

- *vector* model (or *object* model), and
- *raster* model (or *field* model)

These models use as basic building block the *vector* and the *pixel*, respectively. Specifically, the *vector* model considers the geographic space as a set of *objects* (entities), which are represented by points, lines, polygons, surfaces, and solids. Each object is then described by the corresponding thematic, temporal, and/or quality data.

On the other hand, the *raster* model considers reality as a single continuous field (mosaic), which is fragmented into individual elementary portions, called units, *cells*, or pixels (picture elements). Each cell hosts a value that describes a theme of the corresponding region in space. Note that depending on the area of study and application, a cell corresponds to a space of two, three, or more dimensions, it may be of variable size, and its form is of regular (e.g., square in a space of two-dimensions) or irregular shape.

Although either model can be applied to encode both discrete entities and continuous fields, their definition reveals the close relationship between the vector model with the representation of space as a set of discrete entities; and the raster model with the representation of space as a continuous field.

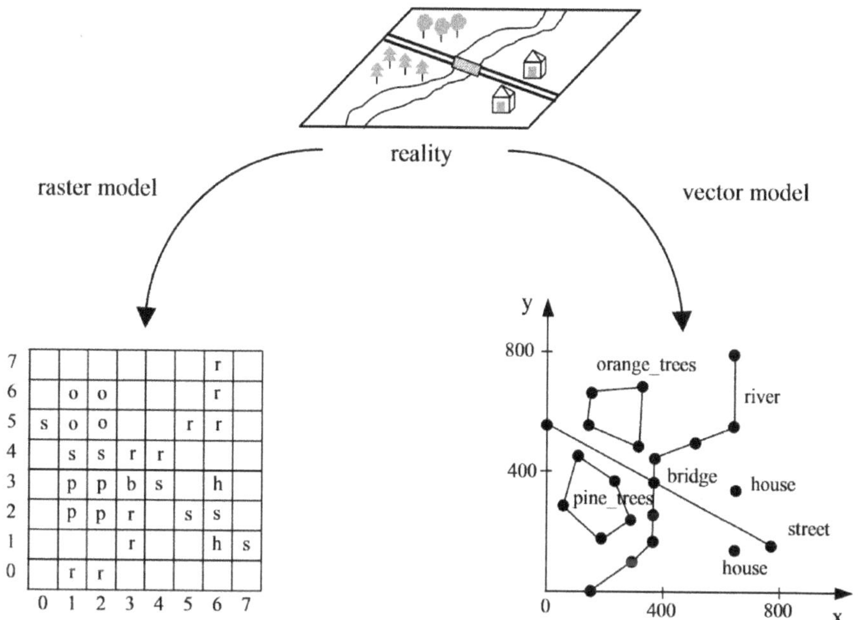


**Fig. 10.1** Spatial data models.

## *10.2.1 Vector Data Model*

The *vector* data model is used for the representation of space as a set of discrete entities or objects. In this model, the study area is considered void, except for the positions occupied by those entities. Each entity is identified by a unique code and is accompanied by the values of thematic attributes of interest to the application. Following, the discussion is limited to the vector space of two dimensions. Note that the concepts presented can be easily extended to spaces of higher-dimensions.

The vector model introduces three basic *geometric data types*: (a) *point*, (b) *line*, and (c) *polygon*. The spatial dimension (geometry) of a particular geographic entity can be described by one or more of these types. If the entity is described by a single type, it is called *simple* vector entity and is of point, line or polygon geometry. Examples of simple vector entities are a lighthouse (point), the seashore (line), and a land parcel (polygon). On the other hand, there are geographic entities, whose geometry is described as a composition of the basic geometric types. These entities are called *composite* vector entities or *collections*. Examples of composite vector entities are the hydrologic and road network of a county in a small-scale map. The hydrological network is composed of sources (point entities), streams (line entities), and lakes (polygon entities). The road network is composed of a set of intersections (point entities) and another of road segments (line entities). Next, the simple vector entities are described.

### 10.2.1.1 Simple Vector Entities

*Simple vector entities* are classified into point, line, and polygon entities. *Point entities* (or points) represent the location of entities, whose shape and size do not matter for the application or size is very small in the scale of representation (e.g., villages and monasteries in a small-scale map, 1:100,000). Each point entity in the two dimensional space is represented by a pair of coordinates (projective: X,Y; or geographic: longitude,latitude) associated with its centroid.

*Line entities* are used to represent linear networks, such as roads, rivers, etc. A line entity or polyline (Figure 10.2) is defined as a set of segments or edges. Each edge is defined by two points or *vertices*, and each vertex belongs to two edges. Exceptions are the vertices that define the two ends of the line entity, which belong to a single edge. When both ends of the line entity coincide, the entity is a *closed* line. An edge may be *directed* or *non-directed* (Figure 10.3a). In a directed edge, the nodes that define the edge are ordered and labelled as *start* node and *end* node. A polyline may comprise of directed or non-directed edges. In the first case, the directed edges may be of the same or different direction (Figure 10.3b).

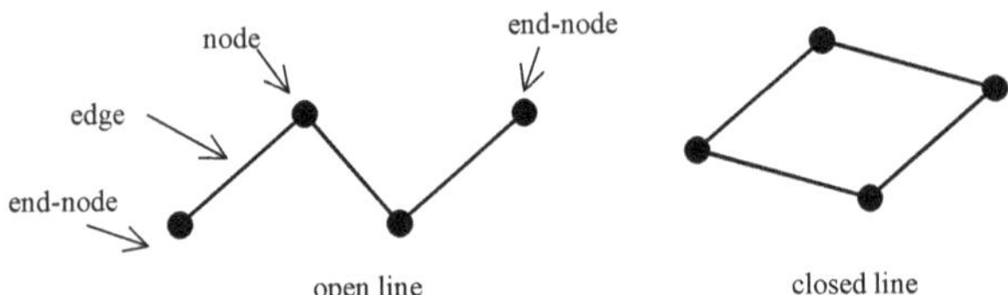


**Fig. 10.2** Simple line entities (polylines).

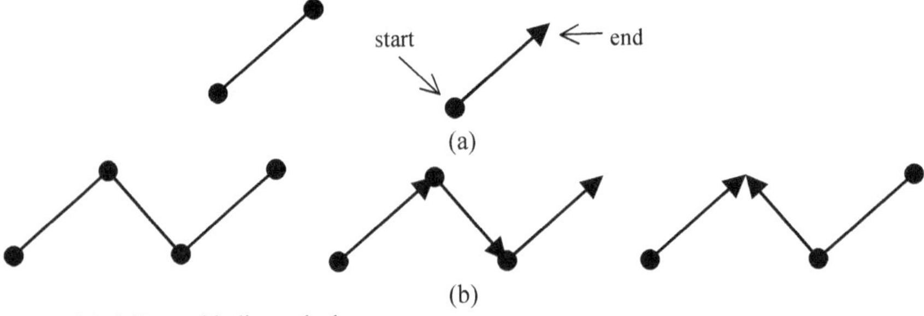


**Fig. 10.3** Polylines with directed edges.

A line entity is called *branched* (or non-planar, see Chapter 8; Figure 8.41) when there are pairs of non-consecutive edges that intersect each other (successive edges intersect at their common point anyway). Next, the discussion will consider *non-branched* lines. The modeling of branched (non-planar) lines can be done by multiple non-branched line entities (Figure 10.4).

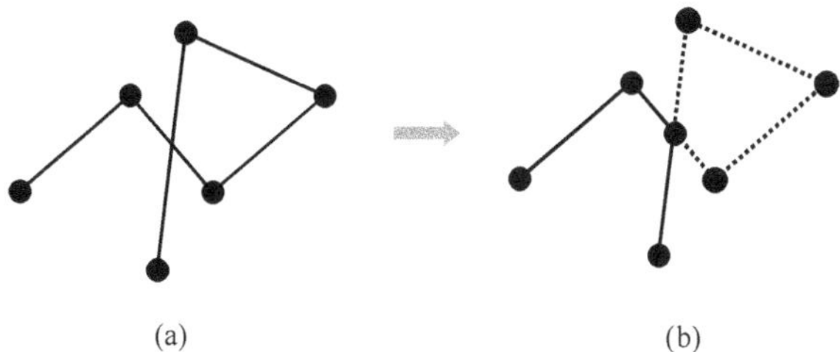

(a)          (b)

**Fig. 10.4** Non-planar polyline (a); representation of non-planar line with two planar (non-branched) polylines (solid and dashed lines) (b).

*Polygon entities* or polygons are applied to represent geographic entities that occupy a discrete area (e.g., administrative units, counties, lakes, etc.). A polygon is defined as a region surrounded by a closed polyline, called the boundary or *outline* of the polygon (Figure 10.5).

A polygon is called polygon without *bottlenecks* when its outline is a non-branched polyline. Otherwise, it is called polygon with bottlenecks. A polygon with bottlenecks can be divided into a set of polygons without bottlenecks. A polygon is called *convex* if the edge connecting any two points of the polygon does not intersect the edges of its outline. Otherwise, it is called *non-convex* polygon. A polygon is called *compact* if every point enclosed by the outline of the polygon belongs to the polygon. Otherwise the polygon is called *polygon with islands* (holes). Simple polygon entities refer to compact polygons without bottlenecks.

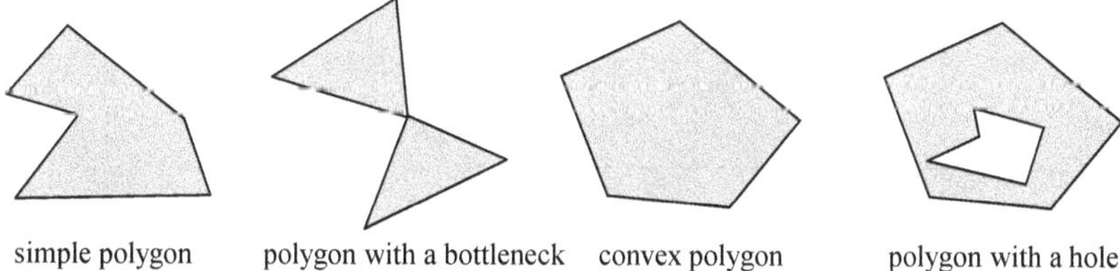

simple polygon    polygon with a bottleneck    convex polygon    polygon with a hole

**Fig. 10.5** Types of polygon entities (polygons).

### 10.2.1.2 Discrete Geometric Bases

The representation of the Euclidean space in a digital environment is accompanied by its *discretization*, due to the limited accuracy in the representation of real numbers and measures in computer systems. The discretization of continuous space can lead to erroneous conclusions in the processing and analysis of the spatial dimension of geographic data.

One example that substantiates the above statement is the following. Given two lines A, B in a plane with equation: $ax + by + c = 0$ (Figure 10.6):

A:  $4x - 7y = 0$
B:  $3x + 7y = 21$

Solving the system of the two lines to locate the point of intersection results to:

$$\begin{bmatrix} 4 & -7 \\ 3 & 7 \end{bmatrix} \cdot \begin{bmatrix} x \\ y \end{bmatrix} = \begin{bmatrix} 0 \\ 21 \end{bmatrix} \Rightarrow \begin{bmatrix} x \\ y \end{bmatrix} = \begin{bmatrix} 0{,}142857 & 0{,}142857 \\ -0{,}06122 & 0{,}081633 \end{bmatrix} \cdot \begin{bmatrix} 0 \\ 21 \end{bmatrix} \Rightarrow \begin{bmatrix} x \\ y \end{bmatrix} = \begin{bmatrix} 3 \\ 1{,}714286 \end{bmatrix}$$

The coordinates of the intersection point K of the two lines are K(3, 1.714286). Obviously, as the following equation is true:

$4x - 7y = 4*3 - 7*1{,}714286 = 0$

the point lies right on line A. The same applies to line B. If, however, a software system represents the real numbers with five decimal digits, then the intersection point coordinates will be stored as $K_5$(3, 1.71429). Obviously, point $K_5$ does not lie on line A as:

$4x - 7y = 4*3 - 7*1{,}71429 = 0{,}00003 \neq 0{,}00000$

The same applies for line B. Figure 10.6 shows the discretization of the space to integers. In this case, the intersection of the two lines has coordinates $K_a$(3,2) and, obviously, does not belong to either of the two lines A, B.

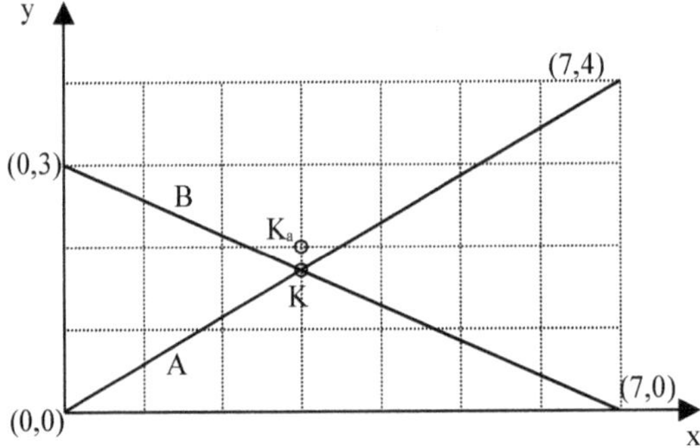


**Fig. 10.6** Intersection of two lines and discretized representation of the intersection point.

Due to the presence of similar geometric problems, the appropriateness of Euclidean geometry in modeling spatial (geometry) data in computer systems has been questioned. One alternative geometry framework, introduced by Gueting and Schneider, is based on the *realms*.

A *realm* is a complete description of the geometry of an application. It is composed of a collection of two basic geometric types: (i) *point*, and (ii) *line segment*. Both the points and line segments are defined on a grid, so that (Figure 10.7): (a) each point can only lie on grid vertices; (b) both start and end points of each line segment are points on the grid; (c) there is no point within a line segment in the realm (in this case the segment is treated as two distinct segments); and (d) line segments of realm neither intersect nor overlap (they can only meet at an end point). Note that the grid implements the discretization of space. The cell size is usually chosen equal to the *spatial resolution* of the representation.

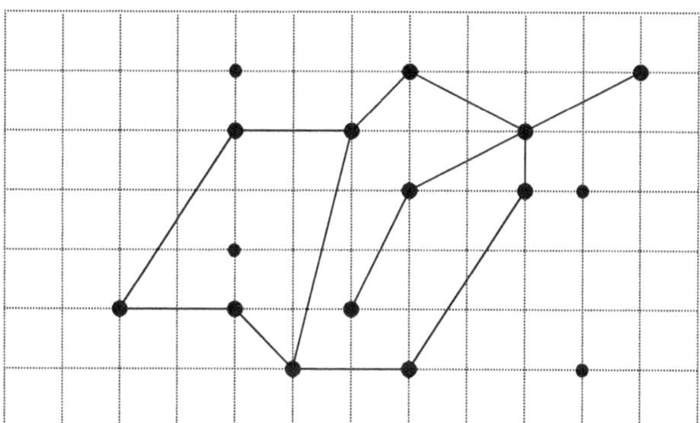

**Fig. 10.7** Example of a realm.

Accordingly, no two segments intersect in a realm. In the event that two line segments intersect, they must define a new point in the realm and possibly be *displaced*, as show in Figure 10.8. The initial segments are represented by dashed lines, while the adjusted ones with solid lines. The new intersection point is placed on the nearest grid vertex.

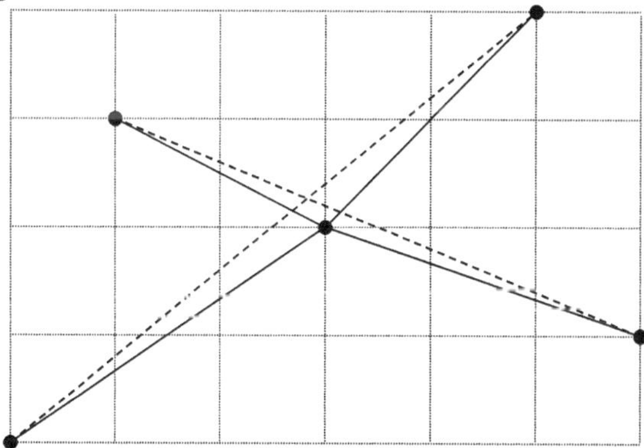

**Fig. 10.8** Intersection of two line segments. Dashed lines violate the geometry of the realm. The solid lines, although displaced, are compliant with the geometry of the realm.

The adjustment of the geometry to comply with the geometry of the realm, can lead to certain topological problems. For example, consider the three segments A, B, C and a point K in Figure 10.9a. In Figure 10.9b the segments A and B were displaced and a new intersection point was defined. In Figure 10.9c, the sections A1 and C were further adjusted. Examining Figure 10.9c, which is compliant with the conditions of a realm, it is discovered that point K lies below line A, while initially it was lying above it (Figure 10.9a).

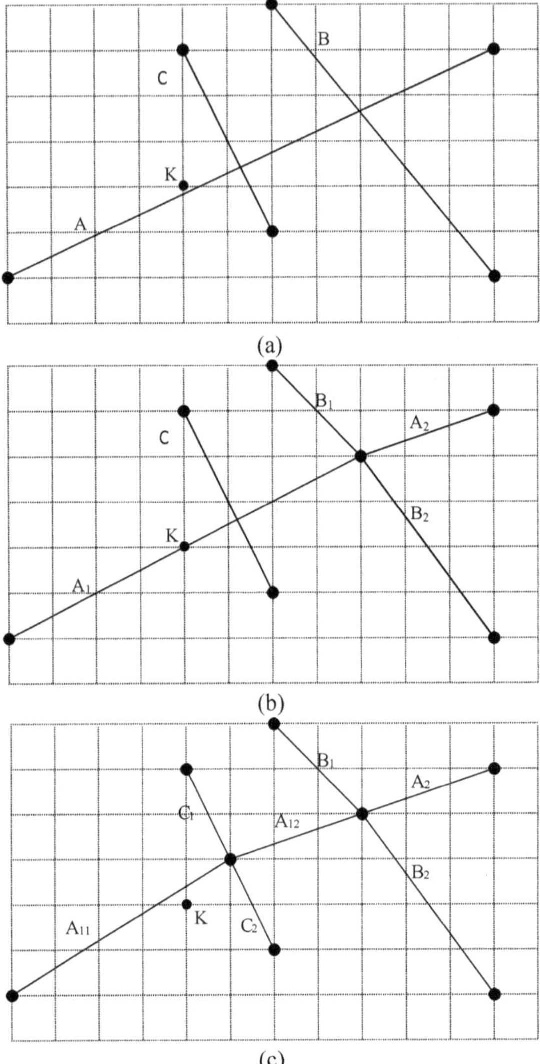

**Fig. 10.9** Example realm adjustments of line segments. The geometry of the realm violates the initial topological relations (point K versus line A).

To avoid any topological problems during the transition from the Euclidean geometry (continuous space) to the representation by realm (discretized space), any displacement of a line segment must result in a new line that entirely falls within its envelope. The *envelope* of a line segment is defined as the area occupied by the cells of the grid intersected by the line (Figure 10.10). As shown in Figure 10.10, the representation of the segment A with the $A_{11}$, $A_{12}$, and $A_2$ in the realm of Figure 10.9, overpasses the envelope of A and this causes a violation of the original topological relations (e.g., case of point K). The algorithm to meet the envelope constraint is not described in this book.

## *10.2.2 Raster Data Model*

The *raster* data model is a representative model for modeling continuous fields in a computer system. In this case, the two-dimensional space is broken into small areas, called *cells* or *pixels*, which form the units for the representation of the spatial dimension of geographic data. The cells are disjoint and their union results the whole space.

Each cell is assigned values for thematic attributes describing the portion of the space occupied by the cell, e.g., vegetation type, land cover, average temperature, etc. The cells are of regular or irregular shape and form a *regular* or *irregular tessellation*, respectively. Next, some basic tessellations are examined.

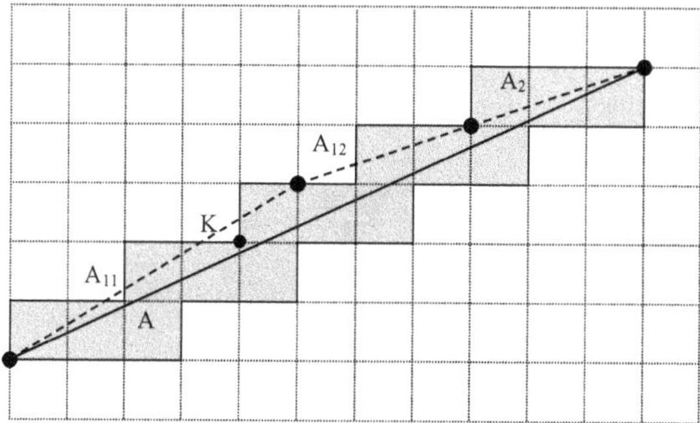


**Fig. 10.10** The envelope of line segment A (Figure 10.9a) in shaded cells and the representation of A in the realm (Figure 10.9c).

### 10.2.2.1 Regular Tessellations

If the space is broken into *cells of regular shape*, the most common geometric shapes for the plane are: square, rectangle, triangle, and hexagon (Figure 10.11). Of these, the most prevalent in the geosciences and computer graphics, due to its simplicity in geometry, is the square, which results to a *square grid* (or *tessellation*).

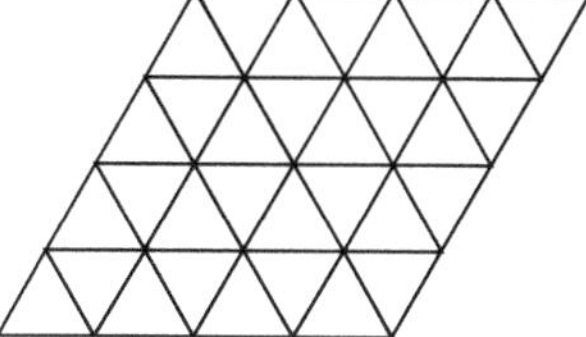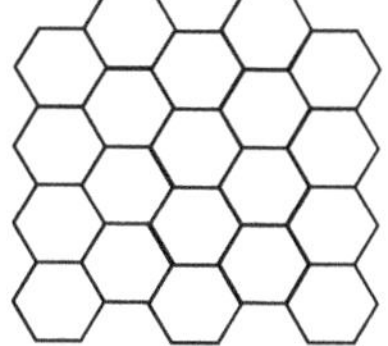

**Fig. 10.11** Regular shaped patterns in raster model.

In the square grid, the two dimensional space (plane) is broken into NxM square cells, which make up a table. Each cell is indexed by the line i (i=1,...,N) and column j (j=1,...,M), to which it belongs. The implementation of a square grid in software systems is straightforward using the data structure of *table* (see Chapter 8). Obviously, the access to a grid cell is direct (by reference to row i and column j). A similar representation is applied when using rectangular cells. The principle of a square grid is extended to spaces of more than two dimensions. For example, in the three-dimensional space each cell is a cubic solid located in: line i, column j, and level k. The implementation of the cubic grid in a software system is done using the data type of a three-dimensional table (see Chapter 8) .

A disadvantage of the square grid is that the distance between adjacent cells is not constant. Specifically, each cell has four *direct* neighbor cells, with which it shares a side, and four *indirect* neighbor cells, with which it shares a vertex (Figure 10.12). The distance from the center of a cell to the centers of its direct neighbors equals to $a$, where $a$ the length of the cell side. On the other hand, the distance from the center of a cell to the centers of its indirect neighbors equals $a\sqrt{2}$. Note that the hexagon grid (Figure 10.11) offers a constant distance from each cell to all its six neighbors (all of them are direct neighbors).

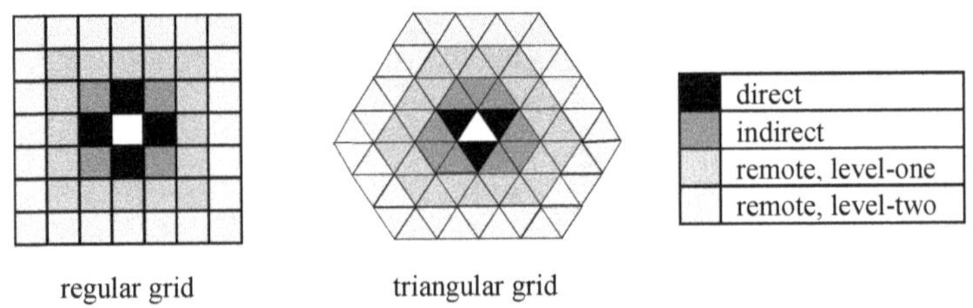


**Fig. 10.12** Types of neighbors for a cell in the square and triangular grids.

The size of a regular cell is called *spatial resolution* and characterizes the raster model. Note that the spatial resolution may be *fixed* or *variable* over the space (Figure 10.13). A variable resolution seeks to selectively retrieve a greater detail in certain regions of space (see Chapter 13).

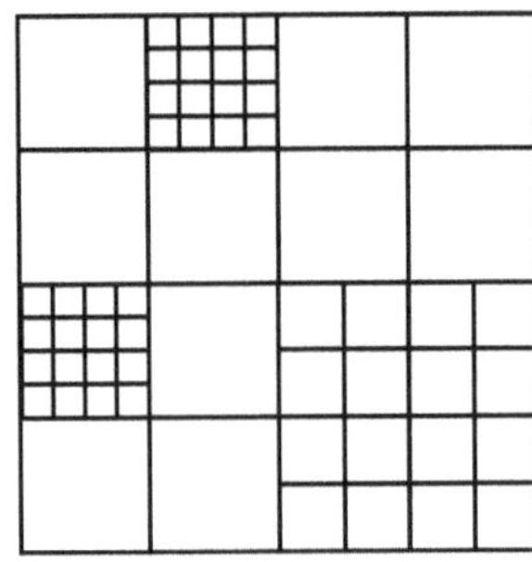

**Fig. 10.13** Variable resolution in a square grid.

## 10.2.2.2 Irregular Tessellations

There are raster data structures, in which the space is broken into cells of irregular shape. The most representative tessellations are: (a) *triangular irregular networks (TIN)*, and (b) *Thiessen polygons*.

### *Triangular Irregular Networks*

In a *Triangular Irregular Networks (TIN)* the space is broken into a set of non-overlapping triangles. Consider a set of two-dimensional points (Figure 10.14). After connecting these points with non-intersecting line segments (edges) results into a tessellation of irregular triangles. Obviously, there are many alternative tessellations (Figure 10.14b,c), depending on the set of edges chosen.

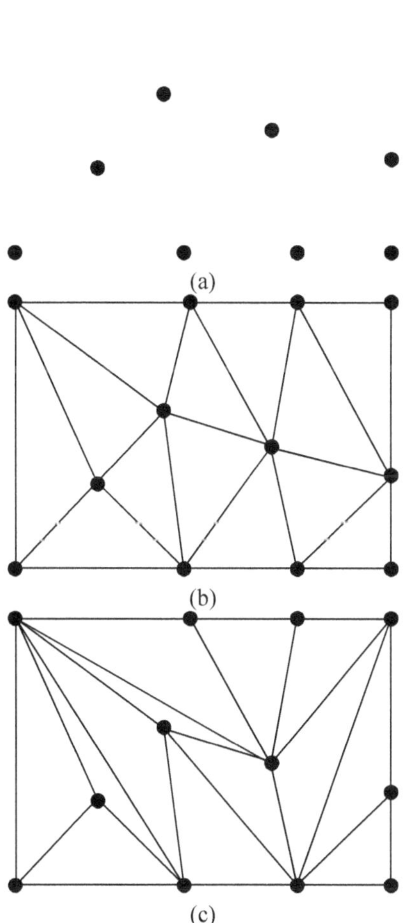

Fig. 10.14 A set of points in 2D space (a) and two alternative partitions of space in irregular triangular networks (b, c).

In geosciences, a common triangular tessellation in two-dimensional space is the *Delaunay triangulation*. This triangulation is defined as follows. Suppose a set S of n points on the plane. Three points k, l, m $\in$ S are vertices of the same triangle T in the triangulation by Delaunay over S, if the circle passing through these three points (circumscribed circle) does not include any other points of S (*Delaunay condition*). Unless there are four points in S that lie on the same circle, the above criterion, when applied, leads to a unique triangulation.

The Delaunay triangulation maximizes the minimum angle of the triangles of the resulted TIN network. In other words, there is no other triangulation over S with the largest smallest angle in the corresponding TIN. This property leads to well-formed (as normal as possible) triangles. Figure 10.14b presents the network by Delaunay for the points in Figure 10.14a.

In the literature there are many algorithms for forming the Delaunay triangulation over a set of points S. Chapter 13 presents one of them.

Triangular Irregular Networks are used as a *model* to represent the terrain in a digital environment. In this case, the vertices of the triangles in the network (set S) are elevation points (points of known altitude). These points are joined by three to form a set of inclining plane triangles in space (three non-collinear points define a plane in space) as shown in Figure 10.15. Note that the network of triangles, after projected on the plane, is independent on the elevation of the vertices (third dimension). The set of triangles can be visualized in a perspective view (Figure 10.16). With an appropriate shading or coloring of the triangles aspect, shading, or slope maps can be generated. Triangular Irregular Networks are widely used in cartography, computer graphics, and geography.

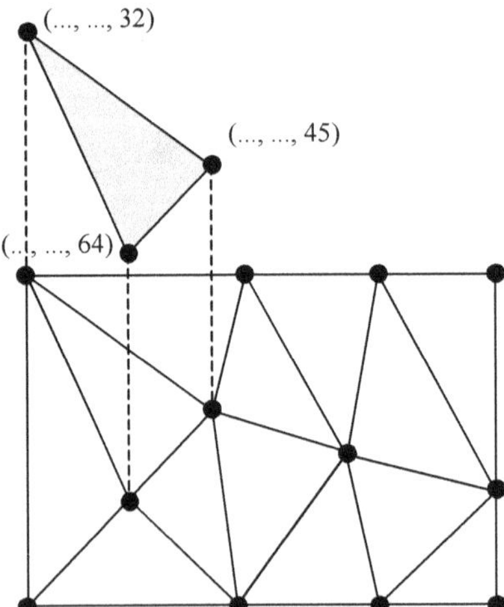


**Fig. 10.15** Example triangular irregular network. All vertices are assigned the terrain elevation.

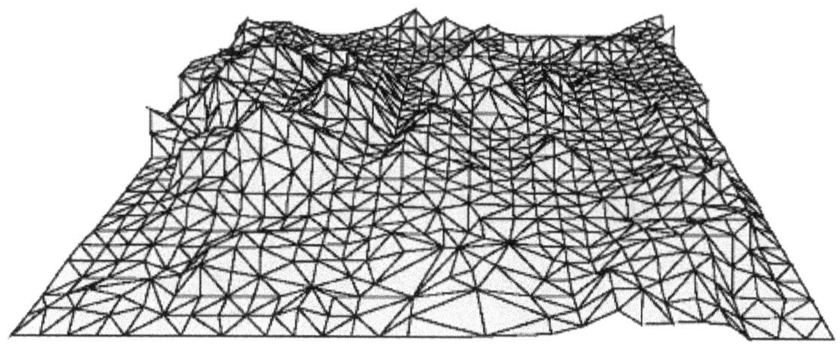

**Fig. 10.16** Perspective view of a triangular irregular network.

## *Thiessen Polygons*

A second tessellation with irregular shaped polygons is the *Thiessen polygons*. Thiessen polygons find a wide application in geography, as they are commonly used in spatial analysis.

These polygons are defined as follows. Consider a set S of N points in two-dimensional space (Figure 10.17). For each location in space it is requested to find the nearest point of S. In other words, space will be partitioned into N polygons (Figure 10.18). Each polygon will enclose a single point of set S and all locations of the space that are closer to this point than any other point of S.

Example application could be the definition of zones (neighborhoods or areas of influence) that are served by ten schools or health centers in a study area. Criterion for the service is proximity. The triangular irregular networks can effectively support the formation of the above zones (Thiessen polygons), as described in Chapter 13.

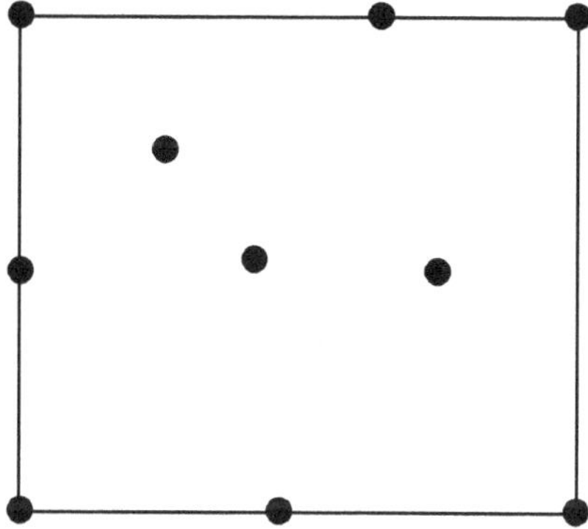

**Fig. 10.17** A set S of points in 2D space.

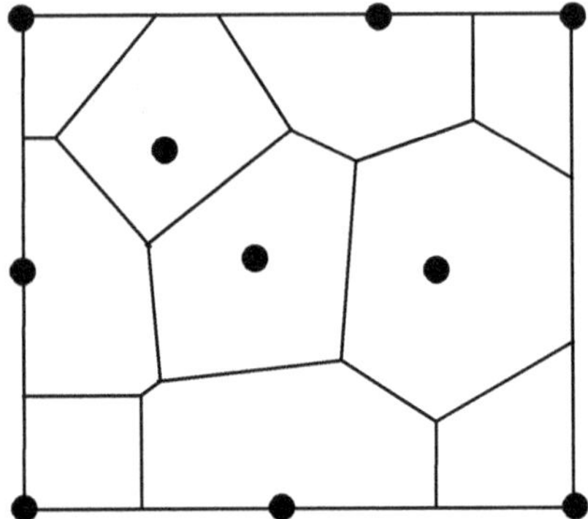

**Fig. 10.18** The Thiessen polygons over the set S of points in Figure 10.17.

### 10.2.2.3 Tessellations for Non-Planar Surfaces

So far the discussion has focused on raster models for the plane surface. Following, two raster models that can be applied to represent a curved geographic surface are briefly discussed: (a) *geographic grid*, and (b) *hierarchical structure of spherical triangles*. These models can be used to represent a larger portion of or even the whole Earth's surface. Although the discussion is limited to the sphere (as a rough approximation of Earth's surface), the principles can easily be extended to the spheroid (ellipsoid surface).

*Geographic Grid*

The raster model of the *geographic grid* divides the earth's surface in curved trapezoidal cells, each of which is delimited by two consecutive meridians and two consecutive parallels. Obviously, the resolution of the geographic grid along parallels and meridians defines the size of the trapezoids and affects the grid size (number of cells).

Figure 10.19 shows an example of a geographic grid with resolution of $30°$ in both latitude and longitude. Obviously, as heading towards the poles the size of the cells (trapezoidal sections) decreases. In addition, their shape also changes significantly (from trapezoids to triangles). This is a drawback of using the geographic grid as a tessellation of the Earth's surface.

*Hierarchical Structure of Spherical Triangles*

The *hierarchical structure of spherical triangles* proposed by Goodchild and Shiren is a raster model for the entire globe. This structure has the advantage that creates cells of similar size and shape. In addition, the space can be partitioned in variable resolutions, depending on the application requirements and the desired level of detail.

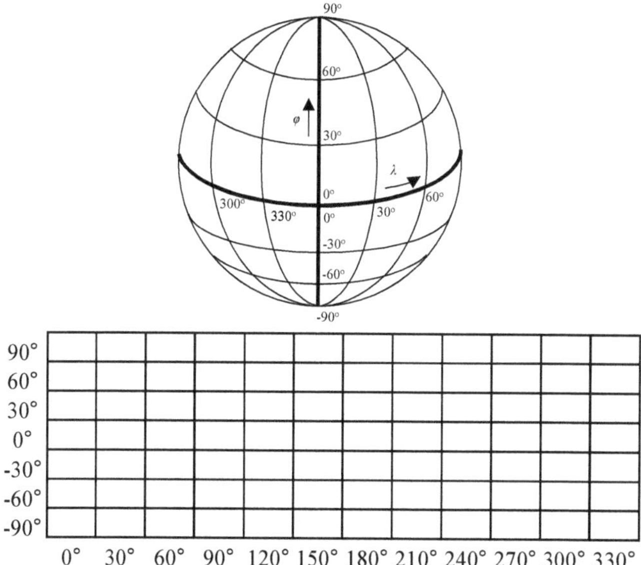


**Fig. 10.19** Example geographic grid.

According to this structure, the globe is divided initially (first level of segmentation) into eight disjoint *spherical triangles*. Four of these are located in the northern hemisphere of the earth and the remaining four in the southern hemisphere. All the triangles of the northern hemisphere share the North Pole as a common vertex. The other two vertices of these triangles lie on the equator and in particular, by triangle in the following pairs of geographic longitudes: 0° and 90°, 90° and 180°, 180° and 270°, 270° and 0°, respectively. In an analogous way, the four triangle of the southern hemisphere are defined. Figure 10.20a highlights one of the eight initial triangles.

Then each of the eight triangles is further divided (second level segmentation) into four new spherical triangles. The new triangles are defined by connecting the medoids of the adjacent edges in the original triangle with great circles (see Chapter 2). Figure 10.20b shows schematically the second level segmentation of one of the eight spherical triangles. The segmentation recursively continues to new triangles until the desired level of resolution is reached (Figure 10.20c).

## References and Further Reading

Burrough, P.A., and McDonnell, R.A., 1998. *Principles of Geographic Information Systems*. Oxford University Press.

de By, R.A. (Ed.), 2000. *Principles of Geographic Information Systems – An Introductory Textbook*. ITC Educational Textbook Series.

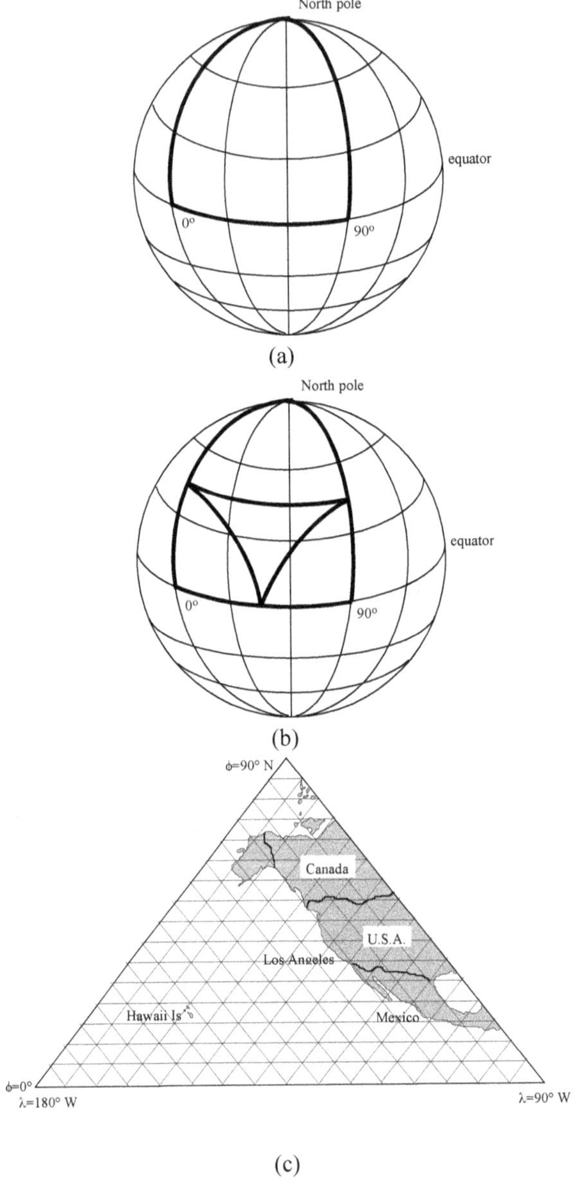


**Fig. 10.20** Schematic representation of the recursive partition of a spherical triangle.

Goodchild, M.F., and Shiren, Y., 1990. A hierarchical spatial data structure for global GIS. In the *Proceedings of the 4th International Symposium on Spatial Data Handling*. pp. 911-917.

Goodchild, M.F., and Shiren, Y., and Dutton, G., 1991. *Spatial data representation and basic operations for a triangular data structure*. Technical Report 91-8. National Center for Geographic Information and Analysis.

Gueting, R.H., and Schneider, M., 1993. Realms: a foundation for spatial data types in database systems. In the *Proceedings of the 3$^{rd}$ Symposium on Spatial Databases*. Singapore, pp. 14-35.

Gueting, R.H., and Schneider, M., 1995. Realm-based spatial data types: the ROSE algebra. *VLDB Journal*. Vol. 4, pp. 243-286.

Longley, P.A., Goodchild, M.F., Maguire, D.J., and Rhind, D.W., 2001. *Geographic Information Systems and Science*. Wiley.

Jones, C., 1997. *Geographical Information Systems and Computer Cartography*. Addison Wesley.

Petrie, G., and Kennie, T.J.M., 1990. *Terrain Modelling in Surveying and Civil Engineering*. Whittles Publishing.

Rigaux, P., Scholl, M., and Voisard, A., 2002. *Spatial Databases with Applications to GIS*. Morgan-Kaufmann.

Shashi, S., and Chawla, S., 2003. *Spatial Databases – A Tour*. Prentice Hall.

van Kreveld, M., 1997. Algorithms for triangulated terrains. In the *Proceedings of the XXIV-th SOFSEM Conference*. Lecture Notes in Computer Science, No. 1338, pp. 19-36. Springer-Verlag.

Worboys, M.F., 1995. *GIS – A Computing Perspective*. Taylor-Francis.

# 11 Geographic Data Analysis

## 11.1 Introduction

Geographic data *analysis operations* process the attributes of geographic entities towards: (a) the creation of new entities; (b) modification of the attributes of existing entities; or (c) extraction of knowledge, measures, and information about geographic data (see Chapter 4). Geographic analysis operations, unlike management operations, cannot easily be standardized, and for this reason there is no officially recognized algebra for analysis of geographic data. Commercial and open source GIS software packages adopt different classifications of analysis operations, which vary depending on the system, the scope, and obviously the underlying data models.

This Chapter provides a classification of geographic analysis operations based on Tomlin's model (Chapter 9) so that there is no dependence on the data model adopted by the software package. Some operations outperform in the vector model, while others in the raster model. Next, the analysis of static geographic data is examined. Then, analysis operations for dynamic geographic data and geographic data with uncertainty are also considered.

## 11.2 Spatial Data Analysis

In the Tomlin's representation model geographic analysis operations are defined in a *layer-by-layer* basis. That is, each operation gets one or more existing layers as input (the operands; see Chapter 9) and generates a new layer as output (the product); which can be used as an operand into subsequent operations (Figure 11.1a).

Analysis operations, also known as *data interpretation operations*, characterize: (a) individual locations, (b) locations within neighborhoods, and (c) locations within zones; and constitute respectively the three classes of operations, i.e., *local*, *focal*, and *zonal* operations (Figure 11.1b,c,d, respectively).

### *11.2.1 Local Operations*

The first class of data-interpretation operations are the *local operations* (Figure 11.1b) and includes those that compute a new value for each *location* on a layer as a function of existing data explicitly associated with that *location* on one or more layers (operands). Local operations include:

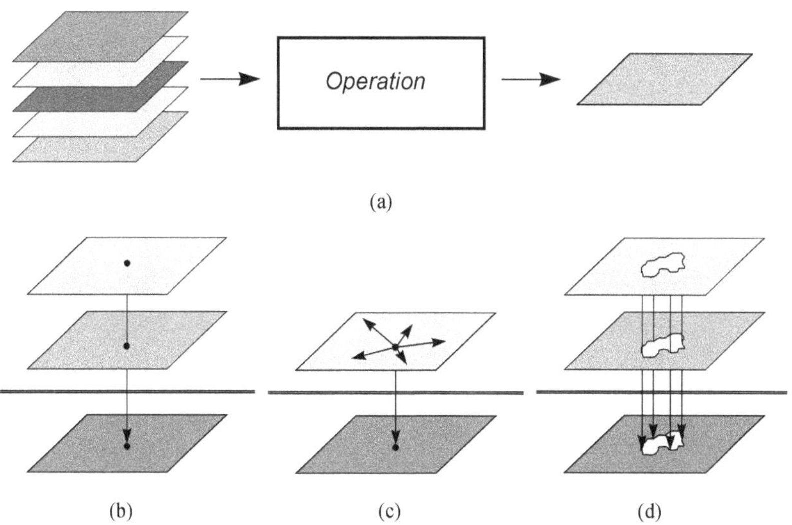

(a)

(b)          (c)          (d)

**Fig. 11.1** Geographic data analysis (or interpretation) operations.

- *Classification and recoding operations*, i.e., assignment of new attribute values to individual locations on a layer based on mathematical transformations or look up tables.
- *Generalization operations*, i.e., reduction of detail associated with individual locations on a layer.
- *Overlay operations*, i.e., assignment of new attribute values to individual locations resulting from the combination of two or more layers. These operations are analogous to join operations in conventional database systems and are usually named as *spatial join operations*.

An example of a local classification operation is given in Figure 11.2. Figure 11.2a shows a thematic layer with the maximum annual temperature by geographic location. The application of the local operation leads to the creation of a new thematic layer (Figure 11.2b), which classifies the locations into two groups. The first, denoted by the value of 0, includes all locations with maximum temperature below $30°$, while the second, denoted by the value of 1, includes all other locations.

## *11.2.2 Focal Operations*

*Focal operations* compute new values for each location as a function of the values in its *neighborhood* (Figure 11.1c). A neighborhood is defined as any set of one or more locations that bear a specific distance and/or topological or directional relationship to a particular location, the *neighborhood focus*. Focal operations include:

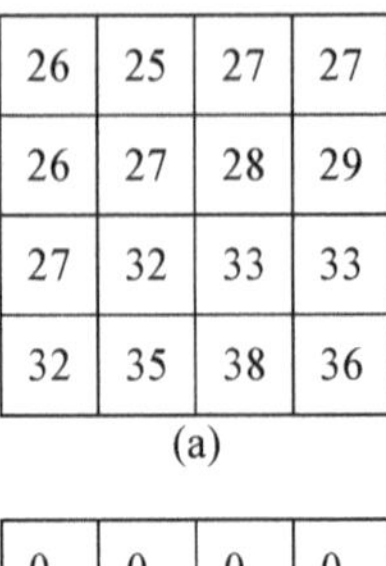

(a)

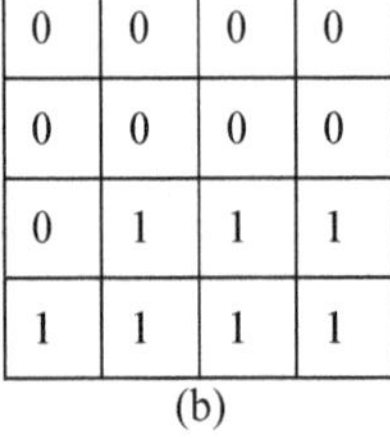

(b)

**Fig. 11.2** Example of a local classification operation: (a) input layer, and (b) output layer. Value 0 or 1 in (b) represents a value less or greater than 30° in (a), respectively.

- *Neighborhood operations*, i.e., assignment of new attribute values to individual locations on a layer, which depict their distance, topology or direction in a neighborhood with respect to the neighborhood focus.
- *Interpolation operations*, i.e., assignment of new attribute values to individual locations on a layer derived by averaging sets of two or more target values associated to selected locations in their immediate or extended vicinity.
- *Surficial operations*, i.e., assignment of new attribute values to individual locations on a layer indicating their surficial characteristics (slope, aspect, volume, etc.).
- *Connectivity operations*, i.e., assignment of new attribute values to individual locations on a layer derived from a running total of the results being retained in a quantitative or qualitative step-by-step fashion and considering the values associated to locations in the immediate or extended vicinity (optimum path finding, intervisibility, etc.).

The neighborhood is further classified into *floating* or *static*. In the former, the neighborhood of each location on the map is considered. In the latter, each location is characterized by its spatial or topological relation against a static zone.

Figure 11.3 shows an example of a focal operation that implements a *floating neighborhood*. Given the layer in Figure 11.3a, the implementation of a focal operation leads to the creation of a new thematic layer (Figure 11.3b), in which each location hosts the sum of the values of the location and those of the direct neighboring locations (locations with which it shares an edge; Figure 10.12) (Figure 11.3a). This operation is called *FocalSum* in GIS and remote sensing software.

Figure 11.4 shows an example of a focal operation that implements the *static neighborhood*. The input layer (Figure 11.4) defines the static neighborhood (highlighted locations; with a value of 1). The locations in the output layer are assigned a coded value that describes the directional relationship with the neighborhood. Specifically, the following coding scheme is applied: 1=north, 2=north-east, 3=east, 4=south-east, 5=south, 6=south-west, 7=west, and 8=north-west.

| 2 | 1 | 1 | 2 |
|---|---|---|---|
| 2 | 3 | 3 | 3 |
| 1 | 4 | 4 | 2 |
| 3 | 1 | 4 | 2 |

(a)

| 5 | 7 | 7 | 6 |
|---|---|---|---|
| 8 | 13 | 14 | 10 |
| 10 | 13 | 17 | 11 |
| 5 | 12 | 11 | 8 |

(b)

**Fig. 11.3** Example focal operation that implements the floating neighborhood: (a) entry layer, and (b) output layer. Each location (e.g., shaded cell) is assigned the sum of the value in that location in the input layer plus those of its four immediate neighbors (lightly shaded cells), e.g., 3 + (1 +2 +3 +4) = 13 for the dark cell.

| 0 | 0 | 0 | 0 |
|---|---|---|---|
| 0 | 0 | 0 | 0 |
| 0 | 1 | 1 | 0 |
| 0 | 0 | 0 | 0 |

(a)

| 8 | 1 | 1 | 2 |
|---|---|---|---|
| 8 | 1 | 1 | 2 |
| 7 | 0 | 0 | 3 |
| 6 | 5 | 5 | 4 |

(b)

**Fig. 11.4** Example focal operation that implements the static neighborhood: (a) input layer, and (b) output layer. The input layer defines a static neighborhood. The output layer describes the relative direction of each location to the static neighborhood.

## *11.2.3 Zonal Operations*

The third and final class of data-interpretation operations comprises the *zonal operations*. Zonal operations compute a new value for each location as a function of existing values associated with a zone containing that location (Figure 11.1d). Zonal operations include:

- *Search operations*, i.e., retrieval of information characterizing individual locations on a layer that coincide with the zones of another layer (i.e., the mask or filter layer). These operations are analogous to selection operations in conventional database systems and are usually named as *spatial selection* operations.
- *Measurement operations*, i.e., assignment of new attribute values to individual locations on a layer that correspond to a measurement (e.g., area, length) characterizing their zones.

Figure 11.5 shows an example of a zonal operation. Assume the thematic layer in Figure 11.5a, which shows three parcels zones A, B, C (each location in space is characterized by an A, B or C depending on the parcel it lies within). Considering the filter layer in Figure 11.5b, which comprises a single zone that occupies the whole space, by applying the zonal operation the size (areas) of each zone A, B, C in the first layer are computed (Figure 11.5c). The measurements refer to number of locations per zone (and are assigned to all locations of each zone).

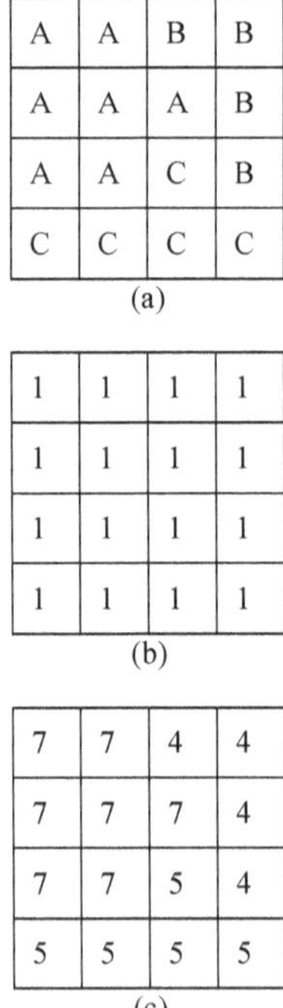

| A | A | B | B |
|---|---|---|---|
| A | A | A | B |
| A | A | C | B |
| C | C | C | C |

(a)

| 1 | 1 | 1 | 1 |
|---|---|---|---|
| 1 | 1 | 1 | 1 |
| 1 | 1 | 1 | 1 |
| 1 | 1 | 1 | 1 |

(b)

| 7 | 7 | 4 | 4 |
|---|---|---|---|
| 7 | 7 | 7 | 4 |
| 7 | 7 | 5 | 4 |
| 5 | 5 | 5 | 5 |

(c)

**Fig. 11.5** Example of zonal operation: (a) input layer, (b) filter layer, and (c) output layer. The output layer reports the size (number of locations) per zone of the input layer (a).

## *11.2.4 Taxonomy of Data Interpretation Operations*

Table 11.1 summarizes the basic *classes* of data interpretation operations accompanied with some representative examples. These operations can be combined to form complex *procedures* and support decision-making in various domains. A simplified example procedure is presented in the following Section.

**Table 11.1** Basic classes of data interpretation operations.

| Classes of Operations | Examples of Operations |
|---|---|
| | |
| ***Local Operations*** | |
| • Classification & recoding | *re-code, re-compute, re-classify* |
| • Generalization | *generalize, abstract* |
| • Overlay (spatial join) | *overlay, superimpose* |
| ***Focal Operations*** | |
| • Neighborhood | |
| – window & point queries | *zoom-in, zoom-out, point-in-polygon* |
| – topological | *disjoint, meet, equal, contains, inside, covers, overlap* |
| – direction | *north, north-east, weak-bounded-north, north-south* |
| – metric (distance) & buffer zones | *near, about, buffer, corridor, thinning* |
| – nearest neighbor | *nearest-neighbor, k-nearest-neighbors* |
| • Interpolation | |
| – location properties | *point-linear, (inverse) distance-weighted* |
| – thiessen polygons | *thiessen-polygons, voronoi-diagrams* |
| • Surficial | |
| – visualization | *contours, TINs* |
| – location properties | *height, slope, aspect, gradient* |
| • Connectivity | |
| – routing & allocation (network) | *optimum-path-finding, optimum-routing, spread, seek* |
| – intervisibility | *visible, light-of-sight, viewshed, perspective, illumination* |
| ***Zonal Operations*** | |
| • Mask queries (spatial selection) | *select-from-where, retrieve* |
| • Measurement | *distance, area, perimeter, volume* |

## *11.2.5 Site Selection Example*

The purpose of this Section is to present a sequence of data-interpretation operations, which may compose one or more procedures to accomplish the task *of site selection for a residential housing development*. The basic approach to this is to create a set of *constraints*, which restrict the planned activity, and a set of *opportunities*, which are conducive to the activity.

The combination of these two sets: constraints and opportunities, is considered in order to find the best locations. In the simplified situation that follows the sets of *constraints* and *opportunities* are as follows:

- vacant area (i.e., no development),
- dry land,
- level and smooth site (e.g., slope $< 10\%$),
- nearness to the existing road network,
- south-facing slope.

In addition all candidate sites should have an adequate size to satisfy the needs of the planning activity (e.g., between 1 and 1.5 sq. km).

Obviously, a wider set of constraints and opportunities could be considered. However, the above is significant to illustrate some basic data-interpretation operations available in GIS.

The whole task requires as input *three layers* of the study area:

1. *hypsography* layer: the three-dimensional surface of the area (elevation values),
2. *development* layer: the existing infrastructure on the area (e.g., roads, buildings, etc.), and
3. *moisture* layer: the soil moisture of the area (e.g., lakes, wet-lands, dry-lands, etc.).

For the purposes of processing, an artificial layer, named dummy, is used. The *dummy layer* covers the whole extent of the study area and consists of individual locations with no attribute values assigned to them (i.e., no zones defined). This layer is commonly used in Zonal(search) operation as an operand (see Figure 11.5); notice that the second operand (b) is the filter layer.

The procedure of site selection, based on the sets of constraints and opportunities determined above, may consist of a sequence of operations. This sequence is presented next. The syntax adopted for an operation is:

*new-layer* = Operation-class(operation-subclass) of *existing-layer* and ...

In addition Figure 11.6 graphically represents the procedure. Rectangles are used to denote the operations. Parallelograms represent the thematic layers. Their names are annotated inside the shapes. Arrows show the operation flow.

1. *Vacant areas:* A new layer of vacant areas is produced from the layer of development by classifying, generalizing and finally performing a selective search on the result (Figure 11.6a).
   - *development-classes* = Local(classification) of *development*
   - *vacant-developed* = Local(generalization) of *development-classes*
   - *vacant* = Zonal(search) of *dummy* and *vacant-developed*

2. *Dry lands:* A new layer of dry lands is produced from the layer of moisture by classifying, reducing detail and performing a selective search on the result (Figure 11.6b).
   - *moisture-classes* = Local(classification) of *moisture*
   - *dry-wet* = Local(generalization) of *moisture-classes*
   - *dry* = Zonal(search) of *dummy* and *dry-wet*

3. *Level sites:* A new layer of level and smooth sites is produced from the layer of hypsography by computing, classifying, generalizing and finally performing a selective search on the result (Figure 11.6c).
   - *slope* = Focal(surfacial) of *hypsography*
   - *slope-classes* = Local(classification) of *slope*
   - *level-steep* = Local(generalization) of *slope-classes*
   - *level* = Zonal(search) of *dummy* and *level-steep*

4. *South-facing areas:* A new layer of south-facing areas is produced from the layer of hypsography by computing, classifying, generalizing and finally performing a selective search on the aspects (Figure 11.6d).
   - *aspect* = Focal(surfacial) of *hypsography*
   - *aspect-classes* = Local(classification) of *aspect*
   - *south-north* = Local(generalization) of *aspect-classes*
   - *south* = Zonal(search) of *dummy* and *south-north*

5. *Accessible areas:* A new layer of accessible sites by the existing road network is produced implicitly from the layer of development by highlighting the road network, computing, classifying, generalizing and finally performing a selective search on the proximities (Figure 11.6e).
   - *roads* = Zonal(search) of *dummy* and *development*
   - *road-proximity* = Focal(neighborhood) of *roads*
   - *road-proximity-classes* = Local(classification) of *road-proximity*
   - *accessible-inaccessible* = Local(generalization) of *road-proximity-classes*
   - *accessible* = Zonal(search) of *dummy* and *accessible-inaccessible*

6. *Good-sites:* A new layer of sites that satisfy the set of constraints and opportunities is produced by the successive overlay of layers produced in the previous steps. Finally, good sites are highlighted by performing a selective search on the result (Figure 11.6f).
   - *vacant-dry* = Local(overlay) of *vacant* and *dry*
   - *vacant-dry-level* = Local(overlay) of *level* and *vacant-dry*
   - *vacant-dry-level-accessible* = Local(overlay) of *accessible* and *vacant-dry-level*
   - *vacant-dry-level-accessible-south* = Local(overlay) of *south* and *vacant-dry-level-accessible*
   - *good-sites* = Zonal(search) of *dummy* and *vacant-dry-level-accessible-south*

7. *Candidate sites:* A new layer of sites that satisfy the set of constraints and opportunities and have adequate size is produced from the layer of good sites by measuring the sizes of zones and highlighting those that are within the predefined size interval (Figure 11.6g).
   - *good-sites-size* = Zonal(measurement) of *good-sites*
   - *candidate-sites* = Zonal(search) of *dummy* and *good-sites-size*

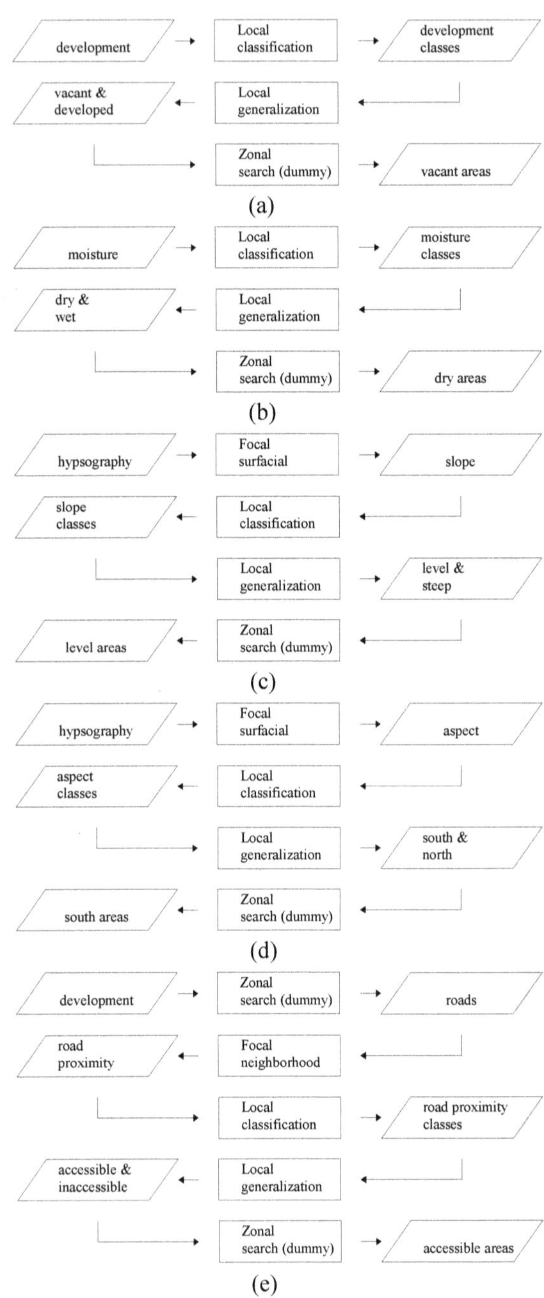
development
Local classification
development classes
vacant & developed
Local generalization
Zonal search (dummy)
vacant areas
(a)
moisture
Local classification
moisture classes
dry & wet
Local generalization
Zonal search (dummy)
dry areas
(b)
hypsography
Focal surfacial
slope
slope classes
Local classification
Local generalization
level & steep
level areas
Zonal search (dummy)
(c)
hypsography
Focal surfacial
aspect
aspect classes
Local classification
Local generalization
south & north
south areas
Zonal search (dummy)
(d)
development
Zonal search (dummy)
roads
road proximity
Focal neighborhood
Local classification
road proximity classes
accessible & inaccessible
Local generalization
Zonal search (dummy)
accessible areas
(e)

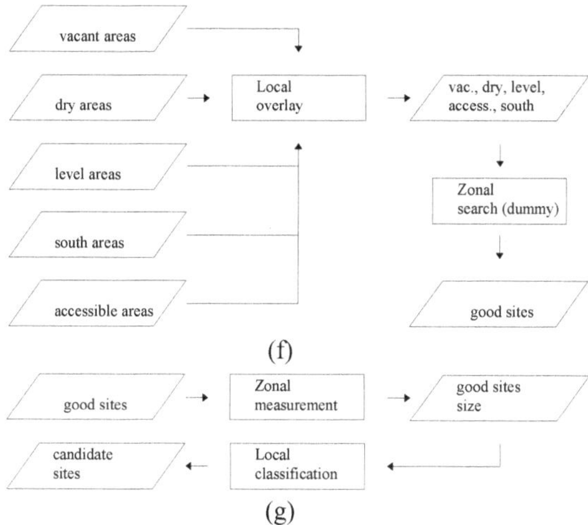



**Fig. 11.6** Site selection for a residential housing development.


## 11.3 Spatio-temporal Data Analysis

In Chapter 9, an extension of the spatial model to accommodate spatio-temporal data was presented. The extended model can also support the *spatio-temporal reasoning*. Specifically, the basic data interpretation operations discussed previously can be extended to handle the lists of *temporal attributes* assigned to individual locations on a layer. Following, the basic classes of data interpretation operations are redefined to include the temporal dimension. Then, one representative operation for each class is examined and accompanied with examples that commonly appear in spatio-temporal reasoning.

Provided that the thematic information associated with individual locations on a layer changes over time and this change is recorded using temporal attributes, the three classes of data interpretation operations are redefined as follows:

- *Temporal local operations:* they compute a new list of temporal attributes for each individual location on a layer as a function of existing data explicitly associated with that location. An example operation is the *spatio-temporal overlay*.
- *Temporal focal operations:* they compute a new list of temporal attributes for each individual location on a layer as a function of its neighborhood. An example operation is the *spatio-temporal distance*.
- *Temporal zonal operations:* they compute either a new list of temporal attributes or single attribute values for each individual location as a function of existing values associated with a temporal zone containing that location. An example operation is the *spatio-temporal select*.

## *11.3.1 Spatio-temporal Overlay Operation*

The spatial overlay operation is commonly applied in GIS. It is analogous to join operation in conventional database systems, and is defined as the assignment of new attribute values to individual locations resulting from the combination of two or more layers. The *spatio-temporal overlay operation* takes a more general form and is defined as the assignment of a new list of temporal attributes to individual locations resulting from the combination of the lists of temporal attributes associated with these locations on two or more layers.

In order to make clear how spatio-temporal overlay operation works, let's consider two layers A and B and an individual location L on them, which is assigned the lists of temporal attributes shown in Table 11.2. Assuming that $A_i$ ($B_i$) occurs before $A_{i+1}$ ($B_{i+1}$) and there are no gaps between successive interval time-stamps (i.e., $t_{(Ai)to} = t_{(Ai+1)from}$ and $t_{(Bi)to} = t_{(Bi+1)from}$) the temporal attributes can be represented as line segments in the one-dimensional space of time. Figure 11.7 illustrates graphically the situation presented in Table 11.2, as well as the result of the overlay operation. Depending on the *duration* and *temporal distribution* of interval time-stamps, the overlay operation results in a new set of interval time-stamps, which are assigned composite attribute values derived from the combination of attribute values of input layers. These composite attribute values with the corresponding interval time-stamps constitute the *temporal attributes* of the new layer.

**Table 11.2** Two lists of temporal attributes associated with an individual location L.

| *layer A* | | | *layer B* | | |
|---|---|---|---|---|---|
| *AV* | $T_{from}$ | $T_{to}$ | *AV* | $T_{from}$ | $T_{to}$ |
| A1 | $t_{(A1)from}$ | $t_{(A1)to}$ | B1 | $t_{(B1)from}$ | $t_{(B1)to}$ |
| A2 | $t_{(A2)from}$ | $t_{(A2)to}$ | B2 | $t_{(B2)from}$ | $t_{(B2)to}$ |
| A3 | $t_{(A3)from}$ | $t_{(A3)to}$ | B3 | $t_{(B3)from}$ | $t_{(B3)to}$ |

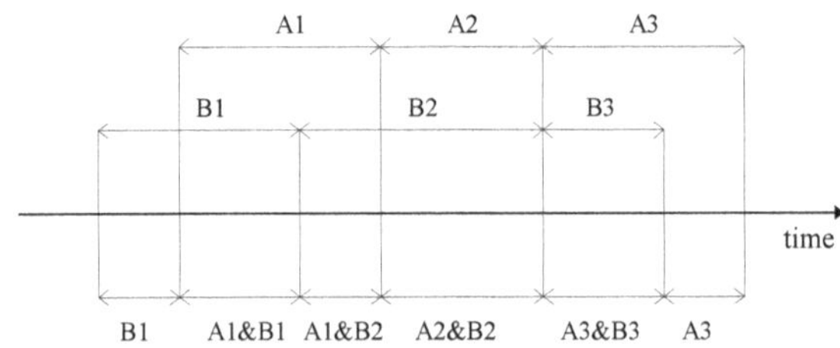


**Fig. 11.7** Temporal overlay operation: the temporal values of location L in layers A, B (top) and the new layer resulted from the overlay operation (bottom).

Figure 11.8 and 11.9 illustrate two examples of temporal overlay operations. The first example (Figure 11.8) shows how a layer of *tobacco farmers* is generated by performing a temporal overlay operation with operands the layer of *farmers* and the layer of *crop types*. The second example (Figure 11.9) shows how the layer of *wind velocity at vessel A* is derived from the layer of *vessel trajectories* and the layer of *wind velocity*. Obviously, only the locations traversed by vessel A are assigned lists of temporal attributes in the new layer.

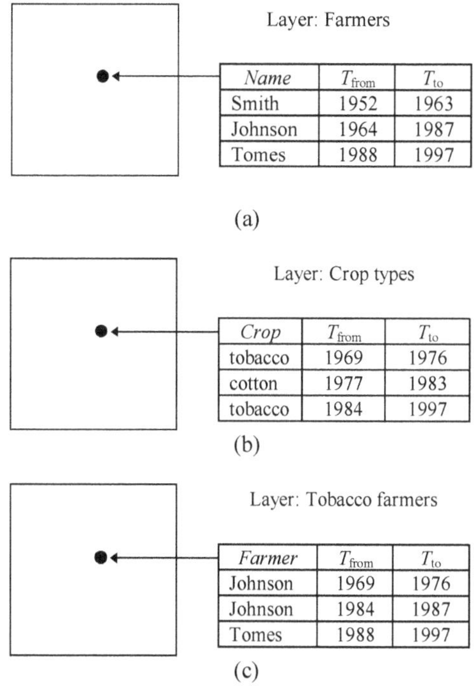



| Name | $T_{from}$ | $T_{to}$ |
|---|---|---|
| Smith | 1952 | 1963 |
| Johnson | 1964 | 1987 |
| Tomes | 1988 | 1997 |

(a)



| Crop | $T_{from}$ | $T_{to}$ |
|---|---|---|
| tobacco | 1969 | 1976 |
| cotton | 1977 | 1983 |
| tobacco | 1984 | 1997 |

(b)



| Farmer | $T_{from}$ | $T_{to}$ |
|---|---|---|
| Johnson | 1969 | 1976 |
| Johnson | 1984 | 1987 |
| Tomes | 1988 | 1997 |

(c)

**Fig. 11.8** Temporal overlay operation: An example of areal units that change over time (tobacco farmers: when and where).

## 11.3.2 Spatio-temporal Distance Operation

A distance metric is usually required in order to analyze the spatial relationships between entities in GIS. Several metrics are available in the literature. The choice depends on the application domain and the requirements posed by decision-makers. In this study, the conventional *Euclidean distance* is considered. For two individual locations i,j the Euclidean distance is given by the formula:

$$d(i, j) = \sqrt{\left(x_i - x_j\right)^2 + \left(y_i - y_j\right)^2}$$

where $(x_i, y_i)$ and $(x_j, y_j)$ denote the coordinates of the two locations i,j, respectively.

What differentiates a *spatio-temporal distance operation* from a conventional distance operation is that, since entities location and shape change over time, the distance metrics referenced to those entities change accordingly. Hence, the distance between two dynamic entities or between a static and a dynamic entity always refers to a particular instant or interval time-stamp. Following, the problem of assigning a list of temporal attributes to an individual location K depicting its distance from a moving point object is examined (Figure 11.10).

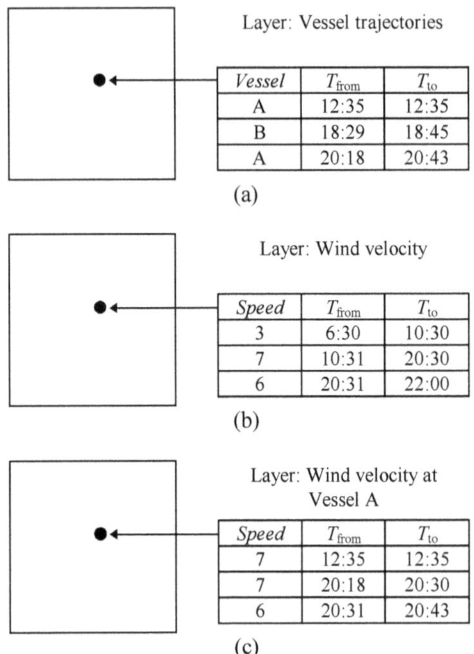



| Vessel | $T_{\text{from}}$ | $T_{\text{to}}$ |
|---|---|---|
| A | 12:35 | 12:35 |
| B | 18:29 | 18:45 |
| A | 20:18 | 20:43 |

| Speed | $T_{\text{from}}$ | $T_{\text{to}}$ |
|---|---|---|
| 3 | 6:30 | 10:30 |
| 7 | 10:31 | 20:30 |
| 6 | 20:31 | 22:00 |

| Speed | $T_{\text{from}}$ | $T_{\text{to}}$ |
|---|---|---|
| 7 | 12:35 | 12:35 |
| 7 | 20:18 | 20:30 |
| 6 | 20:31 | 20:43 |

**Fig. 11.9** Temporal overlay operation: An example with moving point objects (what is the wind velocity at vessel A).

Assume a moving point object P whose track is represented in a layer A. Each individual location L of layer A traversed by P is assigned a set of triplets of the form $(P,T_{\text{from}},T_{\text{to}})$, where $[T_{\text{from}},T_{\text{to}}]$ is the interval time-stamp during which object P has been present at L. In order to compute the list of temporal attributes depicting the distance of an individual location K from P, initially, all these interval time-stamps are retrieved (by performing a temporal select operation) along with the coordinates of the corresponding locations. Hence a list of new triplets of the form: $(L_i,T_{(i)\text{from}},T_{(i)\text{to}})$ is generated. Then, using the equation above, a distance measure $D_i=d(K,L_i)$ is assigned to each interval time-stamp (i) of the list and the list of temporal attributes $(D_i,T_{(i)\text{from}},T_{(i)\text{to}})$ is finally assigned to location K.

More complex problems, such as the computation of temporal distances between a given location and a changing areal unit, or between two dynamic objects, can be solved in a similar manner. Figure 11.11 illustrates an example of generating a new layer which depicts the temporal distance of a moving vessel B from a moving vessel of A (i.e., a moving neighborhood focus). Obviously, only locations traversed by vessel A are assigned lists of temporal attributes in the new layer. Notice that, similarly to temporal distance operation, other focal operations, such as temporal direction (e.g., north, east, south-west, etc.) and temporal topological operations (e.g., disjoint, meet, overlap, etc.) may be defined.

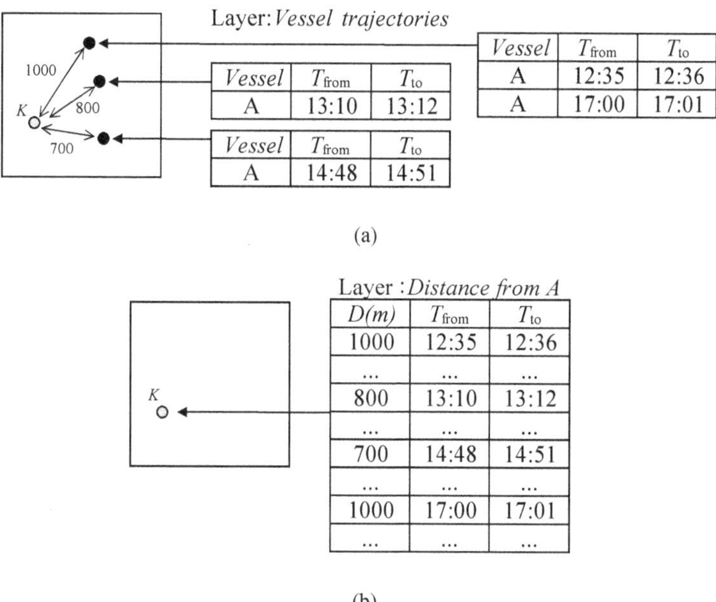

(a)

(b)

**Fig. 11.10** Temporal distance operation: an example with moving point objects (i.e., finding the temporal distance of a location K from a moving vessel).

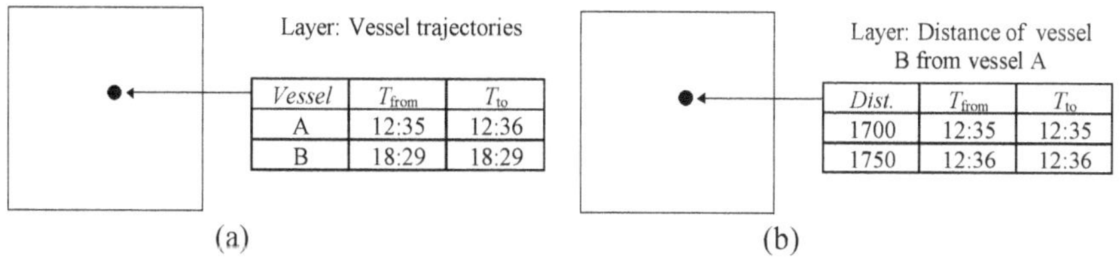

(a)                (b)

**Fig. 11.11** Temporal distance operation: an example for moving point objects (i.e., finding the temporal distance of a moving vessel from another moving vessel).

## 11.3.3 Spatio-temporal Select Operation

The scope of *temporal select operation* is to highlight individual locations on a layer based on the list of temporal attributes assigned to them. The temporal predicate involved in the selection may be a metric or topological relationship on interval time-stamps of temporal attributes (see Chapter 9). Notice that those locations may be filtered by static or temporal zones of other layers as in standard zonal select operation (mask query).

Obviously, the temporal select operation is usually applied in order to generate snapshot images (layers) depicting the situation of single entities or of the entire area under study at specified instant of interval time-stamps. Those static layers are compatible with the standard model for pure spatial data, presented in Chapter 9, and all data interpretation operations for handling the spatial dimension of geographic data can be applied.

## 11.4 Analysis of Geographic Data with Uncertainty

The extension of spatial model to accommodate geographic data with uncertainty was presented in Chapter 9. The extended model can also support *spatial reasoning with fuzziness*.

The three classes of data interpretation operations are redefined as follows in order to incorporate fuzziness:

- *Fuzzy local operations:* they compute new fuzzy values (d.o.m. values) for each individual location on a layer as a fuzzy function of existing fuzzy data explicitly associated with that location; e.g., fuzzy overlay operation.
- *Fuzzy focal operations:* they compute new fuzzy values for every individual location as a fuzzy function of its neighborhood; e.g., fuzzy distance operation.
- *Fuzzy zonal operations:* they compute new fuzzy values for each individual location as a fuzzy function of existing fuzzy values associated with a fuzzy zone containing that location; e.g., fuzzy select operation.

Next, one representative operation for each class is examined and accompanied with examples that commonly appear in spatial decision-making process. Then, the simplified example of site selection for a residential housing development is revisited to highlight some advantages of the fuzzy reasoning versus the conventional one.

### *11.4.1 Fuzzy Overlay Operation*

The *fuzzy overlay operation* is defined as the computation and assignment of an overall measure (d.o.m. value) to each individual location, which is derived from the consideration of the d.o.m. values on two or more layers and the execution of appropriate fuzzy operation(s) (see Chapter 9). The overall measure is also expressed in the fuzzy domain $[0,1]$.

Figure 11.12 illustrates an example of the conventional and fuzzy overlay operation. The scope of this operation is to highlight all individual locations that belong to forest areas with a steep ground slope. In conventional overlay operation the input layers are: (a) "vegetation" layer, and (b) "ground slope" layer. The product is the "forest & steep" layer, whose individual locations are assigned the value of 1, if they satisfy both constraints and 0 otherwise. On the other hand, in fuzzy overlay operation the input layers are: (a) "forest" lexical layer, and (b) "steep ground" lexical layer. In both layers individual locations are assigned the corresponding d.o.m. values for forest and steep ground, respectively. The product is the "forest & steep" layer, whose individual locations are assigned new d.o.m. values, characterizing their membership in the new fuzzy set, and derived by applying the appropriate fuzzy operation (e.g., the energy metric; see Chapter 9) and normalizing the result.

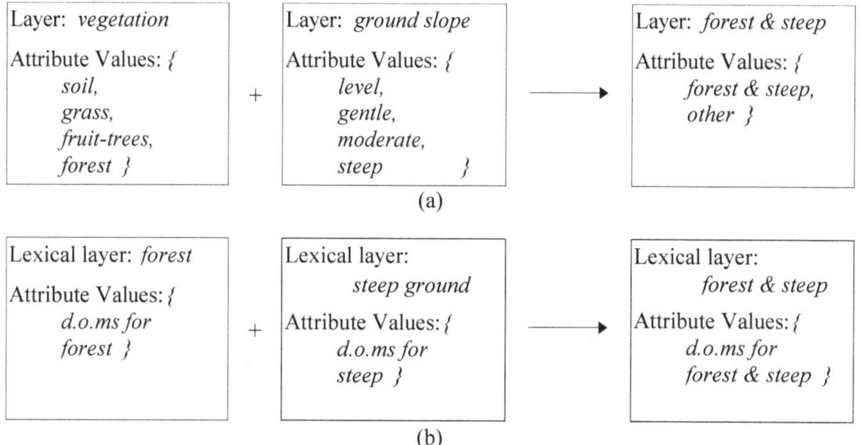

**Fig. 11.12** Example of a classical (a) and a fuzzy (b) overlay operation (input layers on the left and output layer on the right).

## *11.4.2 Fuzzy Distance Operation*

Two cases of *fuzzy distance operation* are examined next. The first shows how individual locations on a layer are classified based on their distance from a given location; while the second shows how individual locations on a layer are classified based on their distance from a given fuzzy zone (Figure 11.13).

In order to characterize an individual location X based on its distance from a given location L (Figure 11.14a) the following procedure is performed (Figure 11.13a). First, the Euclidean distance d from L to X is computed. Then a *membership function* (Chapter 9) is chosen to transform distances into d.o.m. values on the predefined attribute (lexical) values characterizing the theme "nearness" (e.g., close, near, moderate, far, far away). Finally, the distance from L to X is transformed into d.o.m. values. Hence, the product of the fuzzy distance operation consists of a set of layers and each layer accommodates the d.o.m. values regarding a specific attribute value (i.e., close, near, moderate, far, far away) characterizing the theme "nearness to location L".

In order to characterize an individual location X based on its distance from a given fuzzy zone Z, which consists of a set of individual locations $\{L_1, L_2, ..., L_n\}$ with different d.o.m. values in the fuzzy zone (Figure 11.13b), the following procedure is performed (Figure 11.14b). First, the Euclidean distances $d_i$ from all $L_i$ (i=1,2,...,n) to X are computed and transformed into d.o.m. values on the predefined attribute (lexical) values characterizing the theme "nearness" (e.g., close, near, moderate, far, far away). That is, for each attribute value A, the individual location X is assigned a set of pairs $(MF^F_A(X), MF^F_Z(L_i))$, (for i=1,2,...,n), where $MF^F_A(X)$ is the d.o.m. value for A characterizing the theme "nearness", and $MF^F_Z(L_i)$ is the d.o.m. value of location $L_i$ in fuzzy zone Z. Finally, a fuzzy function, chosen by the experts, is applied in order to map the set of pairs into single d.o.m. values (i.e., overall measures) on the predefined attribute (lexical) values characterizing the theme "nearness" (e.g., close, near, moderate, far, far away). Hence, the product of the fuzzy distance operation consists of a set of layers and each layer accommodates the d.o.m. values regarding a specific attribute value (i.e., close, near, moderate, far, far away) characterizing the theme "nearness to fuzzy zone Z".

Some example queries where fuzzy distance operation is applied, in combination with fuzzy select operation, are: "find all areas that are *near* to the existing road network", "find all areas that are *far* from schools", etc. Notice that similarly to fuzzy distance operation, other focal operations, such as a fuzzy direction (with example lexical values: north, east, south, west) or a fuzzy topological operation (with example lexical values: disjoint, overlap), may be defined.


### *11.4.3 Fuzzy Select Operation*


The scope of *fuzzy select operation* is to highlight individual locations on a layer based on their d.o.m. values regarding a single or composite attribute characterizing one or a combination of layers. Hence, depending on the constraints posed by the query, the fuzzy select operation may highlight: (a) those individual locations that have a d.o.m. value within a range of predefined threshold values; and (b) the n–individual locations that are superior to others based on their d.o.m. values (notion of ordering).

Figure 11.15 illustrates an example of fuzzy select operation. Figure 11.15a depicts the d.o.m. values for the attribute "near to highways" assigned to individual locations. Figure 11.15b highlights those locations with a d.o.m. value greater than or equal to 0.8; while Figure 11.15c highlights the 12 most far-off locations from the existing road network. Notice that those locations may be filtered by the zones of other layers as in standard zonal select operation (mask query).

L
d(L,X)
X
(a)
Z
L1
L2
d(L1,X)
Ln
d(L2,X)
X
d(Ln,X)
(b)

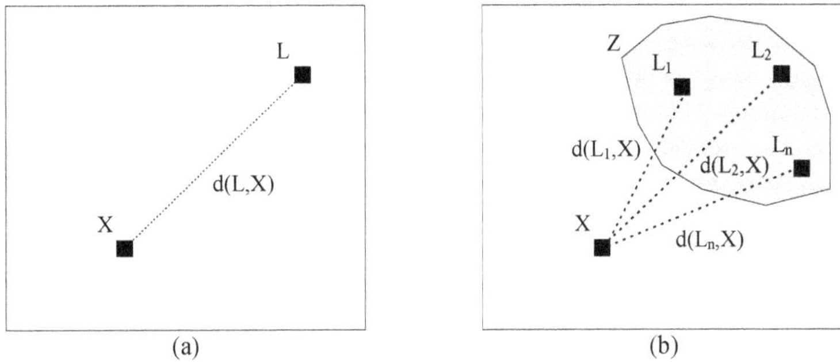

**Fig. 11.13** Example of a fuzzy distance operation between to locations (a) and between a location and a fuzzy zone (b).

Theme: nearness
Lexical values: {
close,
near,
moderate,
far,
far away }
L
X
compute distance
d
membership function
Classification
MF^F close(X)
MF^F near(X)
MF^F moderate(X)
MF^F far(X)
MF^F faraway(X)
(a)
Theme: nearness
Lexical values: {
close,
near,
moderate,
far,
far away }
{ Li }
X
compute distance
{ di }
membership function
Classification
MF^F close(X)
MF^F near(X)
MF^F moderate(X)
MF^F far(X)
MF^F faraway(X)
( MF^F close(X) , MF^F Z(Li) )
( MF^F near(X) , MF^F Z(Li) )
( MF^F moderate(X) , MF^F Z(Li) )
( MF^F far(X) , MF^F Z(Li) )
( MF^F faraway(X) , MF^F Z(Li) )
fuzzy function
(b)

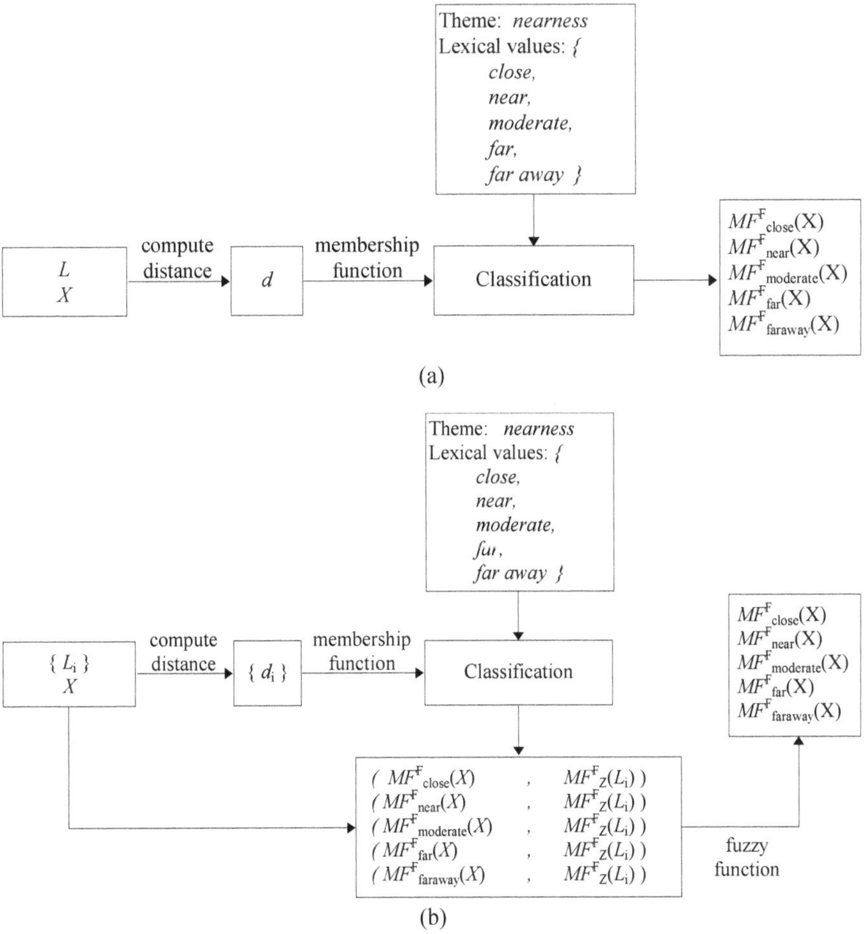

**Fig. 11.14** Procedure to compute the fuzzy distance between two locations (a) and between one location and a fuzzy zone (b).

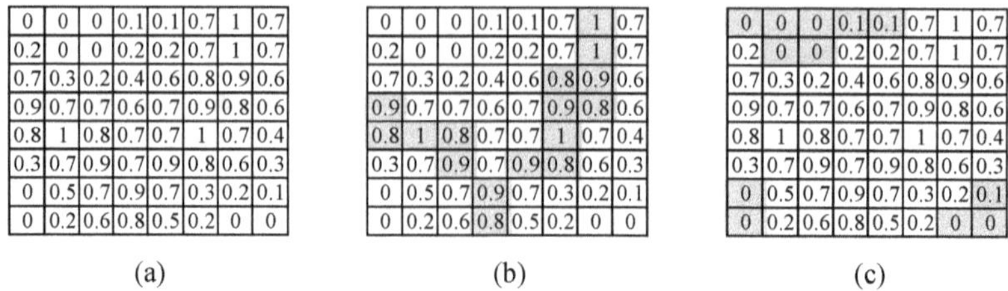

(a)                                    (b)                                    (c)

**Fig. 11.15** Example fuzzy select operation: input lexical layer (a), values > 0.8 selected (b), and locations of the 12 smallest d.o.m. values selected (c).

## 11.4.4 Site Selection and Fuzzy Reasoning Example

Revisiting the simplified example of *site selection for a residential housing development*, presented in a previous Section, some of the advantages provided by the fuzzy data interpretation operations may be highlighted. For the constraints posed by decision-makers the following *lexical values* could be considered:

- ground slope:  { level, gentle, moderate, steep }
- development:  { vacant, semi-developed, developed }
- soil moisture:  { dry, moderate, wet, water }
- accessibility:  { close, near, moderate, far, far away }
- aspect:  { north, east, south, west }

A membership function, like the one shown in Figure 9.18b, will be adopted to map ground measurements to d.o.m. values characterizing individual locations of the area under study. Figure 11.16 illustrates the procedure. After performing a fuzzy classification one layer with d.o.m. values for each lexical value characterizing a theme will be generated (Figure 11.16b). Examining the layers which correspond to the lexical values of interest (i.e., vacant, dry, level, near, south) a fuzzy overlay will provide a new layer which classifies all individual location of the area under study based on the degree they fulfill the constraints posed by decision-makers (Figure 11.16b). Last, a fuzzy select operation will highlight the best sites for the planning activity. As stated in Chapter 9, in this scheme, contrary to conventional way, reasoning is based on a *late and flexible classification* with several advantages.

## 11.5 Analysis of Fuzzy Dynamic Data

Chapter 9 presented a unified model for spatio-temporal and fuzzy data, which extends the spatial model by Tomlin. In the previous Sections, the three categories of data interpretation operations (local, focal, and zonal), were also extended to meet the needs of either a spatio-temporal or a fuzzy reasoning.

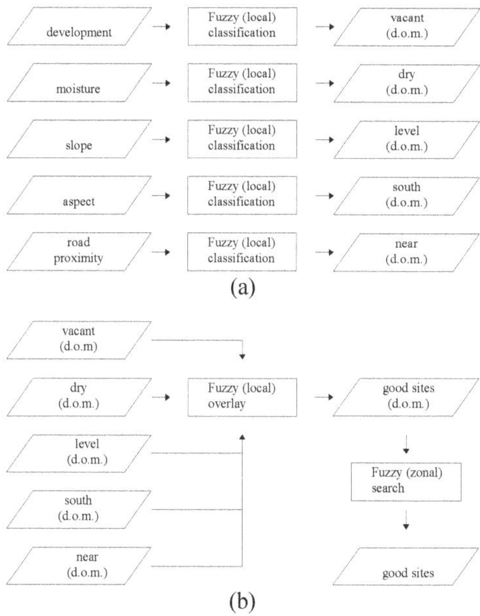


**Fig. 11.16** Site selection for a residential housing development (Figure 11.6) with fuzzy reasoning.

A *unified model* and the corresponding operations would be able to record and analyze both the fuzziness and temporal variation of thematic data associated with individual locations in space. Operations would consider these locations and characterize them depending on whether they meet three criteria (Figure 11.17): (a) *spatial*, (b) *temporal*, and (c) *thematic*. The spatial criterion refers to the spatial reference (location; where) of each location. The temporal criterion refers to the time reference (when it occurs) of each location. Finally, the thematic criterion refers to the d.o.m. values assigned to each location in a lexical cube (what; see Chapter 9).

An example *selection operation* that utilizes the fuzzy spatio-temporal reasoning is the following: "find all forest areas in the Prefecture of Heraklion, Crete during the decade 1961-1970" (Figure 9.24). This function takes as input the lexical cube "forest " for the area of Crete and generates as output a new lexical cube, which hosts the locations in Crete that (Figure 11.17): (a) lie in the spatial area specified in outline of Heraklion, (b) are valid over the temporal interval of the decade 1961-1970, and (c) are assigned a d.o.m. value greater than 0.7 for the lexical value "forest" (0.7 is the threshold specified for the definition of a forest land).

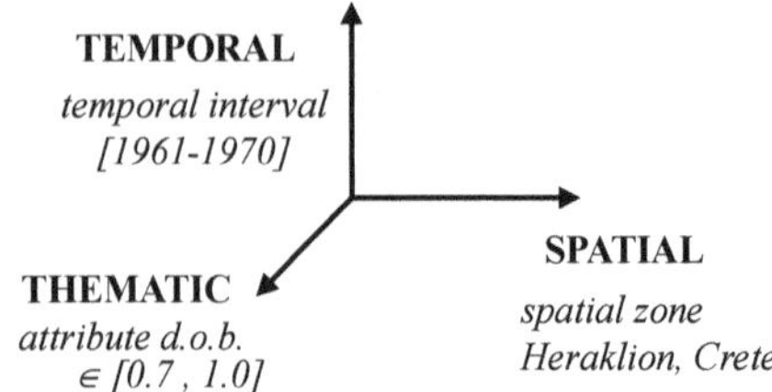


**Fig. 11.17** Criteria for a fuzzy spatio-temporal reasoning.

# References and Further Reading


Burrough, P.A., and McDonnell, R.A., 1998. *Principles of Geographic Information Systems*. Oxford University Press.

de By, R.A. (Ed.), 2000. P*rinciples of Geographic Information Systems – An Introductory Textbook*. ITC Educational Textbook Series.

Frank A., Raper, J., and Cheylan, J.P. (Eds), 2001. *Life and Motion of Socio-Economic Units.* GISDATA Series, No. 8. Taylor and Francis.

Gueting, R.H., Böhlen, M.H., et al., 2000. A foundation for representing and quering moving objects. *ACM Transactions on Database Systems (TODS)*. Vol. 25, pp. 1-42.

Longley, P.A., Goodchild, M.F., Maguire, D.J., and Rhind, D.W., 2001. *Geographic Information Systems and Science.* Wiley.

Rigaux, P., Scholl, M., and Voisard, A., 2002. *Spatial Databases with Applications to GIS*. Morgan-Kaufmann.

Shashi, S., and Chawla, S., 2003. *Spatial Databases – A Tour*. Prentice Hall.

Stefanakis, E., 2001. A unified framework for fuzzy spatio-temporal representation and reasoning. In *Proceedings of the 20th International Cartographic Conference*. Beijing, China, pp. 2678-2687.

Stefanakis, E., and Sellis, T., 1998. Enhancing operations with spatial access methods in a database management system for GIS. *Cartography and Geographic Information Systems* (Journal). Vol.25, pp.16-32.

Stefanakis, E., and Sellis, T., 2000. *Towards the design of a DBMS repository for temporal GIS.* In ESF- GISDATA Series Book: Life & Motion of Socio-economic Units (Eds A.U. Frank, J. Raper and J.-P. Cheylan). Taylor & Francis.

Stefanakis, E., Vazirgiannis, M., and Sellis, T., 1999. Incorporating fuzzy set methodologies in a DBMS repository for the application domain of GIS. *International Journal of Geographical Information Science*. Vol.13, pp.657-675. Taylor-Francis.

Tomlin, C.D., 1990. *Geographic Information Systems and Cartographic Modeling*. Prentice Hall.

Walford, N., 1995. *Geographical Data Analysis*. Wiley.

Worboys, M.F., 1995. *GIS – A Computing Perspective*. Taylor-Francis.

# 12 Computational Geometry

## 12.1 Introduction

*Computational geometry* is the study of the algorithms that solve geometric problems. In recent decades, the wealth of applications that involve calculations on spatial objects (such as GIS, graphical computing, robotics, automated design, etc.) has pushed the development of efficient geometric algorithms.

The aim of this Chapter is to present some of them, which have been implemented by modern GIS software packages and are often applied in processing geographic data. Note that the collection of algorithms that follows is by no means exhaustive. It simply comprises some representative computational geometry algorithms as well as algorithms which other Chapters refer to.

## 12.2 Area of a Polygon

The determination of the *area of a polygon* is a basic problem of Cartometry. Among various methodologies (e.g., use of planimeter, splitting the polygon into regular triangles, etc.) the analytical computation of the area can be applied.

The area of a polygon can be computed based on the coordinates of the vertices forming its outline. Consider the octagon O{1,2,3,4,5,6,7,8} in Figure 12.1, with two dimensional vertices in the Cartesian coordinate system XY. The octagon can be considered to consist of eight trapezoids with bases $(x_1, x_2)$, $(x_2, x_3)$, $(x_3, x_4)$, ..., $(x_8, x_1)$ and heights $(y_2-y_1)$, $(y_3-y_2)$,..., $(y_8-y_1)$, respectively. The area A of the polygon O is given by the formula (adding/subtracting trapezoid areas):

$$A = \sum\left(\frac{x_i + x_{i+1}}{2}(y_{i+1} - y_i)\right) - \frac{1}{2}\sum\left(y_i(x_{i-1} - x_{i+1})\right)$$

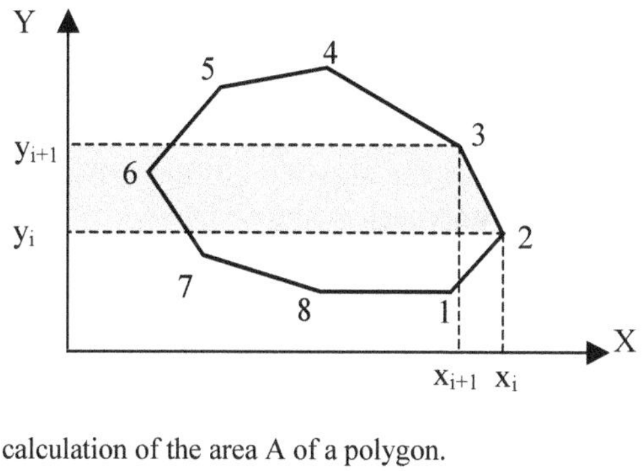

**Fig. 12.1** Analytical calculation of the area A of a polygon.

This method is fast, reliable and applicable to digital systems for determining the area of any polygon on a plane.


## 12.3 Intersection of Line Segments


The *intersection of two line segments* on a plane is a fundamental problem of computational geometry (Figure 12.2). Each segment is part of a straight line, which is described by the equation: y=ax+b. So, for the two underlying lines in Figure 12.2, the following formulas apply:

line 1:   $y = a_1x+b_1$
line 2:   $y = a_2x+b_2$

Each segment is defined by two points. Hence, for each of the above equations there are two unknowns (the coefficients $a_i$, $b_i$) and two equations for their calculation. After the coefficients $a_i$ and $b_i$ are calculated, the system of the two lines can be solved to determine the point of intersection. If this point lies within the line segments, the two segments intersect at this point (Figure 12.2a). Otherwise, they do not intersect (Figure 12.2b).

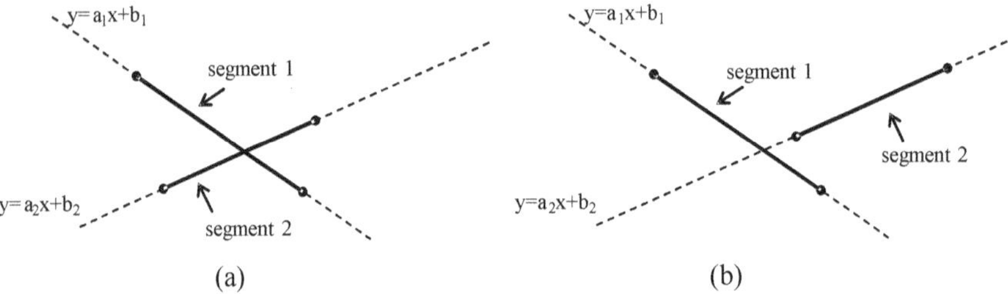


**Fig. 12.2** Intersection of two line segments. The lines as well as the segments intersect (a); the lines intersect, while the segments do not (b).


## 12.4 Intersection of Polylines


Consider a set S of N segments on a plane (Figure 12.3). To test whether any pair of *segments intersect* each other, the algorithm of the previous Section can be applied. The complexity of this approach would be $O(N^2)$.

More effective algorithms have been proposed to solve this problem. These algorithms (not examined here) apply the notion of the *sweep line*. Their complexity equals to $O(NlogN+K)$, where K is the number of intersections. These algorithms can also be applied to detect the intersections in a set of polylines (Figure 12.4).

**Fig. 12.3** A set of N line segments.

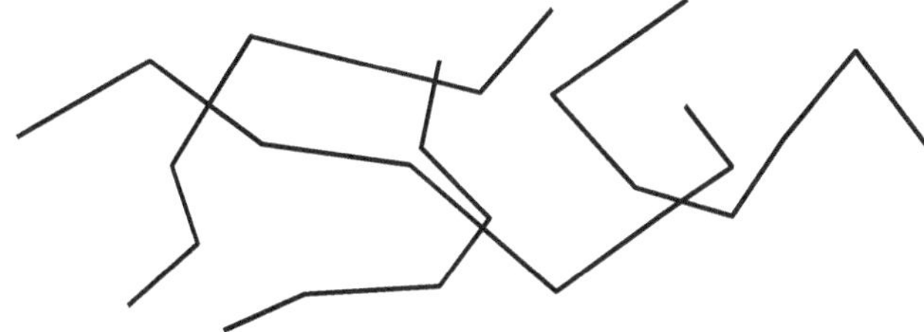

**Fig. 12.4** A set of polylines.

## 12.5 Intersection of Rectangles with Parallel Sides

Assume a rectangle in the two-dimensional space, the sides of which are parallel to the axes X,Y of a Cartesian coordinate system (Figure 12.5). As such, the rectangle can be defined by two points, i.e., the lower left corner with coordinates $(X_{min}, Y_{min})$, and the upper right corner with coordinates $(X_{max}, Y_{max})$.

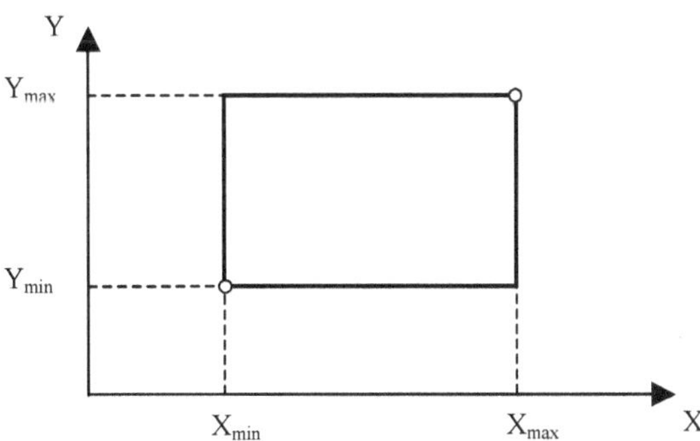

**Fig. 12.5** Rectangle with sides parallel to the axes of the coordinate system.

Any pair of rectangles A, B, defined by the following sets of coordinates:

A: { $(XA_{min}, YA_{min})$, $(XA_{max}, YA_{max})$ }
B: { $(XB_{min}, YB_{min})$, $(XB_{max}, YB_{max})$ }

can intersect each other or not (Figure 12.6). The condition for the detection of an intersection (overlap) between A and B is:

$$\text{IF} \quad (\quad XB_{max} < XA_{min}$$
$$\text{OR} \quad XB_{min} > XA_{max}$$
$$\text{OR} \quad YB_{max} < YA_{min}$$
$$\text{OR} \quad YB_{min} > YA_{max} \text{ ) THEN FALSE}$$
$$\text{ELSE TRUE}$$

A special case of the problem is when one of the two rectangles, e.g., B, degenerates into point (XB,YB). Then, the above condition of intersection, i.e., point B, inside or outside of A, is simplified to the following:

$$\text{IF} \quad (\quad XB < XA_{min}$$
$$\text{OR} \quad XB > XA_{max}$$
$$\text{OR} \quad YB < YA_{min}$$
$$\text{OR} \quad YB > YA_{max} \text{ ) THEN B OUTSIDE A}$$
$$\text{ELSE B INSIDE A}$$

Similarly,

$$\text{IF} \quad (\quad XA_{min} \leq XB \leq XA_{max}$$
$$\text{AND} \quad YA_{min} \leq YB \leq YA_{max} \text{ ) THEN B INSIDE A}$$
$$\text{ELSE B OUTSIDE A}$$

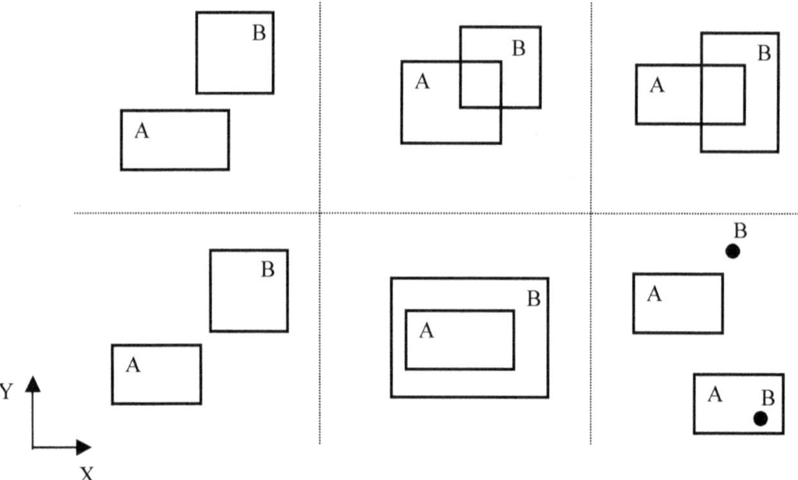


**Fig. 12.6** Examples pairs of intersecting and non-intersecting rectangles.

Note that the rectangles with sides which are parallel to the axes of the coordinate system, find wide application in spatial data structures; as they are used as a rough approximation of the shape of complex polygons or polylines (see Figures 12.8 and 13.13). These rectangles are called *minimum bounding rectangles (MBR)* of the corresponding spatial objects (see Chapter 13).

## 12.6 Point-in-Polygon

A classical algorithm of computational geometry is that of detecting whether a point A lies inside or outside a polygon P. The algorithm is called *point-in-polygon algorithm* and is frequently applied in GIS software packages, computer aided design systems, and digital cartography. Specifically, when a country is selected with the mouse from a digital map of Europe (e.g., Figure 12.9a), this algorithm must be executed so that the country (polygon) chosen is identified.

Consider the polygon P and the points A , B , C and D in Figure 12.7. To find out whether each point lies inside or outside the polygon P, the following algorithm must be applied. From each point a semi-line starting from the point towards any direction, e.g., parallel to the abscissa and heading to $+\infty$ is drawn. Then, the number of intersections of the semi-line with the edges of the polygon is counted. If this number is even, then the point lies outside the polygon (e.g., point D). Else, if the number is odd, then the point lies within the polygon (e.g., point B).

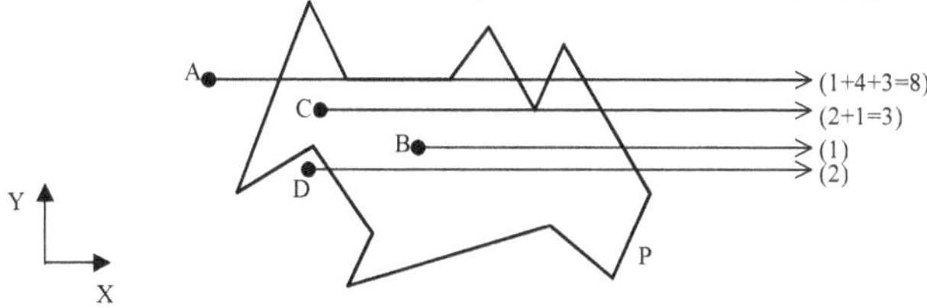


**Fig. 12.7** Point-in-polygon algorithm example.

To apply the above condition (i.e., even=out, odd=in) in special cases where the semi-line passes through the common vertex of two edges (e.g., point C) or is collinear with an edge (e.g., point A), two (2) and four (4) intersections are counted, respectively.

Note that the complexity of the algorithm is O(N) , where N is the number of polygon edges. To accelerate the process, a filter can be applied to detect in O(1) whether the point lies definitely outside the polygon. Specifically, the representation of the polygon with its minimum bounding rectangle (MBR) (Figure 12.8) enables a rapid response for the majority of points lying outside the polygon (see previous section). For example, point A can be labeled as out in O(1), provided that it is not included in the MBR of the polygon. Points B, C, and D, should be considered for the point-in-polygon algorithm against the actual polygon, as they fall inside the MBR. Obviously, the MBR includes the polygon and other areas of the space that are disjoint to the polygon (*dead zones*, as defined in Chapter 13; Figure 13.17).

Consider the map of European countries, as shown in Figure 12.9. This map consists of M polygons, each of which has an average of N edges. Identifying the country selected by the mouse cursor has a complexity equal to O(MxN) as the point-in-polygon check should be applied for each polygon (country) in the worst case. If countries are also represented by their minimum bounding rectangle

(MBR) in a filter structure (Figure 12.9b), then in a very low complexity, namely O(M), a small set of countries that possibly contain the cursor can be identified. These qualified countries (polygons) are usually more than one; because the MBRs are not disjoint (they may overlap). For the qualified countries the point-in-polygon algorithm against their actual geometries is performed. In the example of Figure 12.9, these countries are France (FR) and Germany (GE) (see MBRs gainst the points in Figure 12.9b). After running the point-in-polygon algorithm Germany will be eliminated and France will be reported.

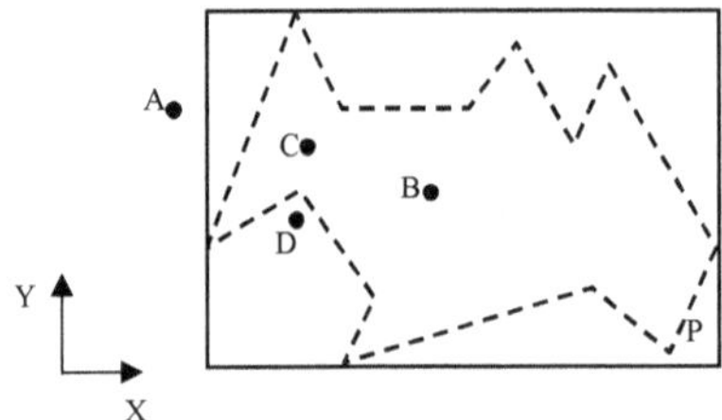

**Fig. 12.8** Point-in-polygon algorithm. The polygon (dashed line) is represented with its Minimum Bounding Rectanlge (MBR; solid line) in a filter.

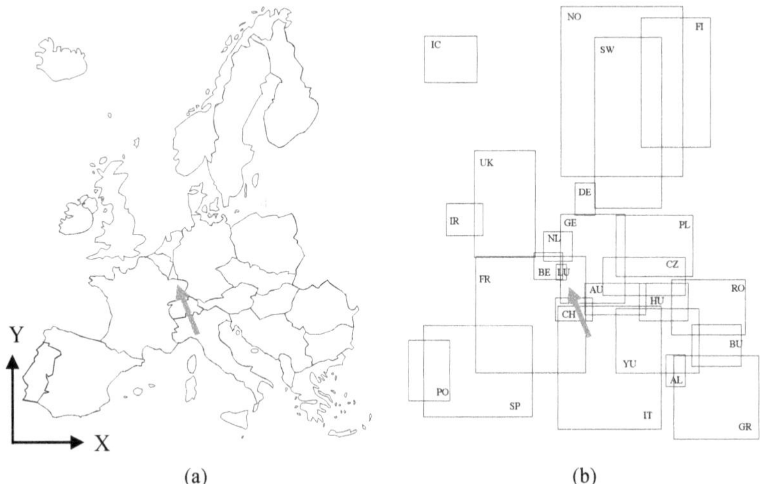

(a)                                                      (b)

**Fig. 12.9** (a) Point-in-polygon algorithm on the map of Europe; (b) representation of European countries with their minimum bounding rectangles (MBR).

## 12.7 Intersection of Polygons

The detection of an *intersection between two polygons* is a common problem in GIS. Two polygons $P_1$, $P_2$ intersect when at least one of the following conditions applies:

- an edge of $P_1$ intersects an edge of $P_2$
- a vertex of $P_1$ lies inside $P_2$
- a vertex of $P_2$ lies inside $P_1$

The algorithms described in the previous Sections can be applied in combination to check whether any of the above conditions applies to the extraction of a relevant conclusion.

## 12.8 Convex Hull of a Set of Points

In the analysis of geographic data the delineation of a polygon that delimits a set of objects is often required. Consider the simple case of N points on a plane (Figure 12.10). The smallest *convex polygon* (*convex hull*; see Chapter 10) containing these points is sought.

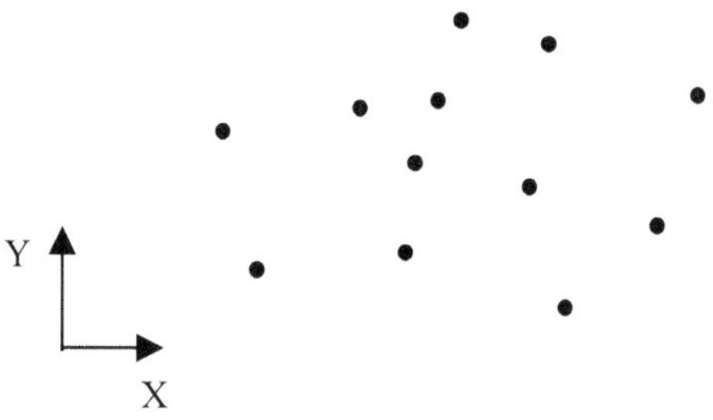


**Fig. 12.10** A set of points in the two dimensional space.

A simple algorithm of finding the convex hull is the following. Starting from the point with the smallest ordinate (this point is definitely a vertex of the polygon) a semi-line parallel to the axis of abscissas (axis X) and towards $+\infty$ is drawn. Obviously, all other points lie above this line. Then, the semi-line is rotated counterclockwise until it encounters the first point. This point is also a vertex of the convex polygon. With fixed point the new vertex the semi-line is rotated counterclockwise until the next point is reached, and so on. The process is repeated until the initial point is reached again (Figure 12.11).

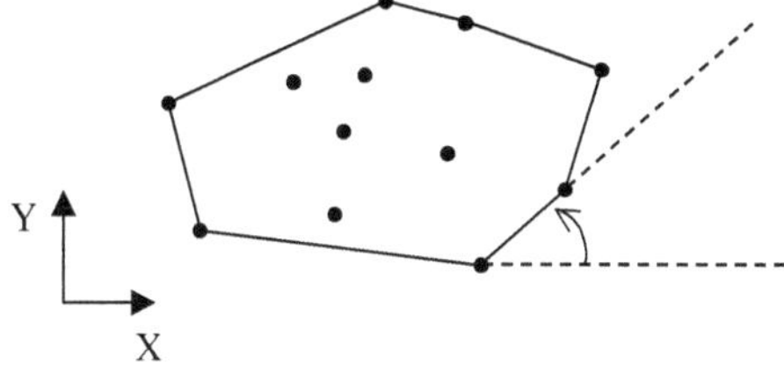


**Fig. 12.11** Delineation of convex hull for a set of points in plane.

The complexity of the algorithm is $O(N^2)$, and corresponds to the worst case that all points are vertices of the convex hull. Several alternative algorithms have been proposed in the past (e.g., Graham, Floyd-Eddy, etc.) to solve the problem with better performance, such as $O(N\log N)$. The description of these algorithms is not included in this book.

## 12.9 Simplification of Polylines

*Simplifying a polyline* is one of the key processes in *cartographic generalization* and is partially supported in a digital environment by appropriate algorithms. The simplification of a line involves the selection and removal of vertices and leads to a new (generalized) line with fewer vertices.

Figure 12.12 shows an example of simplifying the outline of a polygon (prefecture of Chania, Greece). The polygon is represented in Figure 12.12b with fewer points (than in Figure 12.12a), which are enough to represent its form in a smaller scale map.

(a)        (b)

**Fig. 12.12** Simplification of a polygon outline.

### *12.9.1 Methods for Line Simplification*

There are two basic methods for reducing the number of vertices of a polyline. The *first method* systematically retains the first and every $n^{th}$ vertex of the line. In other words, if the vertices of the polygonal line are stored in a data file sequentially by records, the algorithm creates a new file that copies the first and every $n^{th}$ record of the initial file (e.g., for $n = 10$: the $1^{st}$, $11^{th}$, $21^{st}$, ... records). Obviously, the larger the n, the coarser is the simplification of the polyline.

To choose the value of n, *Topfer's rule* is often applied. This rule connects the map scale with the number of elements depicted therein, as follows:

$$n_t = n_s \cdot \sqrt{\frac{S_t}{S_s}}$$

where $n_t$ is the number of elements in the generalized (target) map in scale $S_t$ and $n_s$ the corresponding number on the source map in scale $S_s$. For example, if $n_s$=1,000; $S_t$=1:50,000; and $S_s$=1:10,000, then $n_t$=447.

Often cartographers apply a modified version of the method by assigning *weight values* to the vertices of the line to simplify. In this case, with the support of a classification algorithm, line vertices are removed systematically, based on their significance. For example, a vertex with big weight (i.e., vertex of high importance) will be retained even if it is not the $n^{th}$ in the sequence.

The *second method* is based on random selection of $1/n$ of the polyline vertices. Obviously, the larger the n, the coarser is the simplification of the line.

Both methods above neglect any geographic / cartographic characteristics of the line. For this reason they offer a "blind and random" downsizing of bulky digital files unless weights are assigned to them. For example, suppose that a polyline was digitized in a resolution of 0.0001 inch and it is to be printed on a plotter in a resolution of 0.001 inch. Obviously, the volume of data is 10 times greater than what can be attributed to the graphic plotting. In this case, the application of one of the above methods is suitable for the preparation of the line for plotting.


## *12.9.2 Douglas-Peucker Algorithm*


Simplifying a polyline cannot be a fully automated (algorithmic) process, as it must take into consideration the form (nature or "character") of the line. Cartographers usually eliminate vertices manually, based on the (subjective) perception of the visual significance of these vertices. Obviously, such a tactic requires time and high experience from the cartographer in simplifying lines.

In practice, the cartographer applies some line simplification algorithms, which aim at preserving the form of the line while deleting vertices. The result is supervised and corrected by the cartographer. One of the most popular and effective algorithms was developed by Douglas and Peucker and is available in all modern computer cartography and GIS systems.

A parameter of the *Douglas-Peucker algorithm* is the *threshold segment*, which determines the degree of simplification. The length of this threshold is usually associated with visual acuity (0.25mm or 0.5mm) on the target scale. For example, if the target scale is 1:50,000, the threshold is: 0.5mm x 50,000 = 25m.

Figure 12.13 shows graphically the algorithm in steps. The polyline to simplify has initially 10 vertices. Vertices 1 and 10 (the two extreme vertices) are connected through a straight line segment (Figure 12.13b). Then, the polyline vertex located farthest from the segment 1-10 is detected (vertex 7). The distance of the most distant vertex to the segment is compared against the threshold. If this distance exceeds the threshold (as in Figure 12.13b), the vertex is retained and the algorithm is recursively executed for the two line segments either side of this vertex. Otherwise, it is ignored and the algorithm stops (termination condition). An iterative process for the lines 1-7 and 7-10 follows (Figure 12.13c,d). Figure 11.13e shows the simplified line for the predetermined threshold. This line consists of four (4) vertices (including the end points 1 and 10).

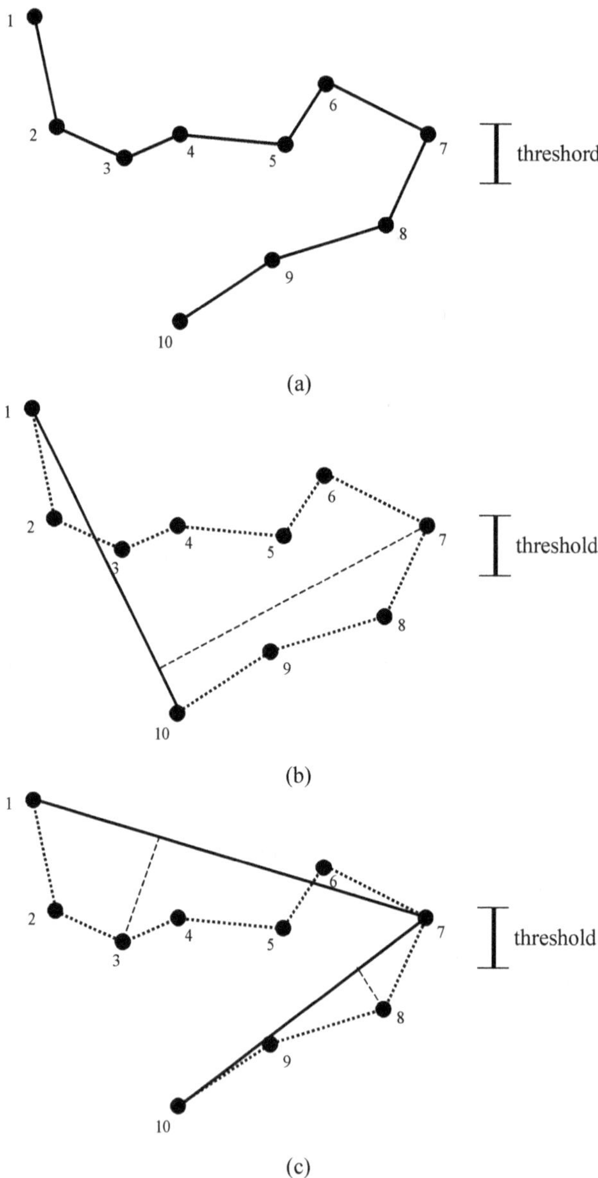



(a)



(b)



(c)

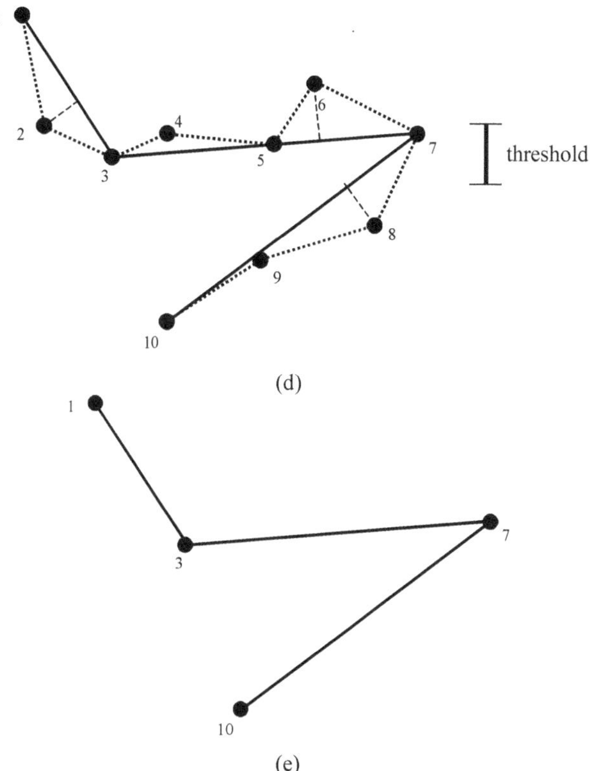


**Fig. 12.13** Douglas-Peucker algorithm (example).

Douglas-Peucker algorithm is not capable of considering the temporal dimension which might be associated with a linear entity. If the linear entity represents the trajectory of a moving object (e.g., a sea vessel; Figure 12.14a), then the simplification needs to also consider the time. Meratnia and de By have proposed an extended version of the Douglas-Peucker algorithm by introducing the notion of the *Synchronous Euclidean Distance (SED)* to consider the temporal dimension associated with the trajectory.

In a trajectory, each point K is assigned a temporal stamp ($t_K$), which determines the time that the moving object crossed that point. Let A, B and K three spatiotemporal locations recorded for a trajectory T, with $t_A < t_K < t_B$ (Figure 12.14a). Then, the SED for the point K is equal to the Euclidean distance KK', where the location K' is calculated with respect to the average velocity vector $U_{AB}$ (Figure 12.14b). In other words, the distance of point K from the straight line approximation AB is equal to KK', where K' is the spatiotemporal trace of K on that approximation at time $t_K$. In general, the SED KK' (dashed line) is not perpendicular to the straight line approximation (AB) and hence its length is greater or equal to the perpendicular distance applied in the original Douglas-Peucker algorithm.

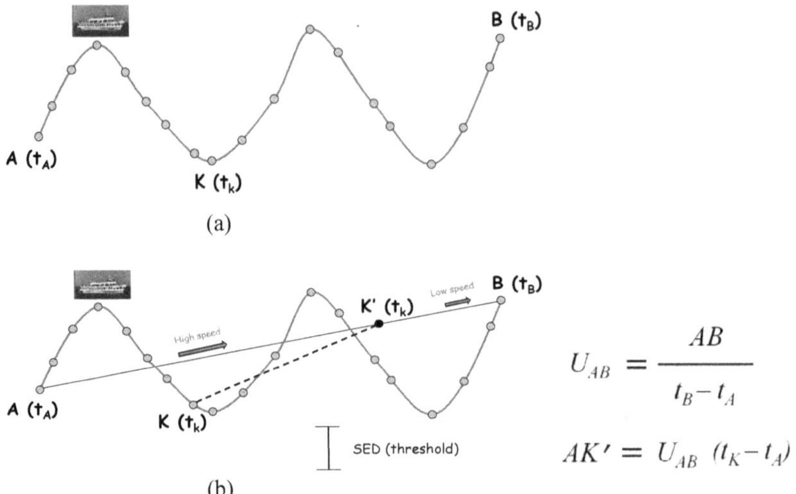


$$U_{AB} = \frac{AB}{t_B - t_A}$$

$$AK' = U_{AB}\,(t_K - t_A)$$

**Fig. 12.14** An example vessel trajectory (a); and its simplification using the Synchronized Euclidean Distance (b). Point K needs to be retained as KK′ >SED.

## 12.10 Delaunay Triangulation

The *triangular irregular network (TIN)* was introduced in Chapter 10. As mentioned there, the Delaunay triangulation is a network of triangles over a set of points on the plane surface. Suppose a set S of N points on a plane. Three points k,l,m ∈ S form a triangle in the Delaunay triangulation of S, if the circle passing through these points (circumscribed circle) encloses no other points of S (Delaunay condition). Unless there are four points in S that lie on the same circle, the above criterion leads to a unique triangulation (Figure 12.18; see next).

There are many algorithms for building the Delaunay triangulation over a set of points S. The following paragraphs describe a simple and computationally efficient algorithm (complexity: O(NlogN), where N is the number of points in S), which is based on an incremental building of the network. This algorithm leads to the Delaunay triangulation regardless the distribution of the points in space or the order of their insertion into the network.

The points of S are added sequentially one after the other, leading to the construction of the triangulation. As the algorithm is incremental, after the insertion of point $K_i$, triangulation $T_i$ is built for all points $K_1$, $K_2$, ... $K_i$ of S. The construction of the triangulation $T_{i+1}$ is obtained by adding point $K_{i+1}$ from S into triangulation $T_i$.

The insertion algorithm of a new point $K_{i+1}$ in the triangulation $T_i$ is as follows. Initially, the triangle T of $T_i$ containing point $K_{i+1}$ is detected. Then, point $K_{i+1}$ is connected to the three vertices of triangle T with temporal edges (Figure 12.15). A check follows to find out whether the three newly formed triangles (sharing vertex

$K_{i+1}$) satisfy Delaunay's condition (Figure 12.16). If the circle passing through $K_{i+1}$ and any other two vertices of the triangle T contains any other vertices of triangulation $T_i$, the edge between the two vertices of T is replaced by a new edge from point $K_{i+1}$ to the opposite vertex of the adjacent triangle which was sharing the replaced edge with triangle T (Figure 12.16). This happens with two edges in the example (the highlighted ones), which have been replaced as shown in Figure 12.16b (the highlighted lines represent the corrections that replaced the temporary dashed edges).

Figure 12.17 shows the adjusted triangulation $T_{i+1}$ after the insertion of point $K_{i+1}$. Figure 12.17b shows that triangulation $T_{i+1}$ is compliant with the Delaunay's condition.

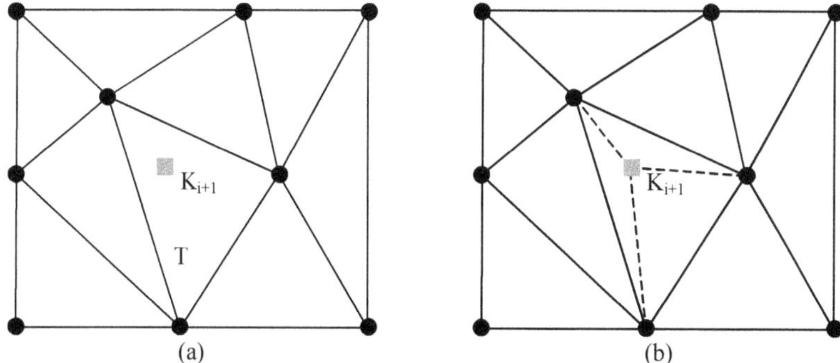


**Fig. 12.15** The new point Ki+1 lies inside triangle T (a). Three new (temporal) edges/triangles (dashed lines) (b).

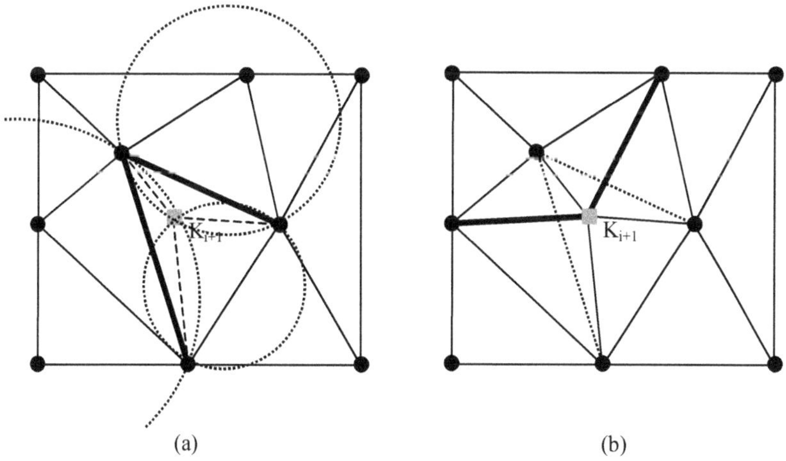


**Fig. 12.16** Replacement of two the highlighted edges in (a) with the solid lines in (b).

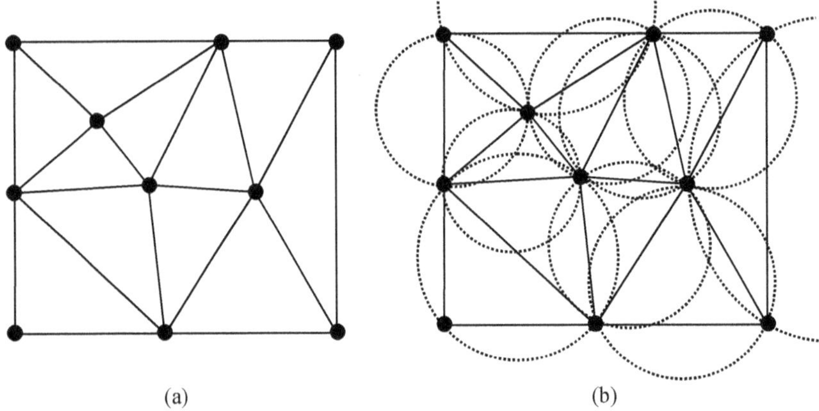

(a)                                                    (b)

**Fig. 12.17** Insertion of a new point (Ki+1) into a Triangular Irregular Network (a), and checking Delaunay condition (b).

The vertices of the two lower-left triangles in Figure 12.18a happen to lie on the same circle. As mentioned above, this case leads to more than one triangulation that all satisfy the Delaunay's condition. Figure 12.18b illustrates the alternative triangulation (with bold segment; while the dotted one represents the original as of Figure 12.18a).

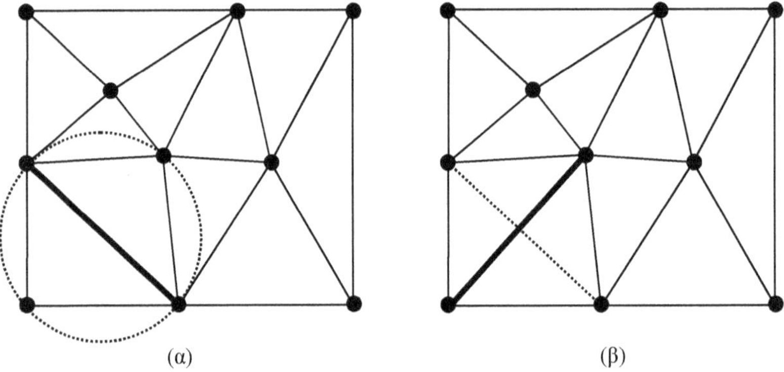

(α)                                                    (β)

**Fig. 12.18** Example of four vertices on the same circle (a), and the dual triangulation (b).

## 12.11 Thiessen Polygons

The triangular irregular networks find application in the creation of *polygons by Thiessen* (Chapter 10). Given a set S of N two-dimensional points on a plane surface (Figure 12.19a), a set P of N Thiessen polygons can be defined (Figure 10.20b). These polygons are disjoint, they occupy all space, and each polygon encloses one point K of S along with all locations in space that lie closer to point K than any other point of S.

The process of creating the Thiessen polygons is the following. Initially, a triangular irregular network is built over the N points of S (Figure 12.19b). Then, the perpendicular bisector to the sides of the triangles of the network are drawn (Figure 12.20a). After applying appropriate cuts to line bisectors the Thiessen polygons are formed (Figure 12.20b).

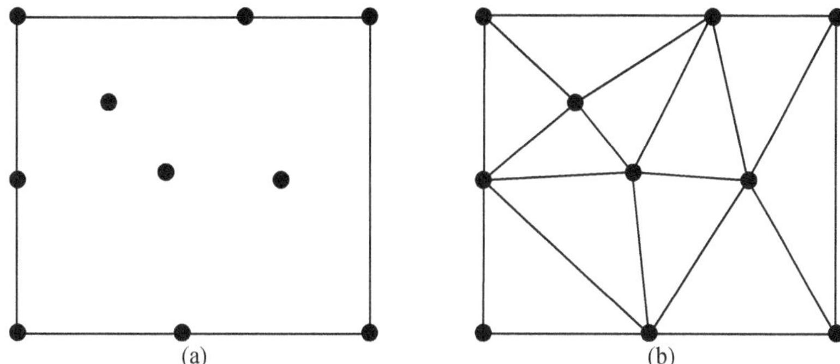

(a)　　　　　　　　　　(b)

**Fig. 12.19** A set of two-dimensional points (a) and the triangular irregular network by Delaunay (b).

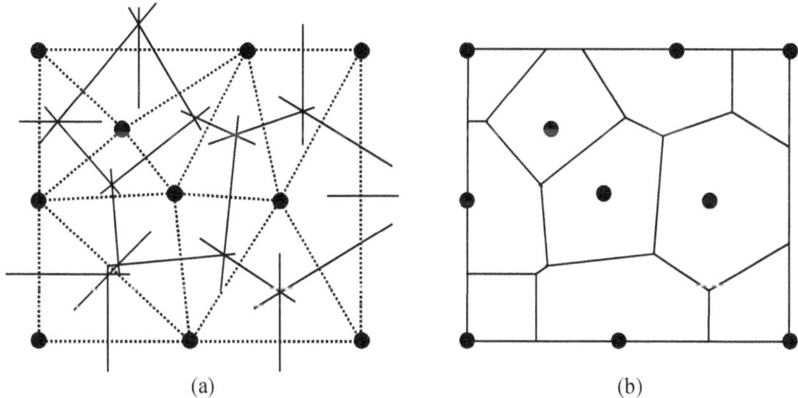

(a)　　　　　　　　　　(b)

**Fig. 12.20** The perpendicular bisectors to triangle edges (a), and the Thiessen polygons (b).

## 12.12 Elevations from a Triangular Irregular Network

Triangular irregular networks find application in the representation of the terrain (Figure 10.15). In this case, the vertices of the triangles are the elevation points (points of known elevation), while the triangles are disjoint ramps on site (three non-collincar points define a plane in space).

To compute the elevation of any point (Figure 12.21), the triangle including the point should be detected first. This can be done either sequentially (point in polygon algorithm) or with the support of a spatial data structure in the role of the filter, such as an R-tree (see Chapter 13).

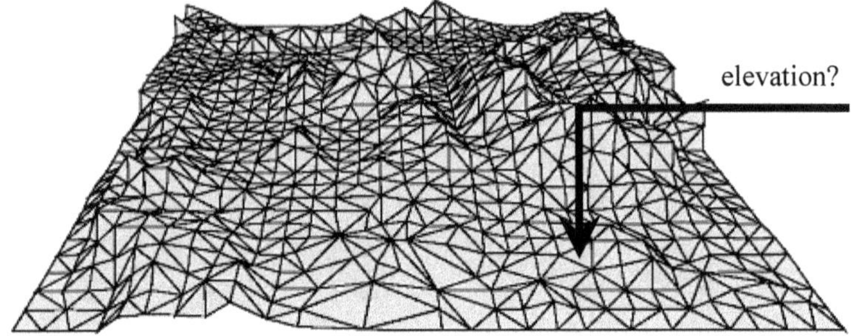


**Fig. 12.21** Computing the elevation of a point.

Then, the problem is reduced to the following: Suppose a triangle 123 in space and a point K within the triangle. Given the coordinates of the triangle vertices $\{(x_i, y_i, z_i); i=1,2,3\}$ and the horizontal coordinates of the point K $(x_k, y_k)$, the elevation $z_k$ of point K is sought (Fig. 12.22). Three solutions to the problem of finding the elevation of point K are described next. From these solutions, the first is the most accurate. Note that neither of these solutions is correct, as they all make the assumption that the land within the triangle 123 is flat.

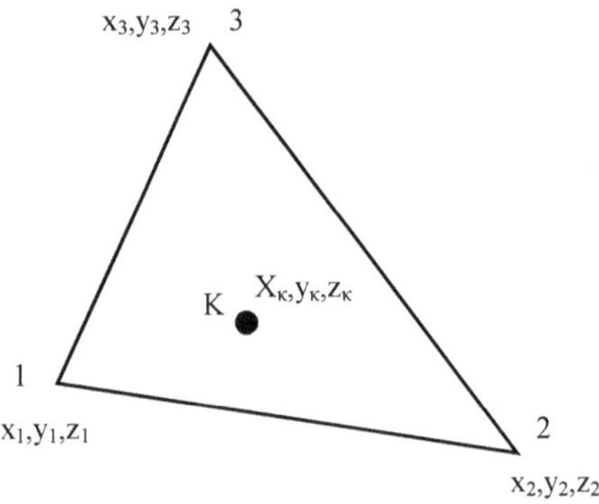


**Fig. 12.22** Finding the elevation of a point falling within a sloping triangle.

*1ˢᵗ Solution*

The first solution to the problem is based on the determination of the coefficients of an equation describing a plane surface in the three-dimensional space:

$$Ax + By + Cz = 1$$

The plane is defined by points 1,2,3. Having determined the coefficients A, B and C (by solving a linear system of three equations with three unknowns) the elevation of any point K on this plane can be computed.

## $2^{nd}$ Solution

The second solution proposes the calculation of the horizontal areas $E_{12K}$, $E_{23K}$, $E_{13K}$, formed by drawing the edges from K to the vertices of the triangle (Figure 12.23). Then, Heron's formula (Figure 12.24) is applied to compute the areas.

Next, three weight values are computed (where $E_{123}$ is the horizontal area of the triangle 123):

$$w_{12K} = E_{12K} / E_{123}$$
$$w_{23K} = E_{23K} / E_{123}$$
$$w_{13K} = E_{13K} / E_{123}$$

The elevation of point K is derived from the following formula:

$$z_K = [ z_1 w_{23K} + z_2 w_{13K} + z_3 w_{12K} ] / [ w_{23K} + w_{13K} + w_{12K} ]$$

Obviously, if point K lies on a side of the triangle, a zero weight occurs and thus a linear interpolation will take place. If point K coincides to a vertex, two weights will turn to null and the elevation will be equal to that of the vertex.

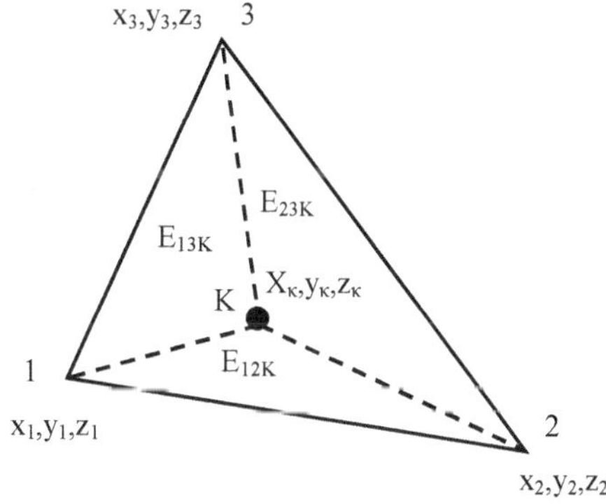


**Fig. 12.23** Edges from point K to the three vertices of the triangle 123.

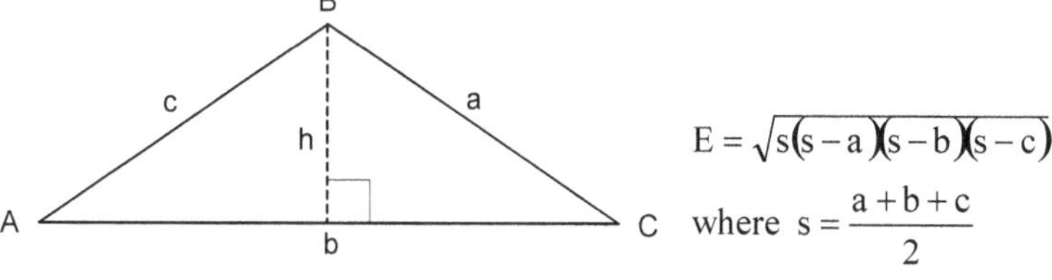


**Fig. 12.24** Heron's formula.

*3$^{rd}$ Solution*

The third solution to the problem applies the *inverse distance weight* interpolation function. Figure 12.25 schematically shows the method and the type of interpolation. Given n points (where n=3; the three vertices of the triangle) each point i affects the elevation of point K inversely the distance $D_{ik}$ in a power p (usually p=2).

To determine the elevation of a point K that lies on a triangle 123 (Figure 12.22), the formula in Figure 12.25 will be applied with: n=3 and p=2.

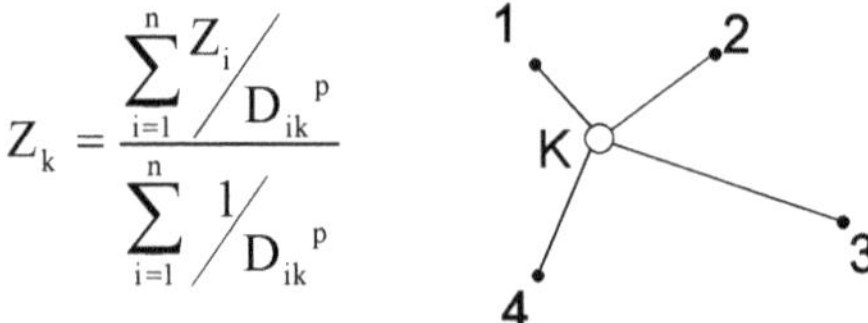


$$Z_k = \frac{\sum_{i=1}^{n} Z_i / D_{ik}^{\,p}}{\sum_{i=1}^{n} 1 / D_{ik}^{\,p}}$$

**Fig. 12.25** The inverse distance weight interpolation function.

## 12.13 Clustering Point Objects

In the literature, many algorithms have been proposed for grouping objects into classes (groups), so that members of a class are as identical as possible between them, while those in different classes differ as much as possible between them. These algorithms are widely used in mining large databases. Next two representative clustering methods are presented: K-Means and DBSCAN.

### *12.13.1 K-Means*

A simple and widely used clustering method is the *K-Means*. There are several variants of the K-Means method for different geometry types and space dimensions. Next, the problem of clustering a set of point objects on a two-dimensional Euclidian space is examined. Given a set S of N points on a plane surface, the K-Means method clusters these points into K clusters based on their location and the notion of proximity. The following algorithm implements the method:

- *Step 1:* Randomly select K points in space. Each of these points will initially represent the center of a cluster (*cluster mean*).
- *Step 2:* Assign each point of S into the nearest cluster, based on the shortest distance to the corresponding cluster mean.
- *Step 3:* After all points in S are assigned to the K clusters, calculate a new mean for each cluster. The new cluster mean will have an abscissa and ordinate equal to the average abscissa and ordinate of the points in the cluster, respectively.

- *Step 4:* Repeat steps 2 and 3 until no change occurs.

Figure 12.26 shows these steps through an example of clustering a set of points into three clusters. The K-Means method works well when the number of clusters K is known in advance (e.g., the Government wants to allocate three schools in a new residential area to better serve the people in that area). On the other hand, it is very sensitive to noise as all points must be assigned to a cluster and any outlier points would distort the clusters shape. The method presented in the next Section does not rely on a predefined number of clusters and can handle noise efficiently.

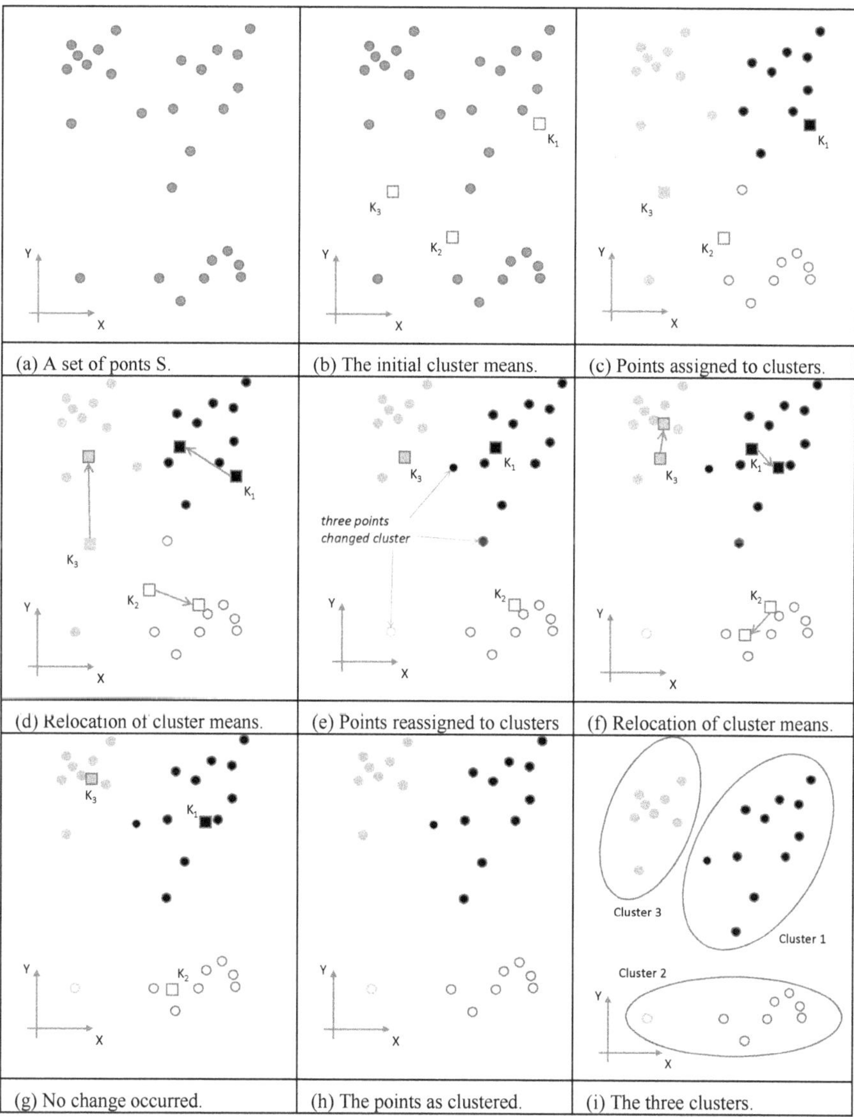


| | | |
|---|---|---|
| (a) A set of ponts S. | (b) The initial cluster means. | (c) Points assigned to clusters. |
| (d) Relocation of cluster means. | (e) Points reassigned to clusters | (f) Relocation of cluster means. |
| (g) No change occurred. | (h) The points as clustered. | (i) The three clusters. |

**Fig. 12.26** Steps of the K-Means algorithm (K=3).

## *12.13.2 DBSCAN*

*Density Based Spatial Clustering of Applications with Noise (DBSCAN)* is a densi-ty-based clustering method initially developed to cluster point objects. The associ-ated algorithm requires two input parameters: e and MinPts. The *neighborhood* of a given point p is examined and judged to be sufficiently dense, if the *number* of points within a distance e from p is greater than MinPts. If so, p is called a *core point* and forms an initial cluster for the data set. The neighborhood within radius e from the point p is called the *e-neighborhood* of p. All point objects of the data set that lie within the e-neighborhood of p are called *non-core points* of the initial cluster of p. The DBSCAN clustering algorithm for a set of N points consists of the following steps:

- *Step 1:* Specify the parameter values for e and MinPts.
- *Step 2:* Form the initial clusters by applying the following rule: For each point p of the data set, examine if it is a core point. If so, generate a new cluster with core point the point p and non-core points all other points of the set that lie within the e-neighborhood of p.
- *Step 3:* Compare repetitively the initial clusters in pairs; if the core point in the one cluster is a non-core point in the other cluster, merge the two clusters into one.
- *Step 4:* If a point is assigned as a non-core point to more than one cluster, re-move it so that it is assigned to only one of them.

At the end of this process each data point is assigned to one or none cluster. In the latter case, the point is considered to be *noise.*

Figure 12.27 presents an example of the DBSCAN algorithm for a small set of points. As shown in Figure 12.27g, MinPts is assigned the value of 3 and e the ra-dius of the circle. Figure 12.27a shows the point data set. In Figure 12.27b, a circle is drawn for each point and all core points are shown in dark black dots. Notice that the circle from each dark dot contains at least 3 (MinPts) other points of the set. Figure 12.27c shows how two initial clusters are compared and finally merged as of Figure 12.27d. Figure 12.27e presents the result of merging the initial clus-ters. It consists of two clusters, and two isolated points, which are considered to be noise. Notice that the two clusters in Figure 12.27e share one point. This point is decided to being removed from the top cluster. The decision to which cluster the point will be assigned is arbitrary. However, the application domain may impose a rule. Figure 12.27f shows the final two clusters discovered by DBSCAN algo-rithm.

A major drawback of the algorithm is the need to specify the values of the pa-rameters e and minPts, which decisively affect the result of clustering. Efficient practices for the choice of these values can be found in the literature. These are not considered here.

Finally, DBSCAN can also be applied to group linear or polygonal objects, after these objects are represented by a set of points (e.g., their centroid or MBR vertices). The presentation of these methods is not included in this book.

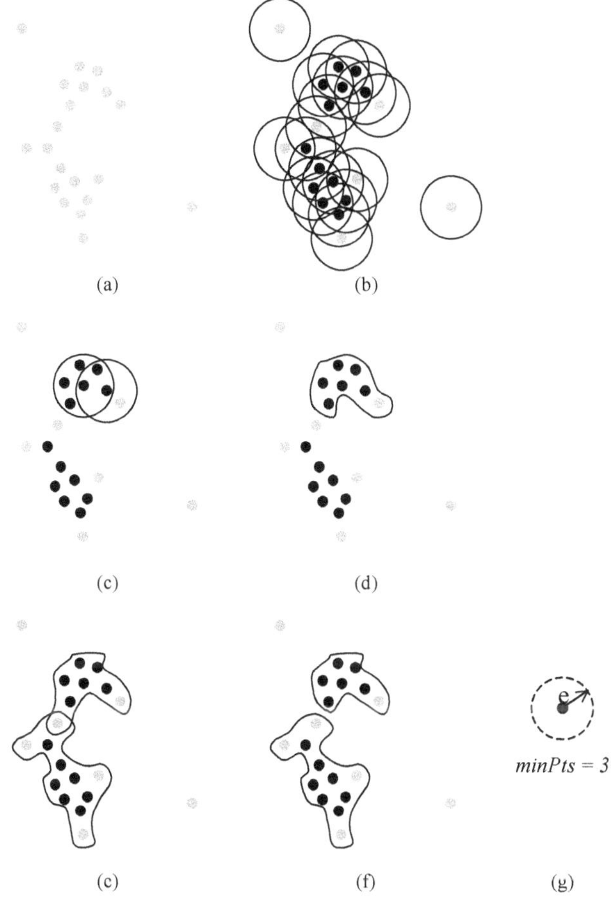



**Fig. 12.27** Steps of the DBSCAN algorithm.

## 12.14 Optimum Path in Space

The determination of the *optimum path* between two physical locations is a very common problem in human navigation and appears very often in applications such as Cartography, Robotics, and GIS. *Optimum* in this context refers to a minimal accumulation of what amounts to incremental travel costs associated with different media. It may be the *shortest, fastest, least-expensive,* or *least-risky* path, to name a few.

When movement is restricted to the chains of a linear network, like a road or airway network, a weighted graph can be used as a model and associated algorithms (see Chapter 8) may be applied for the determination of the optimum path.

Moving in space is a far more complex problem. Finding the optimum path in space is not only used for traveling or movement. It may also be applied to other planning situations, such as building highways, railways, pipelines, and other transport systems. Some representative examples of optimum path finding in space are illustrated in Figure 12.28. These refer to the determination of the fastest path between two villages (Figure 12.28a); the shortest sea course between two ports (Figure 12.28b); the smoothest path over a mountainous terrain (Figure 12.28c); the least-risky path in a hostile environment, for instance, the path with the maximum concealment time vis-à-vis an enemy or an observer (Figure 12.28d).

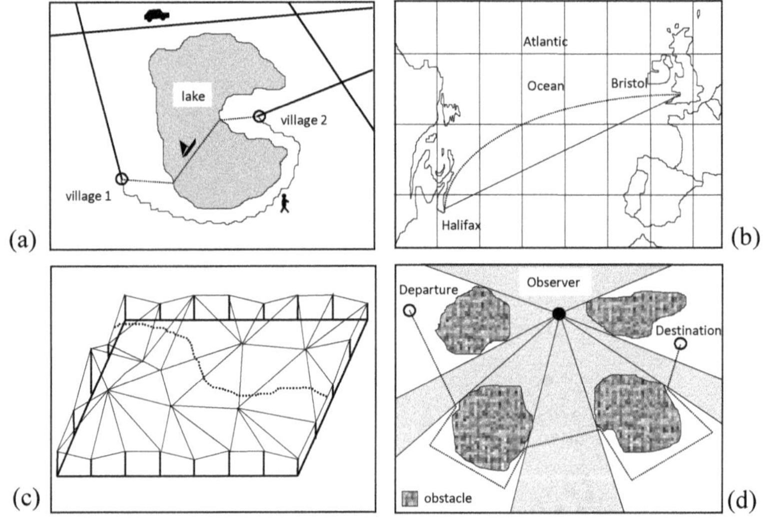


**Fig. 12.28** Examples of optimum paths in space.

Clearly, the space under study has its own peculiarities. For instance, it may be composed of a number of regions or volumes with different travel cost values assigned to them (e.g., walking on grass or sand); it may involve various means of travel (e.g., walking or driving); the direction of movement may introduce a variable cost value (e.g., moving with or against the wind).

The solutions to the problem of finding the optimum path in space are mainly confined to movement on a plane surface (i.e., cross country movement) consisting of zones characterized by different travel cost values. The corresponding methods may be classified into raster-based, and vector-based approaches. Raster-based algorithms have several advantages since they are easy to implement and perform reasonably well. However, an exact solution can only be found when the sampling resolution in the area under consideration is very fine. On the other hand, vector-based algorithms provide exact solutions to the problem. However, both theory and experiments show that their complexity is worse.

In the sequel, a general and effective graph-based approach to the determination of the optimum path in space introduced by Stefanakis and Kavouras is presented.

## *12.14.1 Methodology*

The concept behind this approach is to establish a network connecting a finite number of locations (including departure and destination spots) in space (Figure 12.29), so that effective algorithms of graph theory (Chapter 8) and artificial intelligence can be adopted to find the optimum path(s) for the desired trip.

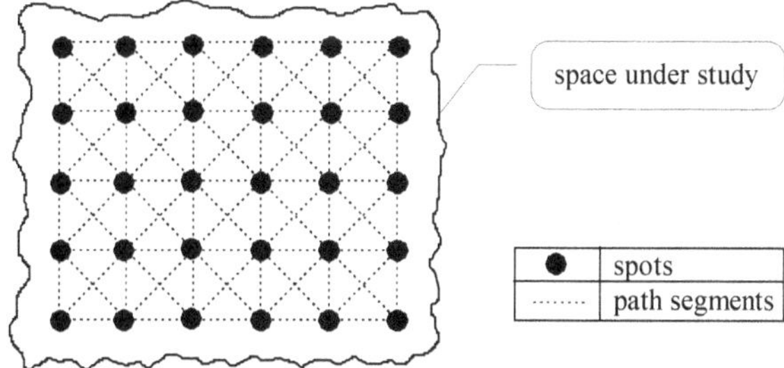


**Fig. 12.29** The concept of the approach on a plane surface.

The approach consists of four steps, which are elaborated in the following Subsections:

1. Determination of a finite number of spots in space.
2. Establishment of a network connecting these spots.
3. Formation of the travel cost model.
4. Determination of the optimum path(s) from the spot of reference (i.e., the departure) to the destination spot.

### 12.14.1.1 Determination of a Finite Number of Spots in Space

Obviously, movement in space involves an infinite number of *spots*. Hence, the space needs to be discretized. *Discretization* (Chapter 10) is the process of partitioning the continuous space into a finite number of disjoint areas or volumes (cells), whose union results into the space. By representing each of these cells with one spot (e.g., its center point), a finite set of spots is generated. These spots will be involved in the process of the determination of the optimum path(s).

A wide variety of tessellations are available for obtaining the desired partitioning of the space (Chapter 10). Tessellations may consist of regular or irregular cells. In the former case, the space is partitioned by a repeatable pattern of regular polyhedra (regular polygons on plane surface); while in the latter case, the space is partitioned by a mesh of polyhedra with variable shape and size.

In order to achieve a uniform distribution of spots over the space a regular tessellation should be adopted. Table 1 lists some possible tessellations for several spaces.

**Table 12.1** Types of Tessellations

| Space | Possible Tessellations |
| --- | --- |
| Plane Surface | rectangular grid, hexagonal grid, triangular grid, etc. |
| 3-d Space | cubic blocks, other platonic solids, etc. |
| n-d Space | n-d blocks |
| Spheroid | geographic grid, polyhedral tessellations, cubic blocks, etc. |

## 12.14.1.2 Establishment of a Network

After the discretization of the space and the determination of the spots (nodes), a network should be established to connect adjacent or non-adjacent spots and indicate the possible paths (finite in number) of movement. For this to occur, each spot (assigned to a cell) is connected through network edges to the spots of the neighboring cells. Independently on the tessellation adopted to partition the space, each cell has three types of neighbor cells: (a) *direct*, i.e., neighbors with shared edges; (b) *indirect*, i.e., neighbors with common vertices; and (c) *remote* neighbors. Remote neighbors are characterized by the level of proximity to the cell of reference. For instance, *level-one* (*level-two*) remote neighbors are the cells which are direct or indirect neighbors of the direct or indirect neighbors of the cell of reference (of the level-one remote neighbors of the cell of reference).

Figure 10.12 illustrates an example for the regular and triangular tessellations on the plane surface. Notice that, by increasing the number of neighbors considered, the directions of movement (i.e., *degrees of freedom*) increase (Figure 12.30). For instance, in a regular tessellation on the plane surface (a chess), the direct neighbors introduce a set of four directions (rook's move is allowed), the indirect neighbors introduce another set of four directions (queen's move is allowed), and the level-one remote neighbors introduce another set of eight directions of movement (queen's+knight's moves are allowed). An exhaustive network would consider all direct, indirect and remote (of any level) neighbors.

## 12.14.1.3 Formation of the Travel Cost Model

The *travel cost model* assigns weights to the edges of the network established in the previous step. Its form depends on both the space under study and the application needs. Some representative examples of travel cost models are:

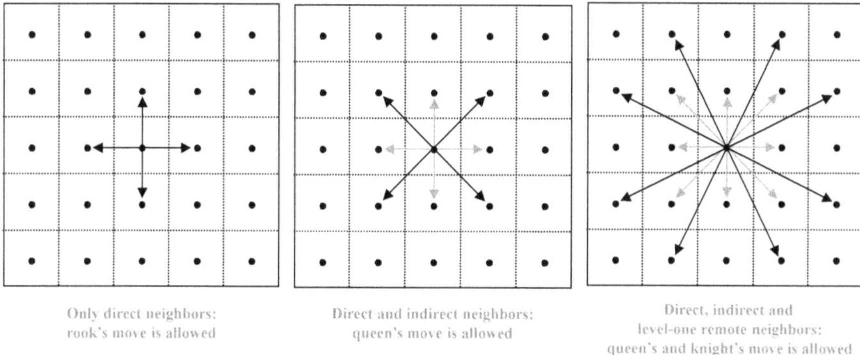



**Fig. 12.30** Various freedoms of movement on a rectangular grid.

- the model of *distance*; e.g., the shortest path (minimize overall distance)
- the model of *time*; e.g., the fastest path (minimize overall time)
- the model of *expenses*; e.g., the least expensive path (minimize the overall expenses)
- the model of *risk*; e.g., the least risky path (minimize the overall risk)

In each case, the space under study consists of areas (or volumes) that are characterized by a weight, which indicates the cost of movement across them per unit of movement; and depends on the travel cost model in use. Figure 12.31a,b illustrates a simplified example of a plane surface consisting of areas with variable travel cost values expressed in seconds per meter. Based on those weight values, the edges of the network are assigned a measure, which is equal to the cost of movement from one spot to another (Figure 12.31c,d). Assuming that the edge j intersects k areas, this measure ($C_j$) is derived from the sum:

$$C_j = \sum_{i=1}^{k} \left[ w_i \cdot d_i \right]$$

where $w_i$ is the weight of area i (i = 1, 2, ..., k) intersected by j and di the length of the edge across that area. For the fast retrieval of areas intersected by a network edge, efficient index structures available in spatial databases, such as R-tree variants (Chapter 13), can be adopted.

The shape of a network edge connecting two spots (network nodes) may vary depending on both the domain and space. Specifically, different line types (straight, curves, splines, etc.) may be considered for a network edge. The most commonly used is the shortest in length line connecting the two spots. On a plane surface the shortest line (i.e., network edge) connecting two spots is the straight line connecting them, while on a spherical surface the shortest arc of the great circle passing through them. Obviously, the line should belong to the space considered. For instance, although the cord is the shortest line connecting two spots lying on a sphere the corresponding arc is considered instead, when movement is restricted on the surface of the sphere.

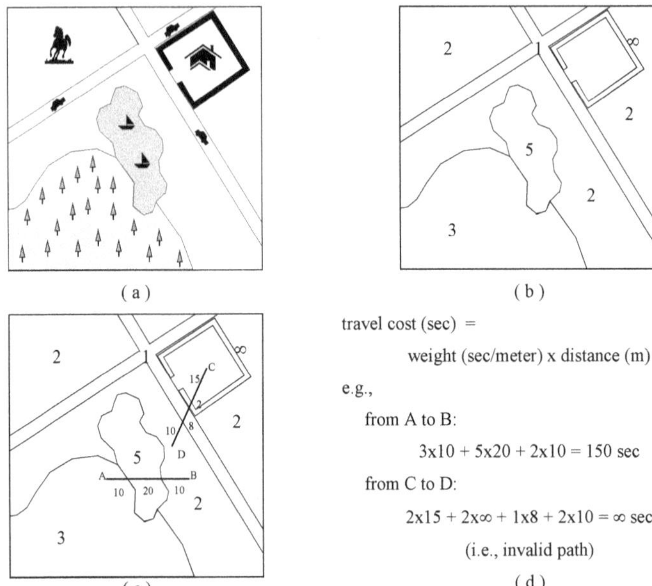



**Fig. 12.31** Example of a plane surface consisting of areas with variable travel cost measures.

There are some applications that involve *dynamic travel cost models*. In a dynamic travel cost model the weight values assigned to the areas or volumes of the space change over time. These applications are characterized as spatio-temporal and with many applications in geography (Chapter 10). Real-world examples that involve dynamic models are the airline routing over the Atlantic and navigation of ships and tankers along the Aegean Sea. In all these applications both obstacles and weight values in general depend on the traffic load and meteorological conditions at specific time intervals.

### 12.14.1.4 Determination of the Optimum Path(s)

After the establishment of the network and the formation of the travel cost model, appropriate algorithms from both the graph theory and artificial intelligence can be applied for the determination of the optimum path(s) from the node(s) of reference to the destination node(s) over the linear network.

As discussed in Chapter 8, a commonly used algorithm available in graph theory is Dijkstra's algorithm, which computes the shortest paths from a reference node to all other nodes of a network. Alternative strategies based on the best-first search are available in the area of artificial intelligence which minimize significantly the computation time and guarantee both completeness and optimality. A[*] algorithm is the most representative example of a best-first search algorithm, which derives the optimum path between two nodes (from reference node to destination node) of a network.

## *12.14.2 Finding the Shortest Path*

Following, it is shown how the four step approach to the determination of the optimum path can be applied in a variety of spaces adopting a simple cost model which is quantitative in nature. Specifically, the problem of shortest path finding is examined in a *variety of spaces*, such as the plane surface, the 3-D space, and the spherical surface as an approximation of the earth surface. In each case, the space is represented by a set of spots, which may be *accessible* or *non-accessible* (i.e., lying on obstacles; usually not considered). For instance, spots lying on the sea are accessible, while those lying on the continents are non-accessible for a ship (Figure 12.28b). The travel cost between two accessible spots is equal to the length of the shortest line connecting them, if they are *intervisible*. Two spots are intervisible, if the shortest line connecting them crosses no obstacle. If this is not the case (i.e., invalid path), the travel cost between the two spots is computed implicitly passing through intermediate spots.

### 12.14.2.1 Plane Surface

The plane surface is the simplest space to study. The finite number of spots can be easily determined by using one of the available two-dimensional tessellations (Table 12.1) and establishing one spot (node) at the center point of each cell. The network is then established by connecting the spots through straight-line segments, which are assigned weight values depending on the nature of the surface and the travel cost model considered.

Figure 12.32 shows an example on the determination of the shortest path on a plane surface with obstacles (shaded areas). The tessellation used is the regular grid with a resolution of 10 units in both X and Y dimensions, while the established network connects each spot (i.e., network node) with its direct and indirect neighbor spots (Figure 12.32b) or direct, indirect and level-one remote neighbor spots (Figure 12.32c) by straight line segments (i.e., eight or sixteen directions of movement are considered). The cells in Figure 12.32b,c depict the accumulated cost values for each network node from the reference (start) node s (e.g., Dijkstra's algorithm product), while solid lines represent the shortest paths for the desired movement. Notice that, by increasing the number of neighbors considered, a more accurate optimum path is derived, for the same set of spots (i.e., resolution of the tessellation) on the plane. Figure 12.33 illustrates two examples of the algorithm on real maps.

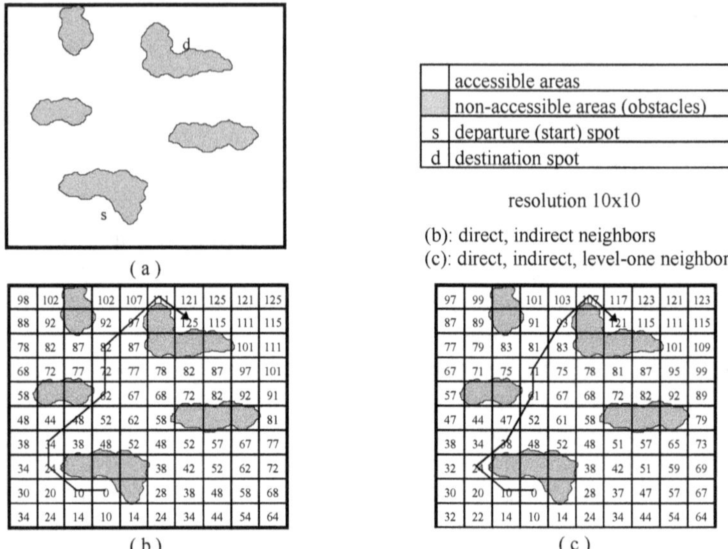



**Fig. 12.32** The shortest path on a plane surface (values assigned to the grid cells refer to the accumulated travel costs of the corresponding spots).

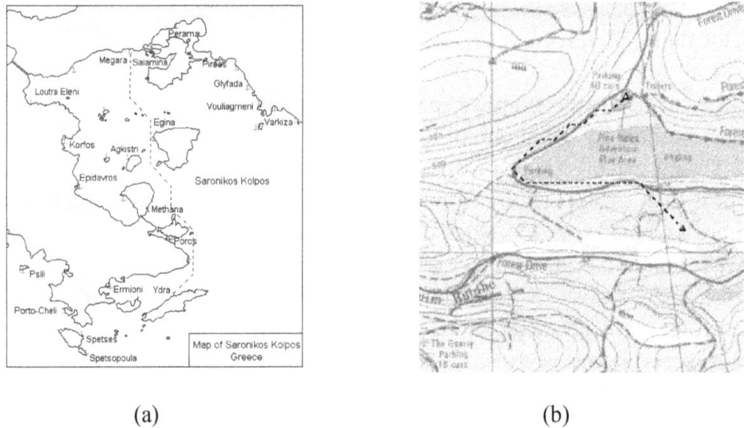



      (a)                           (b)

**Fig. 12.33** Two examples of the algorithm on real maps.

Specifically, Figure 12.33a indicates the shortest sea course between two harbors in the Aegean sea (pixel size is equal to 6 km; total cost: 70,2 km; direct and indirect neighbors considered; $A^*$ algorithm); while Figure 12.33b shows the fastest path from a given location to a parking lot on an orienteering map of Scotland (pixel size is equal to 5 m; total cost: 16 min and 9 sec; direct and indirect neighbors considered; $A^*$ algorithm).

### 12.14.2.2 3-D Space

Moving in the 3-D space is a generalization of movement on a plane surface. The spots are similarly determined using one of the available three dimensional tessellations (Table 12.1) and the network is established by connecting these spots through straight line segments, which are assigned weight values depending on the nature of the space and the travel cost model considered.

Figure 12.34 presents an example on the determination of the shortest path in a 3-D space enclosing obstacles (shaded volumes). All obstacles have their basis at level 0, while their height is denoted in Figure 12.34a over the shaded areas. The maximum height of the space is 40 units, while departure and destination spots have an altitude of 5 and 25 units, respectively. The tessellation used is the regular grid (i.e., cubic blocks) with a resolution of 10 units in all X, Y, and Z dimensions. As for the established network, it connects each spot with its direct and indirect neighbors. The cells in Figure 12.34c,b,e,d depict the accumulated cost values (in layers) for each network node from the reference node s (i.e., Dijkstra's algorithm product), while the solid line represents a shortest path for the desired movement (notice that it jumps across the levels). Figure 12.34f illustrates a perspective view of the 3-D space and the proposed movement.

The extension of the idea to spaces of higher dimensionality (n-D spaces) is considered as a simple generalization of this principle.

### 12.14.2.3 Spherical Surface

The movement on the surface of a sphere is examined separately due to the peculiarities of the underlying spherical geometry. The problem can be solved by examining the movement on a 3-D space (see previous Section), where the interior and exterior of the sphere are non-accessible (i.e., obstacles), while movement is allowed inside a spherical ring (it can be thought as a buffer zone of the spherical surface in space). For instance, if cubic blocks are used for the generation of the spots on the three dimensional space, the accessible spots will be the ones that intersect the spherical surface.

A more adaptive configuration for the spherical surface considers the movement on the surface itself. The determination of the spots is based on the available tessellations for the sphere (Table 12.1). The tessellation, which is closer to cartographers and geo-scientists, in general, is the geographic grid (Figure 10.19, 12.35). Figure 12.35b illustrates an example for the model of distance on a sphere for a geographic grid of 30 degrees resolution (i.e., 62 spots; top and bottom row cells correspond to one cell each). The direct and indirect neighbors are considered for each spot on the established network. The accumulated travel cost values are expressed in radians and refer to a spot (on the shaded grid cell) lying on the equator of the sphere (i.e., Dijkstra's algorithm product). Contrary to the plane and 3-D space, the path connecting two spots is not a straight line segment any more, but the shortest arc of the great circle (i.e., normal section) passing through them. Notice that all cells lying on the top (bottom) row of the matrix are assigned the same value, since they refer to a unique spot lying on the north (south) pole of the sphere.

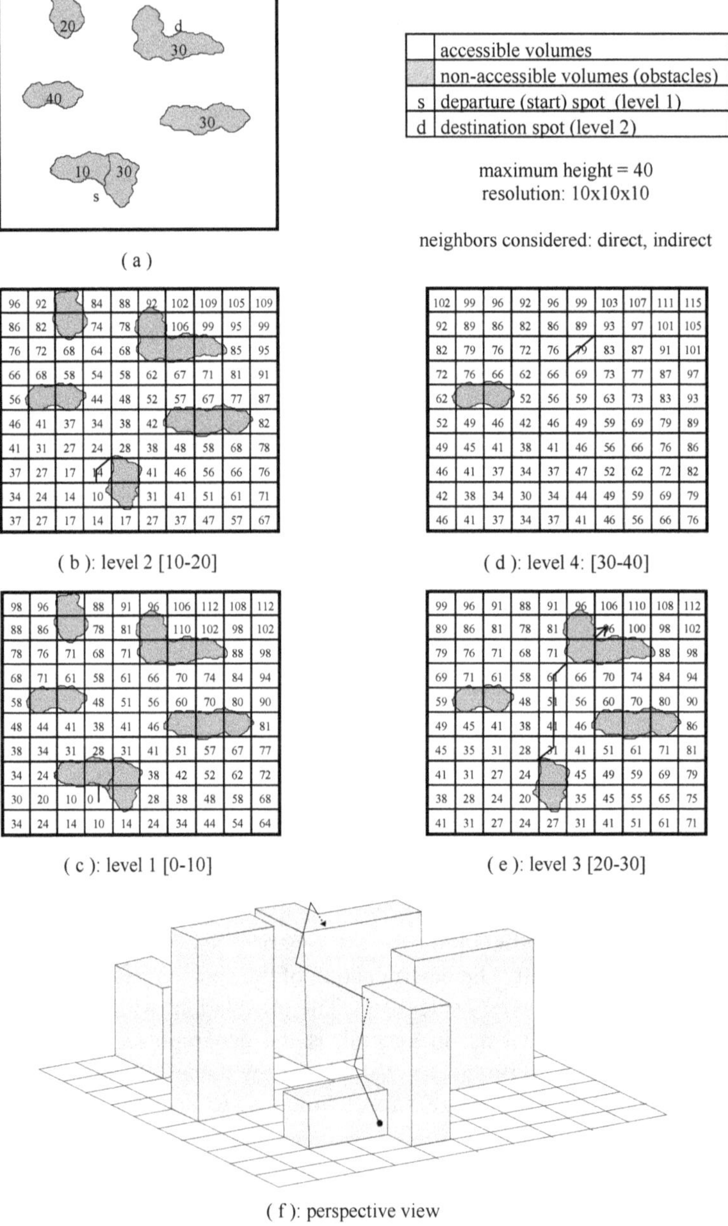



( a )

( b ): level 2 [10-20]

| 96 | 92 |    | 84 | 88 | 92 | 102 | 109 | 105 | 109 |
|----|----|----|----|----|----|-----|-----|-----|-----|
| 86 | 82 |    | 74 | 78 |    | 106 | 99  | 95  | 99  |
| 76 | 72 | 68 | 64 | 68 |    |    |    | 85 | 95 |
| 66 | 68 | 58 | 54 | 58 | 62 | 67 | 71 | 81 | 91 |
| 56 |    |    | 44 | 48 | 52 | 57 | 67 | 77 | 87 |
| 46 | 41 | 37 | 34 | 38 | 42 |    |    |    | 82 |
| 41 | 31 | 27 | 24 | 28 | 38 | 48 | 58 | 68 | 78 |
| 37 | 27 | 17 | 14 |    | 41 | 46 | 56 | 66 | 76 |
| 34 | 24 | 14 | 10 |    | 31 | 41 | 51 | 61 | 71 |
| 37 | 27 | 17 | 14 | 17 | 27 | 37 | 47 | 57 | 67 |

( c ): level 1 [0-10]

| 98 | 96 |    | 88 | 91 | 96 | 106 | 112 | 108 | 112 |
|----|----|----|----|----|----|-----|-----|-----|-----|
| 88 | 86 |    | 78 | 81 |    | 110 | 102 | 98 | 102 |
| 78 | 76 | 71 | 68 | 71 |    |    |    | 88 | 98 |
| 68 | 71 | 61 | 58 | 61 | 66 | 70 | 74 | 84 | 94 |
| 58 |    |    | 48 | 51 | 56 | 60 | 70 | 80 | 90 |
| 48 | 44 | 41 | 38 | 41 | 46 |    |    |    | 81 |
| 38 | 34 | 31 | 28 | 31 | 41 | 51 | 57 | 67 | 77 |
| 34 | 24 |    |    |    | 38 | 42 | 52 | 62 | 72 |
| 30 | 20 | 10 | 0 |    | 28 | 38 | 48 | 58 | 68 |
| 34 | 24 | 14 | 10 | 14 | 24 | 34 | 44 | 54 | 64 |

( d ): level 4: [30-40]

| 102 | 99 | 96 | 92 | 96 | 99 | 103 | 107 | 111 | 115 |
|-----|----|----|----|----|----|-----|-----|-----|-----|
| 92 | 89 | 86 | 82 | 86 | 89 | 93 | 97 | 101 | 105 |
| 82 | 79 | 76 | 72 | 76 | 79 | 83 | 87 | 91 | 101 |
| 72 | 76 | 66 | 62 | 66 | 69 | 73 | 77 | 87 | 97 |
| 62 |    |    | 52 | 56 | 59 | 63 | 73 | 83 | 93 |
| 52 | 49 | 46 | 42 | 46 | 49 | 59 | 69 | 79 | 89 |
| 49 | 45 | 41 | 38 | 41 | 46 | 56 | 66 | 76 | 86 |
| 46 | 41 | 37 | 34 | 37 | 47 | 52 | 62 | 72 | 82 |
| 42 | 38 | 34 | 30 | 34 | 44 | 49 | 59 | 69 | 79 |
| 46 | 41 | 37 | 34 | 37 | 41 | 46 | 56 | 66 | 76 |

( e ): level 3 [20-30]

| 99 | 96 | 91 | 88 | 91 | 96 | 106 | 110 | 108 | 112 |
|----|----|----|----|----|----|-----|-----|-----|-----|
| 89 | 86 | 81 | 78 | 81 |    | 96 | 100 | 98 | 102 |
| 79 | 76 | 71 | 68 | 71 |    |    |    | 88 | 98 |
| 69 | 71 | 61 | 58 | 61 | 66 | 70 | 74 | 84 | 94 |
| 59 |    |    | 48 | 51 | 56 | 60 | 70 | 80 | 90 |
| 49 | 45 | 41 | 38 | 41 | 46 |    |    |    | 86 |
| 45 | 35 | 31 | 28 | 31 | 41 | 51 | 61 | 71 | 81 |
| 41 | 31 | 27 | 24 |    | 45 | 49 | 59 | 69 | 79 |
| 38 | 28 | 24 | 20 |    | 35 | 45 | 55 | 65 | 75 |
| 41 | 31 | 27 | 24 | 27 | 31 | 41 | 51 | 61 | 71 |

( f ): perspective view

**Fig. 12.34** The shortest path in 3-D space (values assigned to the grid cells refer to the accumulated travel costs of the corresponding spots).

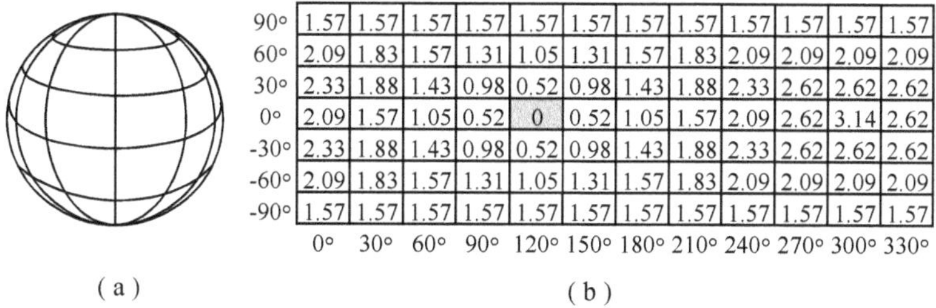

| | 0° | 30° | 60° | 90° | 120° | 150° | 180° | 210° | 240° | 270° | 300° | 330° |
|---|---|---|---|---|---|---|---|---|---|---|---|---|
| 90° | 1.57 | 1.57 | 1.57 | 1.57 | 1.57 | 1.57 | 1.57 | 1.57 | 1.57 | 1.57 | 1.57 | 1.57 |
| 60° | 2.09 | 1.83 | 1.57 | 1.31 | 1.05 | 1.31 | 1.57 | 1.83 | 2.09 | 2.09 | 2.09 | 2.09 |
| 30° | 2.33 | 1.88 | 1.43 | 0.98 | 0.52 | 0.98 | 1.43 | 1.88 | 2.33 | 2.62 | 2.62 | 2.62 |
| 0° | 2.09 | 1.57 | 1.05 | 0.52 | 0 | 0.52 | 1.05 | 1.57 | 2.09 | 2.62 | 3.14 | 2.62 |
| -30° | 2.33 | 1.88 | 1.43 | 0.98 | 0.52 | 0.98 | 1.43 | 1.88 | 2.33 | 2.62 | 2.62 | 2.62 |
| -60° | 2.09 | 1.83 | 1.57 | 1.31 | 1.05 | 1.31 | 1.57 | 1.83 | 2.09 | 2.09 | 2.09 | 2.09 |
| -90° | 1.57 | 1.57 | 1.57 | 1.57 | 1.57 | 1.57 | 1.57 | 1.57 | 1.57 | 1.57 | 1.57 | 1.57 |

( a )  ( b )

**Fig. 12.35** The accumulated cost values on a homogeneous sphere (expressed in rad).

Figure 12.36 illustrates an example on the determination of the shortest sea course on the earth surface approximated by a sphere. For visualization purposes the Plate Carrée projection is adopted to represent the continents and oceans of the earth. The solid line shows a shortest path from Adelaide, Australia to New Orleans, USA. The line does not approximate sufficiently the curve of the great circle on the projection plane due to the low resolution of the grid (10 degrees) and the limited number of neighbors (direct and indirect) considered on the established network.

The major problem faced by this tessellation is that the set of spots generated is non-uniformly distributed over the surface, because of the highly variable shape and size of the geographic grid cells. As a consequence a variable accuracy on the determination of the optimum path is introduced, which depends on the region on the sphere. Clearly, there is a large concentration of spots around the poles which gradually decreases by approaching the equator.

A more uniform distribution of the spots over the spherical surface can be obtained by adopting tessellations that are based on the recursive decomposition of regular polyhedra. A tessellation, which is based on the recursive subdivision of the octahedron, the *hierarchical structure for spherical triangles* is considered here (see Chapter 10 and Figure 10.20). The advantage of this tessellation is that the triangular cells derived have a similar shape and size and as a consequence a uniform distribution of the spots over the spherical surface can be obtained. Algorithms for the conversion of triangular cells to geographic coordinates as well as for the determination of their direct and indirect neighbors are available, and can be adopted for locating the spots over the surface and establishing the network over them.

Figure 12.37 shows a simplified example on the determination of the shortest sea course from Hawaiian Islands to Los Angeles, California. The quadrant 90°W to 180°W of the northern hemisphere has been only considered, while the established network is confined to the three direct neighbors of each spot. The tiling corresponds to the fourth level of subdivision and results in 256 triangular cells with an average resolution of 180 km.

The same technique applies when a more precise determination of a distance on the earth's surface is needed, although its implementation is more complex. In such a case, the earth's surface (geoid; physical reality) is closely approximated by

310

a "spheroid" and the shortest line joining two points is the geodesic curve (Chapter 2). Notice that the spheroidal (ellipsoidal) surface may be approximated by the same tessellations used for the sphere.

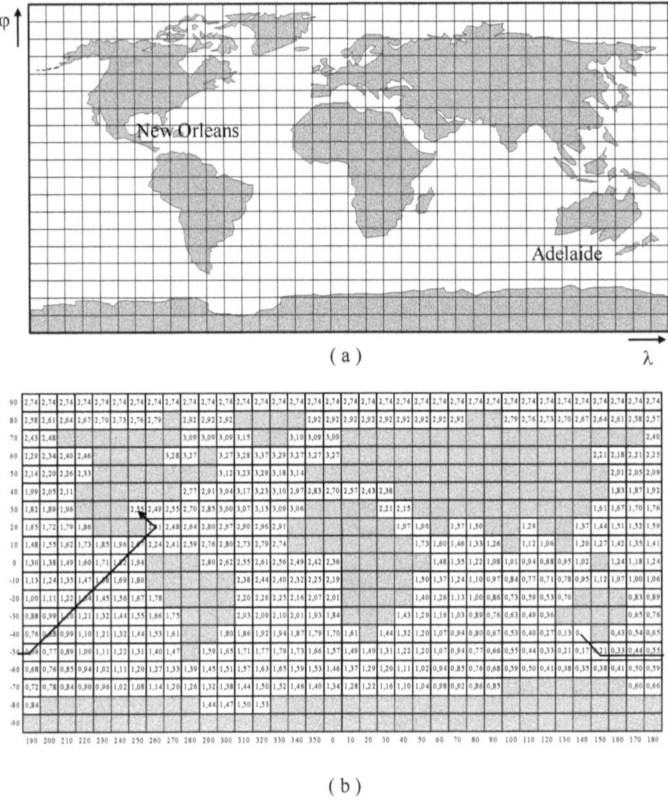


**Fig. 12.36** The shortest sea course (distances expressed in rad; shaded cells in b correspond to non-accessible spots, i.e., land; continents are simplified).

The problem becomes more complex, when one tries to determine the best route not on a ellipsoidal approximation of the earth's surface but on the geoid itself (i.e., an equipotential surface of the earth which is partially known and cannot be described as simply as the ellipsoid). This problem is mostly of theoretical and not practical interest, since there are various other major constraints to navigation which outweigh the fine differences caused by the use of models of the geoid. Even in this case, the general approach introduced in this paper can be applied. The ellipsoid can be represented as a discrete 3-D "relief" (terrain model) with respect to the geoid, and the solution provided in Section 12.14.2.2 can be employed to determine the optimal path on the equipotential surface.

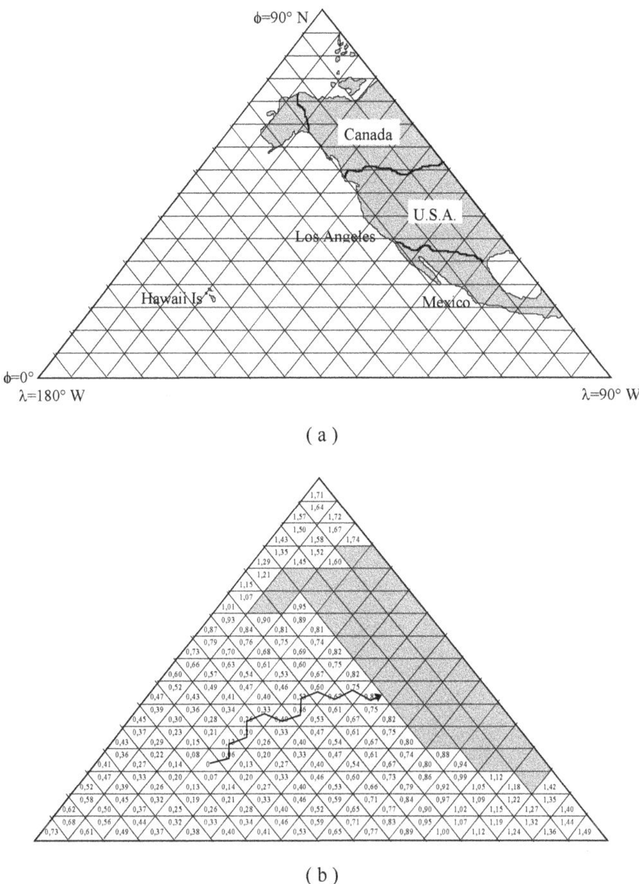



**Fig. 12.37** The shortest sea course (distances expressed in rad; shaded cells in b correspond to non-accessible spots, i.e., land; continents are simplified).

# References and Further Reading


Ester, M., Kriegel, H.P., Sander, J., and Xu, X., 1996. A density based algorithm for discovering clusters in large spatial databases with noise. In the *Proceedings of the 2nd International Conference on Knowledge Discovery and Data Mining*. Portland, Oregon, pp. 226-231.

Malling, D.H., 1989. *Measurements from Maps*. Pergamon Press.

Meratnia, N., and de By, R.A., 2004. Spatiotemporal compression techniques for moving point objects. In the *Proceedings of the International Conference on Extending Database Technology (EDBT)*, Springer, LNCS 2992, pp. 765-782.

Miller, H.J., and Han, J. (Eds), 2001. *Geographic Data Mining and Knowledge Discovery*. Taylor-Francis.

Petrie, G., and Kennie, T.J.M., 1990. *Terrain Modelling in Surveying and Civil Engineering*. Whittles Publishing.

Preparata, F.P., and Shamos, M.I., 1991. *Computational Geometry: An Introduction*. Spinger-Verlag.

Rigaux, P., Scholl, M., and Voisard, A., 2002. *Spatial Databases with Applications to GIS*. Morgan-Kaufmann.

Sedgewick, R., 1990. *Algorithms*. Addison-Wesley.

Shashi, S., and Chawla, S., 2003. *Spatial Databases – A Tour*. Prentice Hall.

Stefanakis, E., 2006. Scheduling trajectories on a planar surface with moving obstacles. *Informatica Journal*, Lith, Acad. Sci., Vol. 17, No. 1, pp. 95-110.

Stefanakis, E. and Kavouras, M., 1995, On the determination of the optimum path in space, In Frank, A., and Kuhn, W., (Ed's.), *Spatial Information Theory: A Theoretical Basis for GIS (COSIT 95)* (Springer-Verlag), pp. 241-257.

van Kreveld, M., 1997. Algorithms for triangulated terrains. In the *Proceedings of the XXIV-th SOFSEM Conference*. Lecture Notes in Computer Science, No. 1338, pp. 19-36. Springer-Verlag.

Worboys, M.F., 1995. *GIS – A Computing Perspective*. Taylor-Francis.

# 13 Spatial Data Structures

## 13.1 Introduction

Spatial data, like alphanumeric data, must be organized properly at the physical level of a computer system, to meet two objectives: (a) allow for *rapid retrieval and processing* by programs, and (b) *minimize the space requirements* for storage and preprocessing.

On the flip side, *spatial data structures*, unlike alphanumeric data structures (Chapter 8), can handle objects with dimensions as well as the complex relations (e.g., topological) that exist among these objects.

In the literature, spatial data structures are divided into different classes depending on the type of structure, type of data, or type of queries they can support. Spatial data structures are usually classified into *vector* and *raster* structures. The first handle vector geometries (e.g., points, lines and polygons), while the second handle raster images. Another important class of spatial data structures is the *topological structures*. These are intended to support the rapid retrieval of topological relations between vectors. This Chapter provides a brief presentation of some representative spatial data structures.

## 13.2 Raster Data Structures

*Raster data structures* organize raster (or image) data to achieve two main purposes: (a) data compression; and/or (b) efficient execution of certain functions in spatial analysis. There is a plethora of structures that fall into this class. Two of these are examined next: *run length encoding* and *quadtree.*

### 13.2.1 Run-length encoding

Raster data, e.g., satellite images, is usually of large size. Hence, the storage requirements of raster data are high. Specifically, a binary (black and white) aerial photo of 4000x3000 pixels requires $4000 \times 3000 \times 1\text{bit} = 1.5 \times 10^6$ bytes (1 byte = 8 bits) = 1.5MB. If the image is a color file of 256 colors per pixel, the size of the file increases drastically.

Various techniques have been developed to reduce the size of a raster file. They all organize the data in appropriate data structures. One of these is the *run-length encoding*. The central idea of this structure is described next through an example.

Suppose the area in Figure 13.1a, which is represented by the raster data in Figure 13.1b. These data can be organized into a two-dimensional table with 16 rows and 16 columns. The storage requirements of this table are [16x16]x2 bits = 64 bytes. The 2 bits correspond to the representation of the four categories (land use), i.e., 00 = sea (s), 01 = urban area (u), 10 = forest (f), and 11 = lake (l).

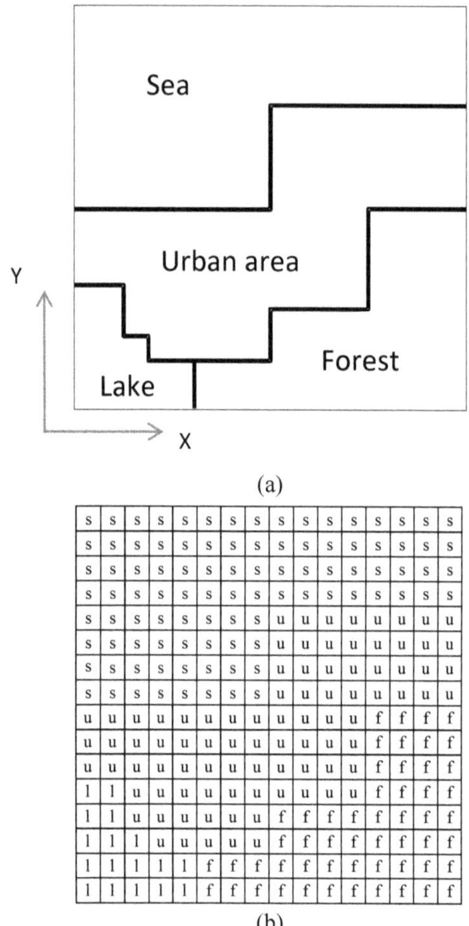


(a)

(b)

**Fig. 13.1** Example of a raster representation.

Data compression can be achieved by applying the coding by lines or *run-length encoding*. A version of this encoding for the representation of the image (table) in Figure 13.1b is 72 pixels for sea followed by 8 for urban, etc., or

72s, 8u, 8s, 8u, 8s, 8u, 8s, 20u, 4f, 12u, 4f, 12u, 6f, 10u, 6f, 6u, 8f, 3l, 5u, 8f, 5l, 11f, 5l, 11f.

The storage requirements for this string are: 24 integers and 24 land use types; hence, 24x1 byte + 24x2 bits = 30 bytes (72 is the maximum integer, so one byte per integer is sufficient).

Note that the more homogeneous the data, the higher compression can be achieved. For example, the first four and a half lines of the image, corresponding to ¼ of the whole image, are represented by a single pair: 72s in the string.

## *13.2.2 Quadtree*

The *quadtree* is a widely used data structure in geosciences. Originally designed for representing raster data, it was evolved to represent vector data as well (see next section). The quadtree structure has been implemented and incorporated as a basic data structure in GIS, remote sensing, and database systems (e.g., Oracle Spatial, see Chapter 15) software.

The quadtree is a tree structure, which structures the space by recursively dividing it into four regions (*quadrants*), named as {NW: north-western, NE: north-eastern, SW: south-western, SE: south-eastern} or {0,1,2,3}, respectively. The recursive splitting stops when either one of the following two conditions is met: (a) the entire area of a quadrant has the same value for the theme attribute (homogeneous area); or (b) the tree has reached its maximum height. The root of the tree corresponds to the entire region. The leaves of the tree are assigned a value for the thematic attribute, which describes the whole or most area of the corresponding region (Figure 13.2). Figure 13.3 shows the recursive partitioning of space and the corresponding quadtree for the area in Figure 13.1.

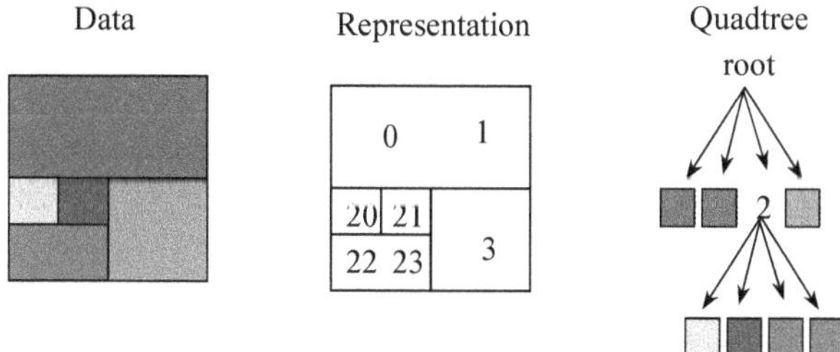


**Fig. 13.2** Example quadtree.

The quadtree structure is very effective in performing certain functions of spatial analysis, such as finding the nearest neighbor, or solving the point in polygon problem (Chapter 12); appropriate algorithms are designed and implemented on top of quatree structure in software systems. Moreover, this structure achieves a high compression of data, wherever there are homogeneous regions. A disadvantage of the structure is that it requires long time in building and updating (maintain) the tree.

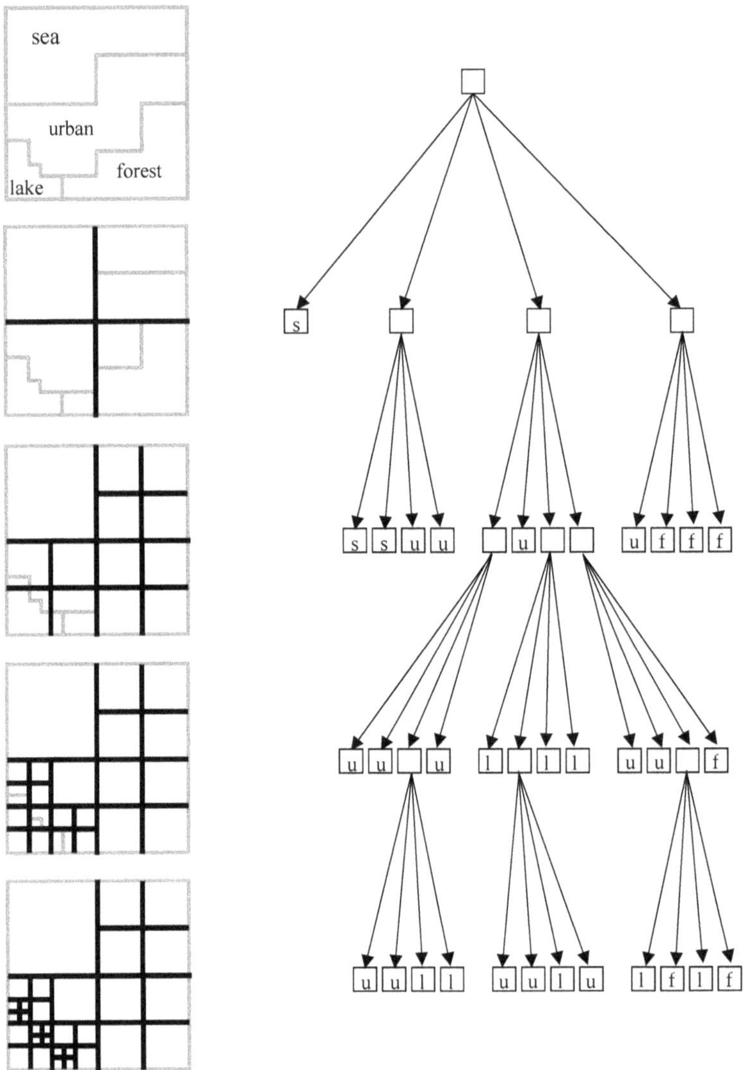



**Fig. 13.3** Quadtree representation of the area in Figure 13.1.

## 13.3 Vector Data Structures

In the literature, there is a plethora of methods for structuring vector data. These are often categorized according to the geometry type of vectors indexed. Specifically, *vector data structures* are divided into (a) *point* structures, (b) *linear* structures, and (c) *polygonal* structures.

*Point data structures* are applied for indexing point objects. Example points are the towns and villages in a small-scale map. Each town is represented by a point or a pair of coordinates (x,y) or (latitude, longitude) in a space of two dimensions. Despite the geographic objects of point geometry, attribute (alphanumeric) data is often treated as points in a space of multiple dimensions. For example, the salary combined with the age of employees in a company can be represented by points in a two-dimensional graph, as show in Figure 13.4; and can be organized in a point data structure. Representative data structures for point objects in a space of n dimensions are: the *k-d-tree, K-D-B-tree, grid file*, and *point-quadtree.*

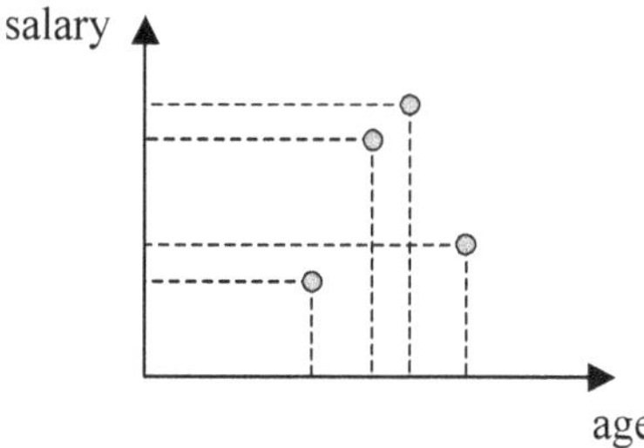


**Fig. 13.4** Representation of two alphanumeric attributes with points in 2D space. Each point corresponds to an employee of a company.

*Linear data structures* have been proposed to support the effective indexing of linear objects, such as road or communication networks. Representative linear structures are the *strip-tree* and the *quadtree variants*. Because of their spatial extent, the linear objects are often treated as polygonal objects and indexed by the polygonal data structures described next.

*Polygonal data structures* were designed for indexing objects with spatial extent in a space of n dimensions. The term "polygon" was borrowed from the two-dimensional space, however it is used in a generic manner. In three dimensions, the "polygonal objects" are solid objects with volume. The most representative structure for indexing polygonal objects is the *R-tree* and its *variants* ($R^+$tree, $R^*$tree, etc.). Other structures are also appropriate, such as the variants of quadtree, the grid-file, etc. These are commonly used to index all three types: point, linear, and polygonal objects. For example, a polygonal data structure can be applied for indexing the counties (as polygons), roads (linear objects) or nodes of the trigonometric network (point-objects) in Canada.

The following Sections briefly describe five vector data structures. The first relates to point objects, while the remaining three to objects of all geometry types. The discussion focuses on the two-dimensional space. However, these structures can easily be extended into spaces of higher dimensions.

These spatial data structures were designed to support two main types of searches: (a) *point queries*; and (b) *region (window) queries*. The former aims at retrieving the objects that contain a point, while the latter at retrieving the objects that intersect a region. Usually, this region is a rectangle with sides parallel to the axes of the coordinate reference system.

Figure 13.5 shows an example point query and an example region query. Both are the most frequent search operations in a digital environment. Specifically, given a map of European countries, when a country is selected with the cursor (which is driven by the mouse), a point search is performed and the country is highlighted. On the other hand, when a zoom in/out operation is executed on the map, a region search takes place and a piece of the map is shown in larger or smaller scale. In reality, both queries relate to a single type as a point query is a degenerated region query on a region of zero spatial extent.

**Fig. 13.5** Example point and region queries.

## *13.3.1 k-d-tree*

The *k-d-tree* is a tree data structure. It is an extension of the *binary search tree (BST)* (see Chapter 8), which indexes k-dimensional point-objects. Figure 13.6 shows an example of indexing a set of points in 2D space with a k-d-tree. Each node in the k-d-tree hosts one point. The point subdivides the corresponding area into two parts: upper/lower or left/right depending on the level of the tree. The subdivision is based on one dimension which alternates from each level to the next.

For example, in the two-dimensional space, the root (level 1) and all the nodes in an odd level ($3^{rd}$, $5^{th}$, ..., etc.) divide the space with an axis perpendicular to the axis of abscissas. On the other hand, the nodes in an even level ($2^{nd}$, $4^{th}$, ..., etc.) divide the space with an axis perpendicular to the axis of ordinates. The main reason for this alternation is to avoid elongated subspaces represented by the nodes of the tree.

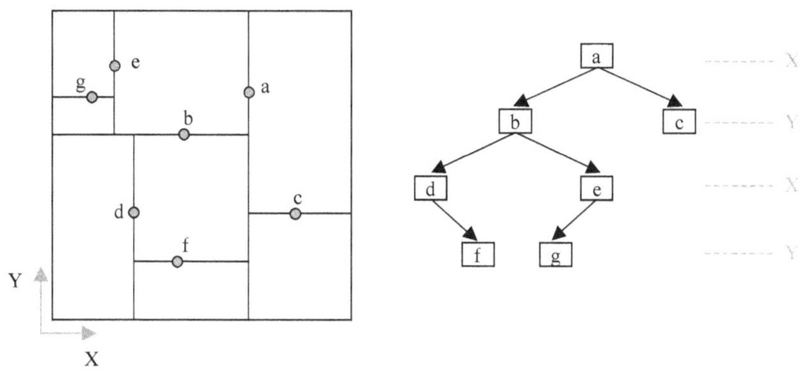

**Fig. 13.6** Example k-d-tree.

Figure 13.7 shows the steps of creating the tree in Figure 13.6 by a sequential insertion of the points in the following order {a, b, c, d, e, f, g}. Each node hosts one point and has two children. The left (right) child node hosts a point with a smaller (greater) abscissa or ordinate than that of the parent node, depending on the tree level.

Figure 13.8 shows schematically the steps of the region search algorithm in the k-d-tree of Figure 13.6. The shaded area represents the search region (range query). First, the point stored in the root of the tree is checked against the search region. Obviously, point a does not fall within the region. As the search region lies totally to the left side of point a (smaller abscissa), the search continues only in the left subtree of the root. Then, point b is checked against the region. As point b falls within the region, point b is reported. As the search region spans both sides of the intercept at point b, the search continues in both subtrees of the node hosting point b. As for the left subtree, point d is checked against the region. As point d does not fall within the region, and the search area lies entirely to the right of the abscissa of d, the search continues only in the right subtree of the node hosting point d. Then, point f is checked against the search region. As point f falls within the region it is reported. Node hosting point f is a leaf node, hence the recursive traversal ends there. Jumping back to the right subtree of b, point e is checked against the search region. As the search region lies entirely to the left of the abscissa of e, only the right branch of the node hosting point e will be searched. As there is no right branch the search ends there too. Eventually, the region search results to points b and f.

A major drawback of the k-d-tree structure – in correspondence with the binary search tree – is that its form is dependent on the order the points are inserted into the tree. This generally results to a non-balanced tree, which degrades the searching performance.

 

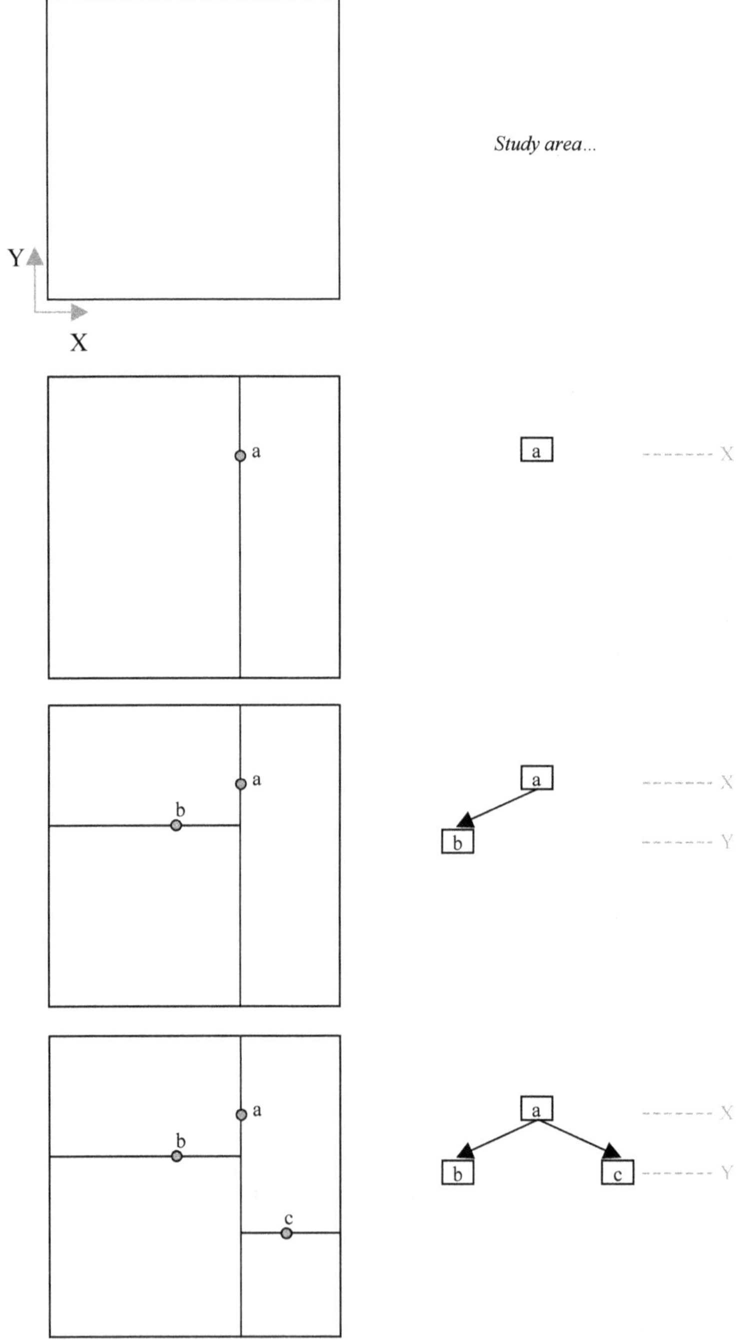
Study area...
Y
X
a
X
a
b
X
Y
a
b
c
X
Y
Study area...

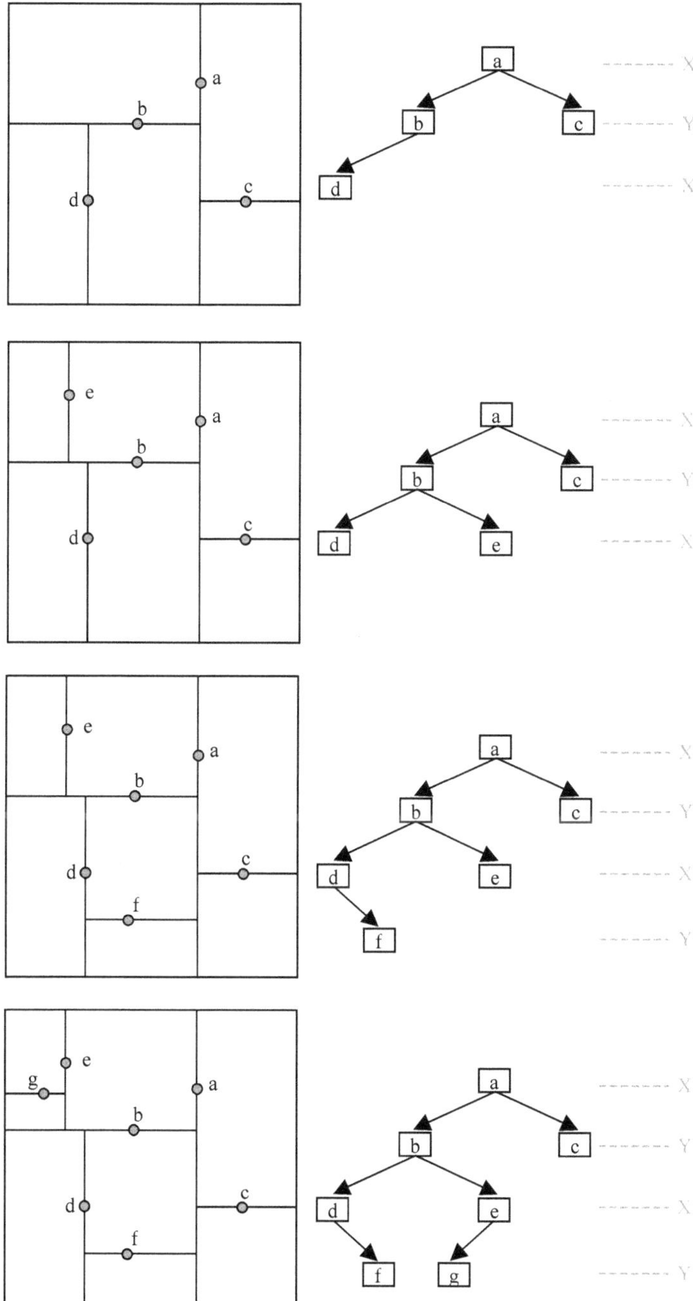

**Fig. 13.7** Building a k-d-tree.

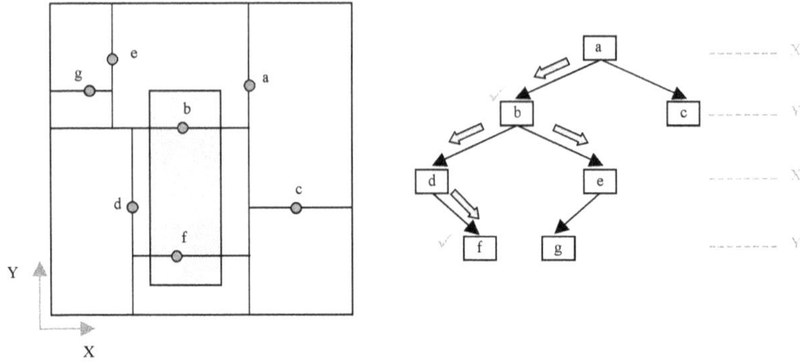

**Fig. 13.8** Region search in a k-d-tree.


## *13.3.2 K-D-B-tree*

The structure of *K-D-B-tree* was proposed as an extension of the k-d-tree to overcome the aforementioned drawbacks. The central idea of this extension is equivalent to the extension of the binary search tree to the $B^+tree$ (Chapter 8). Figure 13.9 describes the technical characteristics of the K-D-B-tree structure.

---

K-D-B-tree is an effective point data structure, which combines the features of the k-d-tree and $B^+$tree. It is a balanced tree, which consists of non-leaf and leaf nodes. A leaf entry is of the form (I, P), where I is the identity of the point object (used as a reference to a point object in the database), and P is an array of the point object coordinates in the n-dimensional space: $(p_1, p_2, ..., p_n)$. An non-leaf entry is of the form: (PTR, R), where PTR is a pointer to a node in the next lower level of the tree, and R is the rectangular area encompassing all entries (areas and points) of that node. Assuming that M is the maximum number of entries in a node (branching factor) , and m is the minimum number of enries in a node (where m $\leq$ M/2), a K-D-B-tree meets all following conditions :
✓   each tree node hosts between m and M entries unless it is the root of the tree,
✓   the root node has at least two children unless it is a leaf of the tree,
✓   all leaf-nodes appear at the same level of the tree
✓   the rectangular areas R of entries in a non-leaf node do not overlap, and
✓   the union of rectangular areas R of all entries in a non-leaf node, excluding the root, is equal to the area assigned to the corresponding entry in the parent node.

---

**Fig. 13.9** Technical characteristics of the K-D-B-tree structure.


Figure 13.10 shows how the lower left corner of the European countries MBRs in Figure 12.9b can be grouped and stored in a K-D-B-tree. Each intermediate node of the tree contains at most four records (M=4) and no fewer than two (m=2). At the bottom level, the lower left corners of the MBRs (denoted by two letter codes) are grouped into seven leaf nodes A, B, C, D, E, F and G. In the next upper level, the seven nodes are grouped into two intermediate nodes X and Y. Those are pointed by the root of the tree. The algorithms for building and maintaining a K-D-B-tree are not examined in this textbook.

The region search algorithm in a K-D-B-tree works as follows. Given a search region Q (Figure 13.10b), the tree is traversed starting from the root. The entries selected by descending the tree are: X and Y (root level), C, D, and F (intermediate level) and end AU, CZ, HU, RO and PL (leaf level) as the corresponding areas intersect with Q. All these entries are displayed as shaded in Figure 13.10.

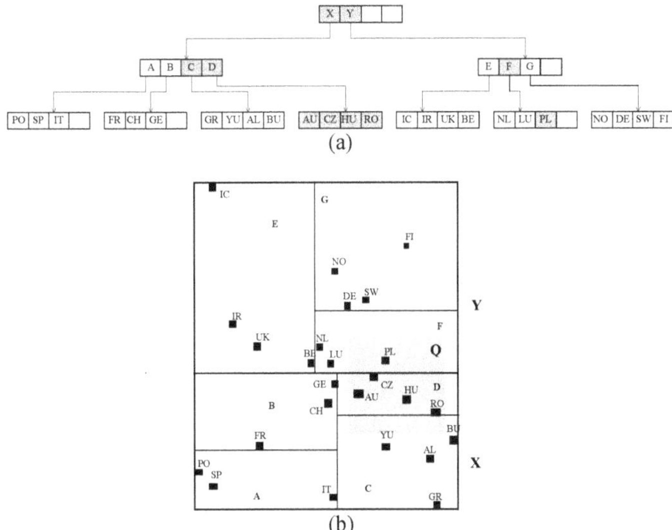

**Fig. 13.10** Example of a K-D-B-tree for the bottom left corners of the MBRs of the European countries (Figure 12.9b). The shaded entries in (a) represent areas or points that intersect the search area Q in (b).


## *13.3.3 Grid Structure*


The structure of the *grid* is the most widely used data structure in paper maps (e.g., city maps). Its digital version was the structure that mainly supported the retrieval of spatial data in the early age GIS and CAD software pachages. In modern software systems it was replaced by more efficient structures, such as the R-tree and quadtree.

In the grid structure, the space is divided into a number of cells by drawing a grid. These cells are generally equal size squares in the two dimensional space. Each cell is assigned a code that indicates its position in the grid, and a linked list which hosts the identifiers of all objects (point, line, polygon) that lie fully or partially in that cell.

Figure 13.11 shows an example of a grid structure in 2D space. It is important to note that some objects (especially the large ones) are assigned to the corresponding lists of more than one cells, e.g., A2, L2.

For a search area Q, the search algorithm works as following (Figure 13.11b). Initially, all cells that are partially or fully covered by Q are detected. This is a simple process, as cell locations and coverage areas are known from the grid definition (partitioning of space). Then, all objects in the linked lists of the fully covered cells are reported in the result, e.g., the objects of cell (3,2): {L2}. The geometries of all objects assigned to the cells that are partially covered by the search area Q are compared against the area Q. If an intersection occurs, the corresponding object is reported to the result; for example: cell (3,1): {A3}. Otherwise, the object is ignored; for example, cell (2,3): {A2}. To avoid unnecessary geometry checks, reported objects are removed from all lists of candidate objects (Figure 13.11b).

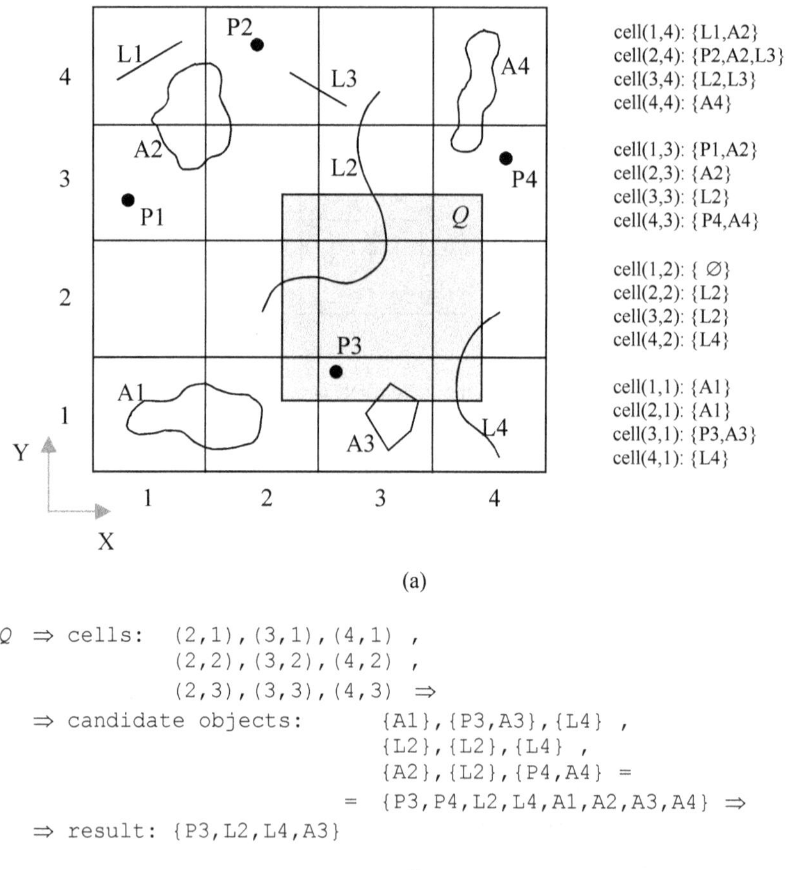


(a)

```
Q  ⇒ cells:    (2,1),(3,1),(4,1) ,
               (2,2),(3,2),(4,2) ,
               (2,3),(3,3),(4,3) ⇒
     ⇒ candidate objects:    {A1},{P3,A3},{L4} ,
                             {L2},{L2},{L4} ,
                             {A2},{L2},{P4,A4} =
                = {P3,P4,L2,L4,A1,A2,A3,A4} ⇒
     ⇒ result: {P3,L2,L4,A3}
```

(b)

**Fig. 13.11** Example grid structure (a) and a region Q search processing (b).

The effectiveness of the grid structure closely depends on the size of the cells. It is obvious that a small cell size minimizes the number of geometry checks caused by partially overlapped cells (although the number of these cells increases for the same search area, the total non-overlapping surface tends to be smaller in area) and hence performance improves. On the other hand, smaller cell size increases the storage requirements, as objects with extent (lines and polygons) will be assigned to a larger number of cells. Therefore, a cell size that improves performance without dramatically increasing the size of the structure should be sought. A parameter that can be considered in choosing the most appropriate cell size is the *density* of objects in space. This parameter is obviously variable and sometimes difficult to predict in dynamic databases. Finally, a variant of grid structure applies the variable cell size (Figure 10.13).

### *13.3.4 Quadtree (for vectors)*

A *variant of the quadtree* that can be used to structure vector data (points, lines, and polygons) has the following properties. The space is recursively divided into four sections (quadrants) of the same size. Each object is assigned to the smallest quadrant of the tree that contains the object entirely (this criterion does not apply to point objects or objects with a very small extent). Therefore, vector objects can be assigned to both non-leaf and leaf nodes of the quadtree.

Generally, large-size objects are placed in nodes near the root of the tree, while small-size objects are usually arranged near the leaves. Of course, there are cases where small objects can be detected near the root, because of their position in space; the early divisions (quadrant at or near the root level) happen to intersect them.

As for the point objects, the partition can be continued indefinitely; actually until the spatial resolution is reached. Hence, a maximum allowable height for the tree (or minimum size quadrant) needs to be initially determined. This height characterizes the quadtree.

Figure 13.12 shows the quadtree indexing for the data in Figure 13.11. It is obvious that each node of the tree can be assigned more than one object, which are hosted in a linked list. The height of this tree is equal to 3 (i.e., number of levels = 3).

The region search algorithm works as follows. For the shaded area Q (Figure 13.12) the tree is traversed starting from the root and following the paths to quadrants, overlapped partially or fully by Q. If the quadrant is fully covered from Q, all entries assigned to the corresponding node and all its descendent nodes are reported. If the quadrant is partially covered by Q, the objects assigned to the node are checked for the geometric intersection against Q (and reported if they intersect). The traversal continues at lower levels of the tree to the nodes whose quadrants intersect the area Q.

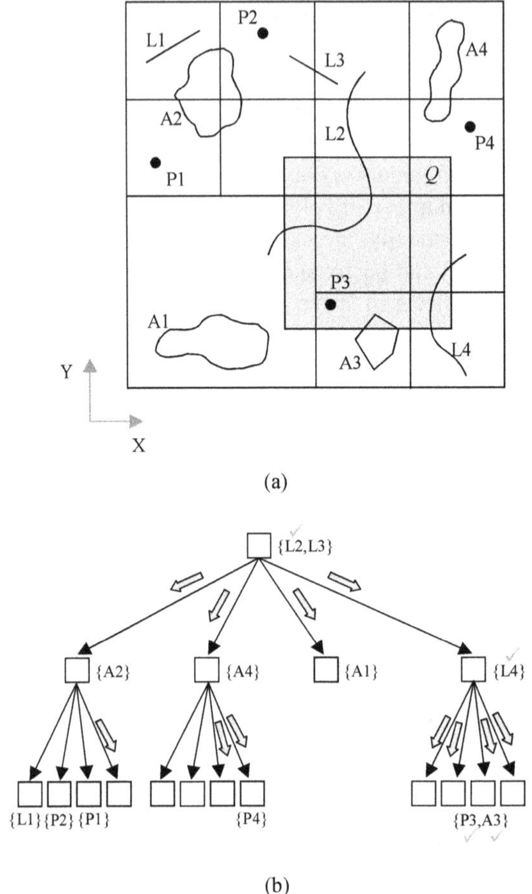



(a)



(b)

**Fig. 13.12** A vector quadtree and the region search over area Q.

## *13.3.5 R-tree*

The *R-tree structure* and its *variants* ($R^+$, $R^*$ trees, etc.) are the most efficient structures for spatial indexes. The R-tree expands the notion of $B^+$ tree to index vector data (points, lines, and polygons). In the Sections that follow, the discussion is limited to 2D space. Nevertheless, the R-tree index structure can be applied to spaces of higher dimensions. Lately, all commercial and open source database systems, geographic information systems (GIS), and computer aided design (CAD) software, make use of this structure as the basic structure for spatial indexing.

The principle of the R-tree structure can be summarized as follows: (a) complex geometric objects are represented in the index with their *minimum bounding rectangle (MBR)*; (b) these MBRs are recursively *grouped* into larger MBRs, which are called *pseudo-MBRs*; and (c) these pseudo-MBRs along with the object MBRs form *hierarchies* by being structured in a tree index.

Figure 13.13 shows the representation of the three types of geometric objects (point, line, polygon) with the corresponding MBRs. Figure 13.14 presents the MBRs of the objects illustrated in Figure 13.11. Figure 13.15 shows the grouping of neighboring MBRs into larger pseudo-MBRs, which are further organized into an R-tree structure.

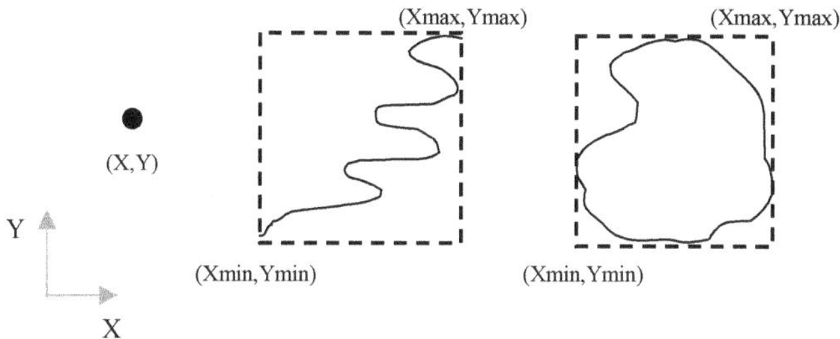


**Fig. 13.13** Representation of 2D geometric objects with MBRs.

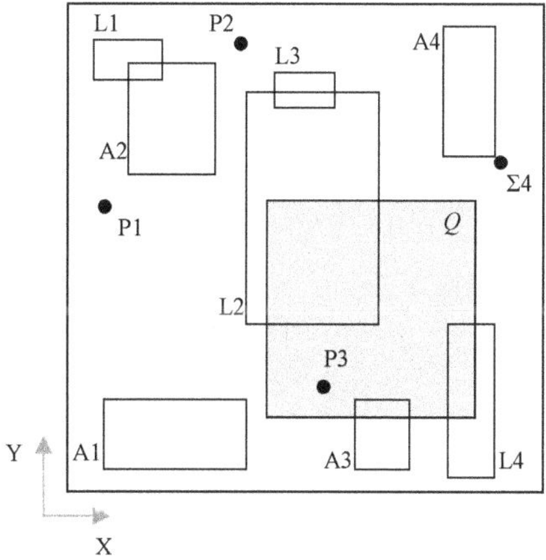


**Fig. 13.14** Representation of 2D geometries (Figure 13.11) with their MBRs.

Figure 13.16 summarizes the technical characteristics of an R-tree. The R-tree is a balanced tree, i.e., all leaf-nodes lie in the same level. The geometric objects are indexed in the leaf nodes of the tree, which accommodate their MBRs. Each node of the tree has a number of children, which varies between a maximum value (M) and a minimum value (m), in accordance with the B$^+$tree. These two values are typical for the R-tree, and called *maximum* and *minimum branching factors*, respectively. The value of m equals to $\lceil$(M/2); with the exception of the root of the tree where m equals to 2. The height of the tree for indexing N objects is less than or equal to $\log_m$(N-1). The proof is similar to that of the B$^+$tree (see Chapter 8). Note that in the implementation of the R-tree in the secondary memory, M is chosen so that each node fits to a disk page. For example, for a page size of 1024 bytes and integer coordinates, the optimal values for M and m are equal to 50 and 25, respectively.

The structure of the R-tree is not unique for the same set of objects, but depends on the order in which these are inserted in or deleted from the tree. Since, the leaf-nodes of the tree host pointers to the objects in the database, all types of geometric objects (points, lines and polygons) can be accommodated in the same structure.

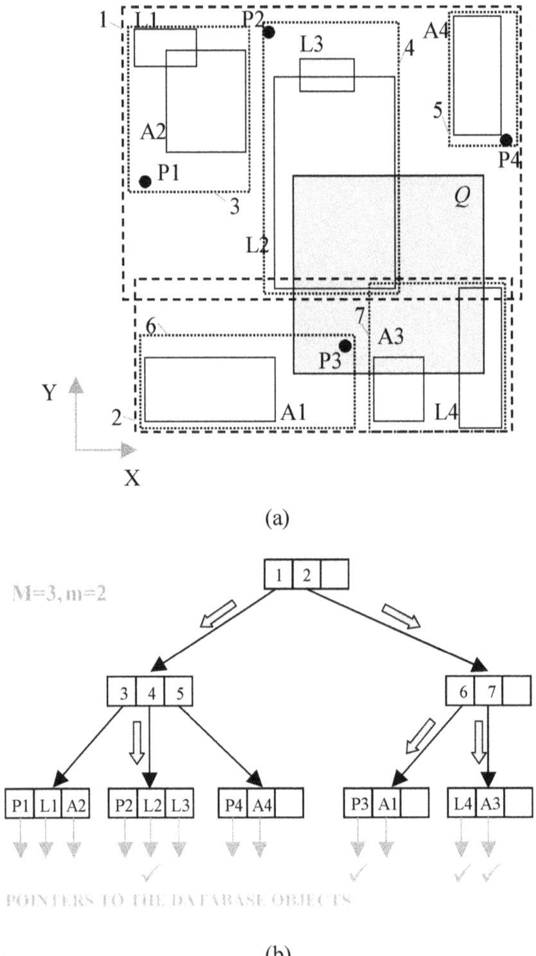



(a)

(b)

**Fig. 13.15** Grouping of MBRs (a) and organized in the corresponding R-tree (b). Example search in R-tree over region Q.

The region search algorithm in an R-tree works as follows. For a search region Q (e.g., Figure 13.15) the tree is traversed starting from the root. At each node, the intersection of its pseudo-MBR against the region Q is checked. If they intersect, the traversal continues recursively to the node descendants. When a leaf node is reached, all objects indexed by the node whose MBR intersects Q, are reported as candidate objects (filter step; see next).

The R-tree was proposed as a direct extension of the $B^+$tree in n-dimensions. It is a balanced-tree structure, comprising of intermediate nodes and leaves. A leaf-node has entries of the form (P, R), where $P$ is the identity of (or pointer to) a database object, and R is the MBR approximation of this object, i.e., it is of the form:$(p_{l-1}, p_{l-2}, ..., p_{l-n}, p_{u-1}, p_{u-2}, ..., p_{u-n})$, which represents the 2n coordinates of the lower left $(p_l)$ corner and the upper right $(p_u)$ corner of an n-dimensional parallelogram $p$. An intermediate node is of the form: (PTR, R), where PTR is a pointer to a node of the next lower level of the tree, and R the MBR of all entries grouped in that node. Assuming that $M$ is the maximum number of entries in a node (i.e., maximum branching factor), and m (where $m \leq M/2$) the minimum number of entries (i.e., minimum branching factor) in a node, an R- tree satisfies the following conditions:
- ✓ each node hosts between m and M entries unless it is the root of the tree,
- ✓ the root node has at least two children unless it is a leaf node,
- ✓ all leaves appear at the same level of the tree,
- ✓ for each entry (P, R) in a leaf node, R is the smallest MBR that encloses the object P, and
- ✓ for each entry (PTR, R) in an intermediate node, R is the smallest MBR that encloses all MBRs of the child node pointed by PTR.

**Fig. 13.16** Technical characteristics of the R-tree structure.

Clearly, the representation of an object with its MBR simplifies the geometry (Figure 13.13). Moreover, the intersection of a search region with the MBR does not assure the intersection with the object itself. An example is given in Figure 13.17. The area of the MBR that is disjoint with the area of the corresponding object is called *dead zone*. Therefore, the R-tree provides a first *filter* to the objects that possibly intersect a search region and not the actual set (Figure 13.18).

The complexity of the search algorithm in R-tree approaches the logarithmic performance. However, it is usually worse for the following reason. The search region may overlap with multiple MBRs or pseudo-MBRs hosted in a single node of the tree; and this overlap will cause a geometry check and/or traversal of more than one subtrees (as opposed to the $B^+$tree). As the overlap between MBRs is minimized, the performance of the structure approaches the logarithmic performance.

Therefore, the performance of the R-tree structure depends on the grouping of the object MBRs into pseudo-MBRs and the recursive grouping of the latter to larger pseudo-MBRs. To make this clear, consider the grouping of the four MBRs in Figure 13.19a into two pairs. Two alternative groupings are shown in Figure 13.19b,c. The first minimizes the dead zones in the pseudo-MBRs, while the second minimizes the overlapping area between them. As shown in Figure 13.19d,e, for two point searches $(e_1,e_2)$ both groupings are ineffective. This is because neither of the two satisfies both conditions: (a) *minimize dead zones*, and (b) *minimize*

*the overlapping area* between MBRs. Moreover, it is clear that these two conditions are often conflicting.

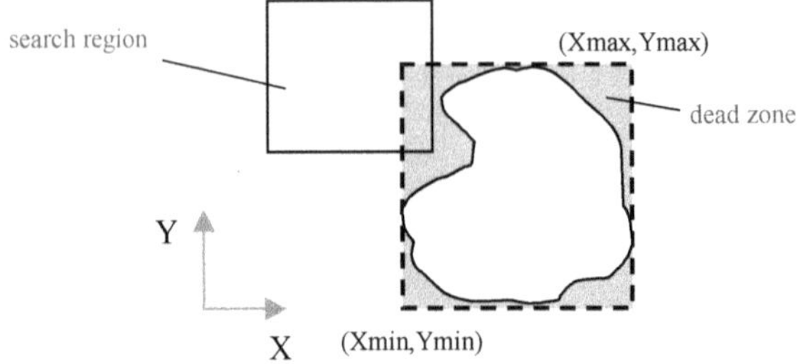



**Fig. 13.17** A polygonal object, its MBR (dashed polygon), and the dead zone (shaded area). The intersection of the search area with the MBR does not assure the intersection with the corresponding object.

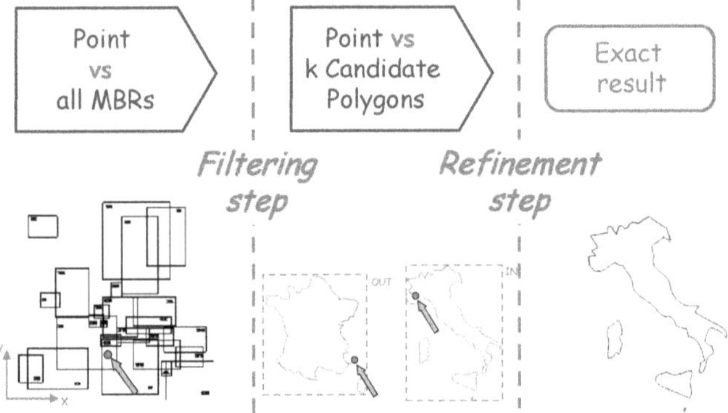



**Fig. 13.18** The two-tier query model.

The organization of MBRs in the various R-tree versions is based on *heuristic rules*, which try to minimize the area of overlap and/or the perimeter of the pseudo-MBRs. The $R^*tree$ is the most efficient version and has been adopted by modern software systems as the default spatial index structure.

Figure 13.20 shows the organization of the European countries MBRs (Figure 12.9b) into an R-tree. Each intermediate node of the tree hosts a maximum of four entries (M=4). At the leaf level the countries MBRs (denoted by pairs of characters) are grouped into seven nodes A, B, C, D, E, F, and G. In the next upper level, the seven nodes are further grouped into two intermediate nodes X and Y, as the first root descendants.

Regarding the search query: "find all the countries whose MBRs intersect the search region Q" (Figure 13.20b), the entries selected by descending the tree are: X and Y (root level), B, C, F, and G (intermediate level) and finally IT, AU, YU, BU, GE, RO, HU, CZ, and PL (leaf-level). All those entries are shaded in the tree of Figure 13.20a.

Continuing the discussion from Chapter 12 (Point-in-Polygon algorithm) it is apparent that the use of the R-tree structure can further accelerate the search performance (in both the region and point searches). The complexity in this case approximates the logarithmic performance in finding the candidate countries. More details on the complexity of the R-tree structure can be found in Chapter 14.

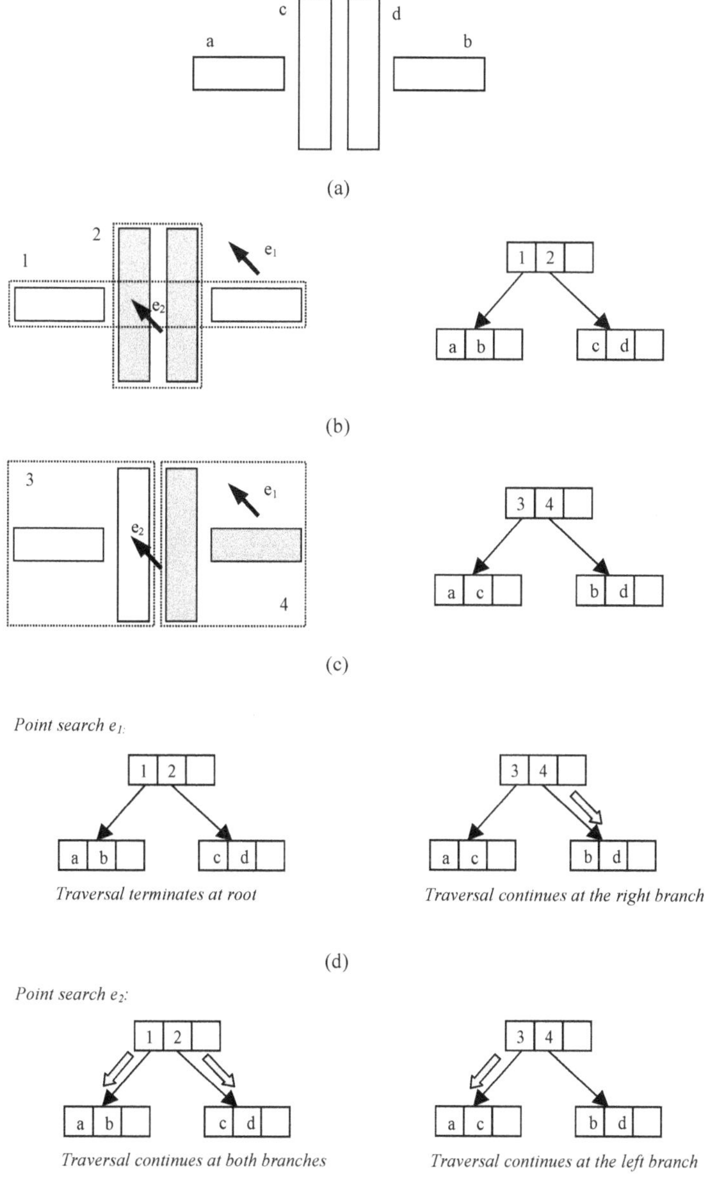



**Fig. 13.19** Alternative groupings of MBRs (a). Minimizing dead zone (b); and minimizing overlap (c). Point search per case (d, e). Grouping in (a) outperforms in point search $e_1$. Grouping in (b) outperforms in point search $e_2$ (b).

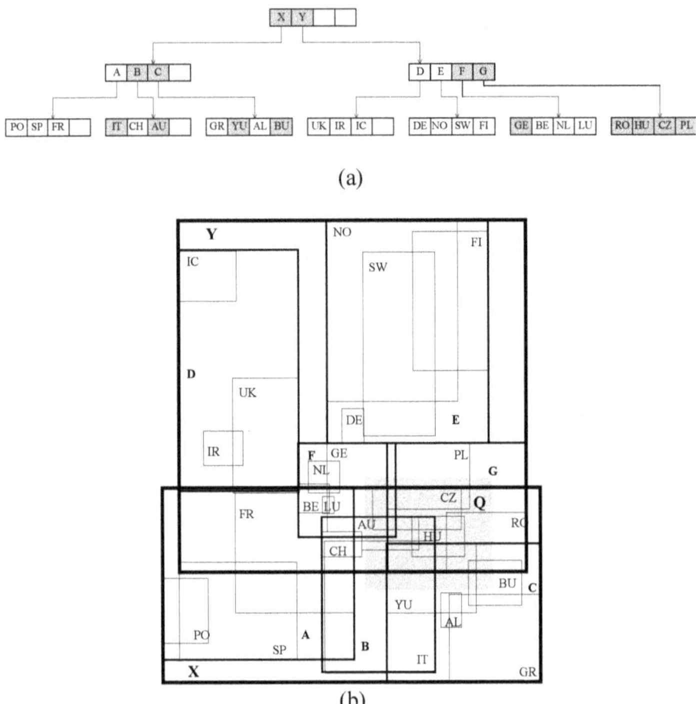

(a)

(b)

**Fig. 13.20** Grouping of the European countries MBRs in an R-tree (a). The highlighted entries correspond to the MBRs that intersect the search area Q (b).

### 13.3.5.1 Spatial Operations and R-tree

As mentioned previously, the vector data structures were originally designed to support the basic geometric region search operation (Figure 13.5; point search is considered as a degenerated case). These structures have also been adopted to support other types of spatial operations. The basic strategy for addressing other operations lies in the mapping of the spatial predicate to a set of region searches and the filtering of candidate objects.

An example is shown in Figure 13.21. The search for a spatial direction relation between polygons, e.g., "find all the north-eastern countries of Switzerland", can be transformed into a region search. Obviously, the transformation is not always as trivial; the spatial relationships may include ambiguities in their definition.

In general, all searches involving spatial relations (i.e., topologic, direction, and metric) can be supported by the R-tree family of structures. Moreover, effective solutions have been proposed for the operation of overlay, or spatial join in general by a synchronized traversal of multiple R-trees. Finally, the R-tree structures have also been applied to support the operation of finding the k-nearest neighbors of an object.

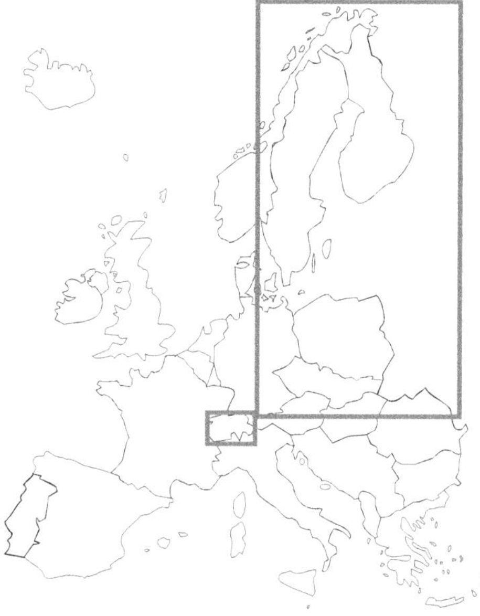

**Fig. 13.21** Example direction relations search (e.g., for countries NE Switzerland) and its transformation to a region search on the large window.


# 13.4 Topological Data Structures

The retrieval of *topological relations* underlying a collection of geometric objects is also a subject of computational geometry (Chapter 12). Although efficient algorithms have been proposed for this purpose, these algorithms are often characterized by a high complexity.

As the search for topological relations is common in a functional environment, computer systems usually *precompute* some of them and store them in the database for use when requested (Chapter 14). The precomputed relations or any relevant ancillary elements are organized into appropriate structures, usually tabular (collection of relational tables), which are called *topological data structures*.

The price of a fast retrieval of topological relations is both the storage requirements for the topological data structure and maintenance cost in a dynamic environment. It is obvious that every geometric update in the database must be reflected in the topological data structures as well.

A couple of typical GIS application queries that retrieve to topological relations between spatial objects would be the following: "what are the neighbors of this polygon?" (e.g., "which parcels border with the park?") or "what arcs meet in this node?" (e.g., "which highways lead to this square?").

The execution of such queries can be supported by appropriate topological data structures. GIS software packages apply different topological data structures to support such types of queries. Figure 13.22 shows a simple topological data structure for the building block of the cadastral example examined in the previous Chapters (Figure 5.5).

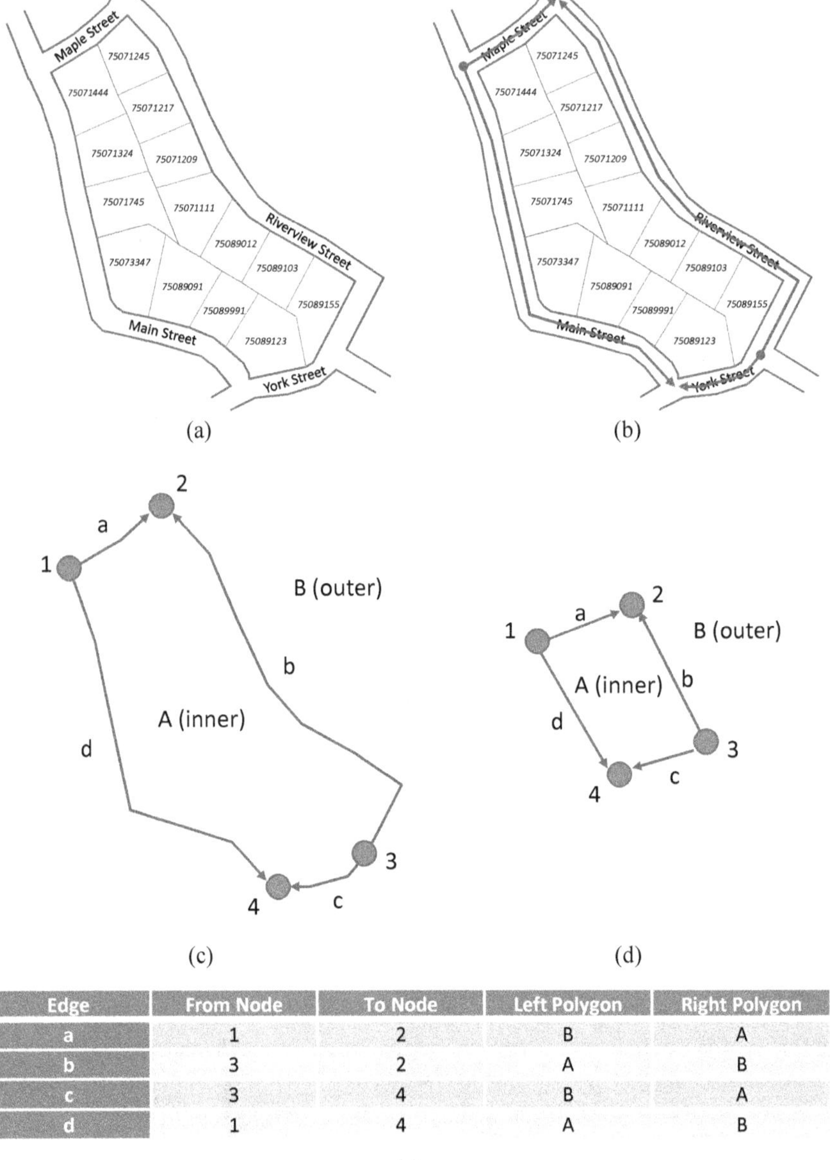

| Edge | From Node | To Node | Left Polygon | Right Polygon |
|------|-----------|---------|--------------|---------------|
| a | 1 | 2 | B | A |
| b | 3 | 2 | A | B |
| c | 3 | 4 | B | A |
| d | 1 | 4 | A | B |

(e)

**Fig. 13.22** Example of a planar graph (c) and its simplified form (d) representing a building block (a),(b). A simple topological data structure (e) built on top of the graph.

# References and Further Reading

Manolopoulos, Y., Theodoridis, Y., Tsotras, V.J., 2000. *Advanced Database Indexing*. Kluwer.

Papadias, D., Theodoridis, Y., and Stefanakis, E., 1997. Multi-dimensional range query processing with spatial relations. *Geographical Systems*. Vol. 4, pp. 343-365.

Preparata, F.P., and Shamos, M.I., 1991. *Computational Geometry: An Introduction*. Spinger-Verlag.

Rigaux, P., Scholl, M., and Voisard, A., 2002. *Spatial Databases with Applications to GIS*. Morgan-Kaufmann.

Samet, H., 1990. *The Design and Analysis of Spatial Data Structures*. Addison-Wesley.

Shashi, S., and Chawla, S., 2003. *Spatial Databases – A Tour*. Prentice Hall.

Stefanakis, E., and Sellis, T., 1998. Enhancing operations with spatial access methods in a database management system for GIS. *Cartography and Geographic Information Systems*. Vol. 25, pp. 16-32.

Theodoridis, Y., Stefanakis, E., and Sellis, T., 2000. Efficient Cost Models for Spatial Queries Using R-trees. *IEEE Transactions on Knowledge and Data Engineering*. Vol.12, IEEE CS Press. pp. 19-32.

# 14 Geographic Operation Execution Strategies

## 14.1 Introduction

Consider a geographic database, which hosts (a) the road network, and (b) the administrative boundaries of the Province of New Brunswick, Canada. A search like "what highways cross York County?" (Figure 14.1), can be executed following different scenarios (Figure 14.2). For example, one scenario would first search for the highways in the Province and then compare the set of highways against the polygon of York County. An alternative scenario would first search for all segments of the road network crossing York County and then select from this set the segments that represent highways. Both scenarios would lead to the same result. But, which one would complete faster? Obviously, an answer to the latter question must take into consideration how data is structured in the database as well as what indices are available.

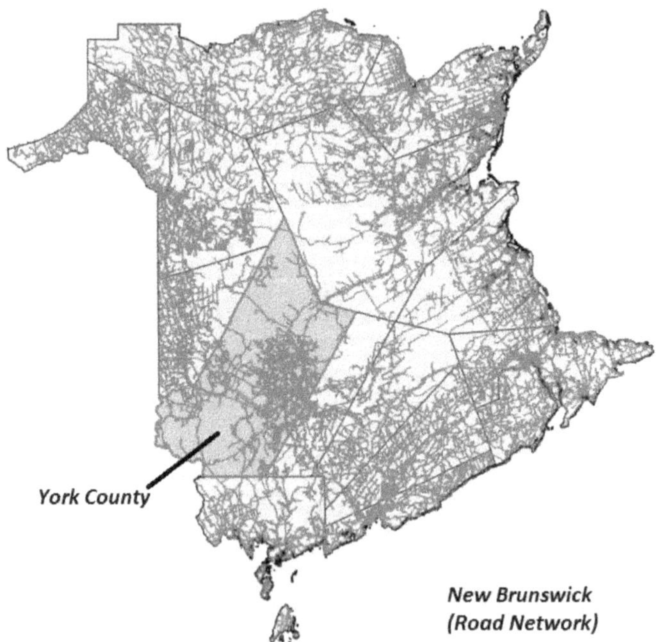


**Fig. 14.1** New Brunswick's road network and counties.

Usually, there are *multiple scenarios* (*strategies*) to execute a complex geographic operation or procedure (i.e., a sequence of simple operations). Each scenario is a separate solution, which leads to the result, but is often characterized by a different complexity. A key role of the database management system (DBMS) is to optimize the execution of complex operations and procedures, in other words, choosing the optimal scenario.

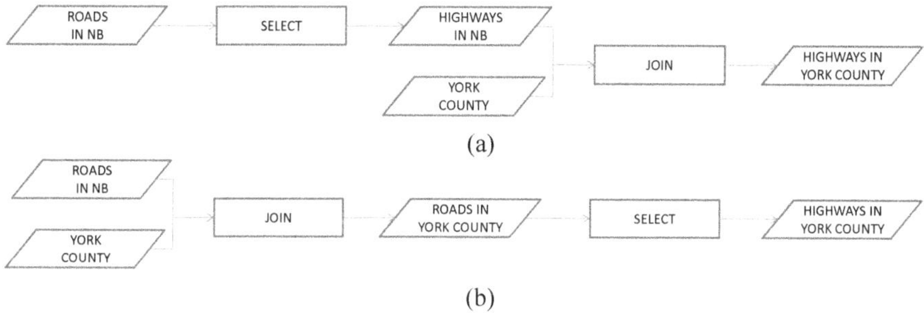


**Fig. 14.2** Two alternative scenarios for finding the highways in York County (Figure 14.1).

The term *optimization*, though used frequently, is misleading especially in non-traditional DBMSs, such as geographic DBMSs. In many cases, the execution strategy as chosen by the DBMS is not necessarily the most effective, but a relatively effective one. In traditional DBMSs, two main optimization techniques are being applied solely or in combination: (a) *heuristic rules*, and (b) *systematic cost estimation*.

The first technique is based on heuristic rules for ordering the operations that make up a procedure. The basic rule that applies to traditional DBMS suggests running first the less-time-consuming operations to reduce the size of intermediate temporary results (e.g., tables). Therefore, the operations of SELECT and PROJECT must always precede JOIN operations (see Chapter 6). Moreover, the most restrictive SELECT and JOIN operations (i.e., those that produce smaller tables) must precede others in the execution of a procedure.

The second technique is based on a systematic estimation of the cost of various scenarios in executing a procedure and the selection of less costly one. This technique requires the presence of reliable cost estimates (*cost models*) for the alternative scenarios, so that it is possible to objectively compare them. Note that the number of execution scenarios considered in each case must be limited, otherwise there is a risk of wasting valuable time.

These two techniques as applied to traditional DBMS can be extended and adopted in GIS, to establish a relatively efficient execution strategy for spatial procedures. However, this is a hard task due to the peculiar nature of geographic data and the underlying data models. Moreover, spatial predicates are more complex and overly dependent on the application domain.

The optimization of spatial queries has been considered in spatial database systems. Heuristic rules along with cost models are available to help design the execution strategy and reduce the cost of complex queries that involve spatial and non-spatial predicates. The following Sections summarize some heuristic rules and show how the technique of systematic cost estimation can be applied in the design of a strategy for execution of spatial procedures. Finally, the practice of pre-computing and storing ancillary spatial attributes and relationships to accelerate the execution of geographic operations is discussed.


## 14.2 Heuristic Rules

*Heuristic rules* define an effective scenario for the execution of the simple operations that make up a procedure. They aim at reducing the total cost of execution in terms of CPU time, number of pages retrieved from the disk (I/O), and/or memory requirements. This is accomplished through the use of indexes, pre-calculated and stored values of spatial attributes and relationships, or retrieval of interim results from non-time consuming and restrictive predicates (spatial and non-spatial).

A set of heuristic rules has been proposed to support the execution of complex procedures in spatial database systems. These rules extend the principles of traditional database systems and are closely related to the architecture of the system, i.e., the spatial data model, the available spatial index structures, etc.

The following list highlights some basic rules for the execution of simple operations that make up a spatial procedure (Figure 14.3).

- *Execute selection and projection operations as early as possible.* This is a standard optimization technique in traditional database systems. Regardless of the underlying data model, both selection and projection are operations that reduce the overall size of a file. On the other hand, the size of the file resulting from a join or other binary operations is usually increasing and depends on the size of the input files. Therefore, before applying the latter functions, it is advantageous to reduce the size of the input files, so that memory requirements, CPU time, and number of disk accesses are minimized.
- *Break up any selection operation with conjunctive conditions (spatial and non-spatial) into a cascade of simple selection operations.* This permits a greater degree of freedom in reordering the execution of individual selections, so that the optimal strategy is achieved.
- *Execute the most restrictive selections first.* This ensures that intermediate temporary files are the smallest possible in both the number of objects (records) and absolute size. The most restrictive selections can be chosen based on a selectivity measure that is usually estimated in the database catalog using metadata items.
- *Execute operations that are supported by indices first.* Both spatial and non-spatial indices may support the execution of selection operations. Operations that are supported by index structures should be performed first, so that intermediate temporary files, used as input to subsequent operations, which are not supported by index structures, accommodate the description of a smaller number of objects. This way, a sequential scan (if any) is less expensive.
- *Group operations that can be executed by a single routine.* When more than one selection operations refer either to the same attribute (spatial or non-

spatial) and can be supported by existing index structures or to the same object and cannot be supported by existing index structures (hence a sequential scan on the database objects in temporary files is imposed), their execution by a single access routine is advantageous.

- *Replace some operations by alternative operations.* This rule is application dependent and can be usually applied in spatial operations. A representative example is the alternative use of nearer versus nearest operation, in order to avoid a repetitive computing of the nearest object computation which meets a series of other conditions in a query.

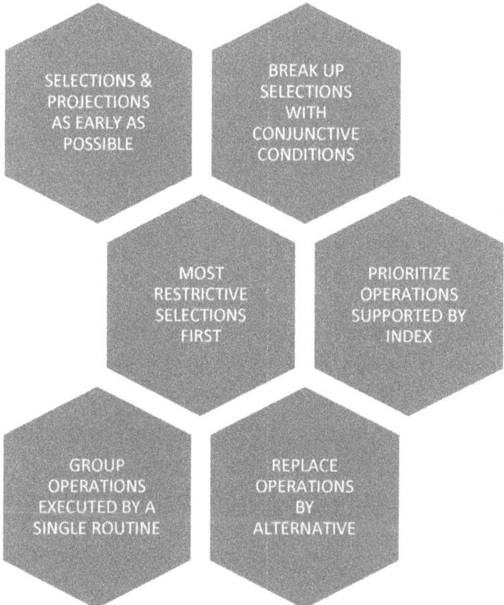


**Fig. 14.3** Common heuristic rules in the execution of procedures in spatial database systems.

## 14.3 Systematic Cost Estimation

In order to choose the execution strategy based on a *systematic cost estimation*, analytical cost models are required, to compute the cost of individual operations. Obviously, this is not the case for all operations due to the complexity of the spatial index structures used, whose performance is heavily dependent on various parameters, such as the distribution of objects in space and their extents (see Chapter 13). This is the reason why GIS packages usually lack optimization techniques, which are based on systematic cost estimation.

On the other hand, efficient models for predicting the cost of the R-tree index structure, which is commonly used to support the execution of common operations such as the range (window) and join queries, are available. The following Subsections briefly present the cost models used in the systematic cost estimate of these queries.


## 14.3.1 Cost Models for the Window Queries

Several analytical models to estimate the performance of R-trees on *window operations (range queries)* have been proposed in the past. Following, a model introduced by Theodoridis and Sellis, is described. This model predicts the R-tree performance using knowledge of the data set properties only (e.g., its distribution in space), and is applicable to point and non-point objects.

According to this model, the expected retrieval cost (i.e., number of disk accesses), for an n-dimensional query window Q with extents ($Q_1$ , $Q_2$ , ... , $Q_n$), along each dimension, is given by the following formula:

$$C(Q) = 1 + \sum_{j=1}^{h-1} \left\{ N_j \cdot \prod_{i=1}^{n} \left( s_j + Q_i \right) \right\}$$

In this formula the parameters involved are: h the height of the tree structure; $N_j$ the expected number of nodes in the tree; and $s_j$ the average node extent along each dimension at level j of the tree (the root is assumed at level j=h, and the leaf-nodes at level j=1). The model assumes that the sides of the nodes are equal along each dimension; this is a simplification that is reasonable for "well structured" R-trees. The expression for computing the height h of an R-tree is:

$$h = 1 + \left\lceil \log_{c \cdot M} \frac{N}{c \cdot M} \right\rceil$$

where N is the number of distinct objects in the database, M is the maximum number of entries in an R-tree node, and c is the average node capacity (typically c=67%; c·M denotes the average number of entries per node). The number $N_j$ of nodes at level j is:

$$N_j = \frac{N}{\left( c \cdot M \right)^j}$$

The following formula expresses the average node extent (MBR or pseudo-MBR) at level j:

$$s_j = \left( D_j \middle/ N_j \right)^{1/n}$$

where $D_j$ the density of the node rectangles at level j, which is computed as a function of the density $D_{j-1}$ of the node rectangles at level j-1 (notice that the density D of a set of N objects is defined as the sum of the object areas $s_j$ divided by the data space):

$$D_j = \left\{ 1 + \frac{\left( D_{j-1} \right)^{1/n} - 1}{\left( c \cdot M \right)^{1/n}} \right\}^n$$

Hence, $D_j$ can be recursively computed using $D_0$, which denotes the density of the actual objects MBRs.

## 14.3.2 Cost Models for Spatial Joins

The cost model presented in the previous Section was extended by Theodoridis, Stefanakis and Sellis to also estimate the cost (in terms of disk accesses) of *spatial join queries* (e.g., overlay operation) between two spatial data sets. Assuming that each data set is indexed by an R-tree, the join query between the two sets can be supported by applying a *synchronized tree traversal* on both R-tree indexes (the algorithm is not included in this book).

According to the extended model the cost formula for join queries is the following:

$$C(R_1, R_2) = \sum_{j=1}^{h-1} \left\{ N_{R_2,j} \cdot N_{R_1,j} \cdot \prod_{k=1}^{n} \left( s_{R_1,j,k} + s_{R_2,j,k} \right) + N_{R_2,j} \cdot N_{R_1,j+1} \cdot \prod_{k=1}^{n} \left( s_{R_1,j+1,k} + s_{R_2,j,k} \right) \right\}$$

where $C(R_1,R_2)$ the average number of disk accesses needed to process a join query between the two data sets with $N_{R1}$, $N_{R2}$ data objects indexed in trees $R_1$ and $R_2$, respectively. $N_{Ri,j}$ denotes the average number of nodes of the tree $R_i$ at level j and is given by the following formula as a function of the actual population $N_{Ri}$ of the data set:

$$N_{Ri,j} = \frac{N_{Ri}}{\left( c \cdot M \right)^j}$$

h denotes the $R_1$-tree height, and is given by the following equation, also as a function of the actual population $N_{Ri}$ of the data set:

$$h = 1 + \left\lceil \log_{c \cdot M} \frac{N_{Ri}}{\left( c \cdot M \right)^i} \right\rceil$$

and $S_{Ri,j,k}$, denotes the average extent of nodes (MBRs and pseudo-MBRs) of the tree $R_i$ at each dimension k at level j and is given by the formula:

$$s_{Ri,j,k} = \left( D_{Ri,j} \middle/ N_{Ri,j} \right)^{1/n}$$

as a function of the amount $N_{Ri,j}$ and the density $D_{Ri,j}$ of the node rectangles at level j, which, in turn, is given by:

$$D_{Ri,j} = \left\{ 1 + \frac{\left( D_{Ri,j-1} \right)^{1/n} - 1}{(c \cdot M)^{1/n}} \right\}^{n}$$

recursively as a function of the actual density $D_{Ri}$ of the data MBRs.


### 14.3.3 Evaluation of the Cost Models

The cost models for window and join queries presented in the previous Sections are based on primitive data properties only and ignore the actual R-tree instance. The analysis assumes uniformity of data in order to express the density of the R-tree nodes at a parent level as a function of the density of the child nodes. This assumption might produce a model that could be efficient for uniform-like data distributions, but hardly applicable to non-uniform ones. To adapt the model in order to efficiently support any type of data sets (uniform or non-uniform ones), the notion of a *density surface* describing the data set has been adopted in place of a single average density value. The density surface of a real data set from the TIGER database of the US Census Bureau is illustrated in Figure 14.4.

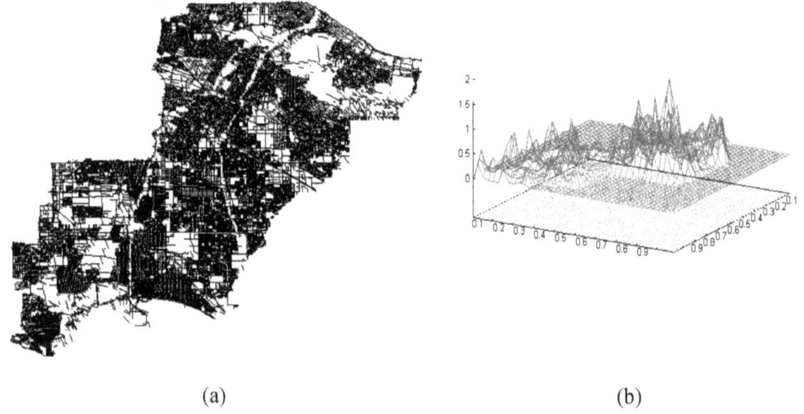

(a)                                                      (b)

**Fig. 14.4** A real data set (a) and the corresponding density surface (b).

An extended experimental analysis for the evaluation of the above models was performed. The analytical estimations of disk accesses were compared with experimental results on synthetic (random and skewed) and real data sets. Table 14.1 summarizes the average relative errors of the experimental results compared to the analytical predictions.

The estimated cost of window (point or range) and join queries is very close to the actual experimental results for both uniform-like and non-uniform data distributions (synthetic and real datasets), with the relative error never exceeding 15% or 20%, respectively.

In conclusion, the above cost models are very useful and can be used in combination with cost models available in traditional databases, such as the logarithmic cost for $B^+$trees, in order to provide systematic cost estimation for operations combining spatial and thematic predicates (Chapter 15).

**Table 14.1** Average deviation of analytical-experimental results: number of disk accesses.

| Data sets | Relative Error | | |
|---|---|---|---|
| | point queries | range queries | join queries |
| *Random data* | 0%-10% | 0%-5% | 0% - 15% |
| *Skewed data* | 0%-15% | 0%-10% | 0%-20% |
| *Real data* | 0%-15% | 0%-20% | 0%-15% |

## 14.4 Pre-computed Spatial Attribute Values and Relationships

In order to facilitate the processing of spatial operations and procedures, it is sometimes beneficial to *pre-compute* and store certain spatial attribute values or relationships. For example, the computations of several spatial attribute values, such as the area and perimeter of a region, which are commonly used in spatial operations, require both large CPU time to compute and many disk accesses to load the region's geometry into the main memory. A trivial and effective solution to this problem is to pre-compute these values while entering (or updating) the spatial objects in the database and store them as thematic attributes along with the objects for later use (Figure 14.5).

On the other hand, *spatial relationships* are seldom stored explicitly due to their large cardinality. Despite the fact that their retrieval can be supported by spatial index structures (see Chapter 13), it is sometimes beneficial to pre-compute and store certain spatial relationships, if they are relatively stable, frequently referenced, and with manageable size. In order to store the pre-computed relationships (e.g., intersect and contain) a special index file can be created and maintained. These files store pairs of object identifiers having the predefined spatial relationship (Figure 14.6). Example indices of this type are the topological data structures discussed in Chapter 13.

## *Store the area explicitly*

**Table: municipalities**

| gid<br>integer | ID<br>double precision | NAME<br>character varying(60) | POP_01<br>double precision | the_geom<br>geometry | area<br>double precision |
|---|---|---|---|---|---|
| 1 | 841 | Irakleio | 137711 | ••• | 108856910.092 |
| 2 | 842 | Agia Varvara | 5310 | ••• | 98473803.8593 |
| 3 | 843 | Arkalochori | 10897 | ••• | 238720576.640 |
| 4 | 844 | Archanes | 4548 | ••• | 31845436.375 |
| 5 | 845 | Asterousia | 6303 | ••• | 204604555.984 |

**Fig. 14.5** The area of polygons (municipalities) is stored explicitly along with the geometry and other attributes (so that any on-the-fly computation is prevented).

## *Store Topological Relations explicitly*

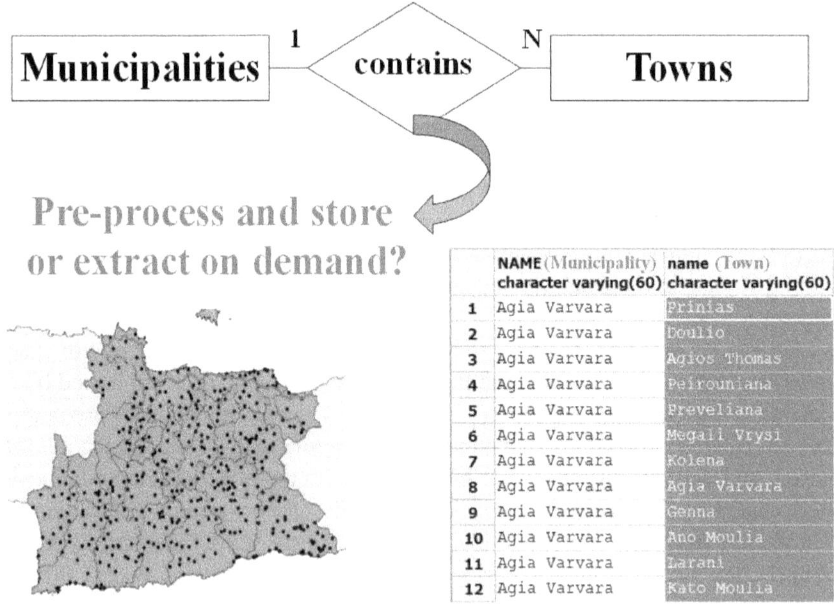


| | NAME (Municipality)<br>character varying(60) | name (Town)<br>character varying(60) |
|---|---|---|
| 1 | Agia Varvara | Prinias |
| 2 | Agia Varvara | Doulio |
| 3 | Agia Varvara | Agios Thomas |
| 4 | Agia Varvara | Peirouniana |
| 5 | Agia Varvara | Preveliana |
| 6 | Agia Varvara | Megali Vrysi |
| 7 | Agia Varvara | Kolena |
| 8 | Agia Varvara | Agia Varvara |
| 9 | Agia Varvara | Genna |
| 10 | Agia Varvara | Ano Moulia |
| 11 | Agia Varvara | Larani |
| 12 | Agia Varvara | Kato Moulia |

**Fig. 14.6** Each town is explicitly associated with a municipality.

Notice that the practice of *pre-computing* spatial attribute values and relationships should not be applied into extreme, because: (a) in most of the cases data is not static and changes quite often; and (b) the database is usually very large and the relationships between objects are many-to-many for each relation. Obviously, it is expensive to re-compute such indices and maintain all data consistent and up-to-date, especially in highly dynamic databases.

## References and Further Reading


Elmasri, R., and Navathe, S.B., 2000. *Fundamentals of Database Systems*. Addison-Wesley.

Garcia Molina, H., Ullman, J.D., and Widom, J., 2000. *Database Systems Implementation*. Prentice Hall.

Ramakrishnan, R, and Gehrke, J., 2002. *Database Management Systems*. McGraw-Hill.

Rigaux, P., Scholl, M., and Voisard, A., 2002. *Spatial Databases with Applications to GIS*. Morgan-Kaufmann.

Shashi, S., and Chawla, S., 2003. *Spatial Databases – A Tour*. Prentice Hall.

Stefanakis, E., and Sellis, T., 1998. Enhancing operations with spatial access methods in a database management system for GIS. *Cartography and Geographic Information Systems*. Vol. 25, pp. 16-32.

Theodoridis, Y., Papadias, D., Stefanakis, E., and Sellis, T., 1998. Direction relations and two-dimensional range queries: optimization techniques. *Data and Knowledge Engineering*. Vol. 27, pp. 313-336.

Theodoridis, Y., and Sellis, T. 1996. A model for the prediction of R-tree performance. In the *Proceedings of the 15$^{th}$ ACM SIGACT-SIGMOD-SIGART Symposium on Principles of Database Systems, PODS'96*, Montreal, Canada, ACM Press.

Theodoridis, Y., Stefanakis, E., and Sellis, T., 1998. Cost models for join queries in spatial databases. In *Proceedings of the 14$^{th}$ IEEE International Conference on Data Engineering (ICDE '98)*, Orlando, Florida.

Theodoridis, Y., Stefanakis, E., and Sellis, T., 2000. Efficient Cost Models for Spatial Queries Using R-trees. *IEEE Transactions on Knowledge and Data Engineering*. Vol.12, IEEE CS Press. pp. 19-32.

# 15 Geographic Database Systems

## 15.1 Introduction

A core component of a GIS as of any Information System is the data manager (Chapter 4). The data manager should be capable to support data modeling, data integrity, concurrent usage, privacy protection, user and program interfaces, data re-organization, and data redundancy control. In practice, most GIS software packages offer limited support in these areas. There are multiple reasons for this:

- Geographic entities are usually complex and the number of relationships that apply among them can be indefinite.
- The spatial dimension of geographic entities is usually represented in long and highly variable in length records.
- A large number and variety of operations is required to manipulate the geographic entities.
- Transactions in GIS usually take very long; up to days or weeks to complete.
- Handling the spatial dimension of geographic entities usually requires complex integrity and consistency checking algorithms.

Next, the GIS architectures that have been adopted over the last few decades are briefly presented. Special focus is given in the architecture of modern systems which adopts the *object-relational technology*. After a brief introduction to this technology, an open source object-relational DBMS along with its spatial model are examined closely. Finally, the steps to define, construct, and manipulate a geographic database are presented.

## 15.2 GIS Architectures

Four architectural approaches have generally been adopted, to implement commercial, open source or prototype systems and satisfy the urgent needs for operational spatial data handling over the last four decades. These approaches are (Figure 15.1):

- Single conventional DBMS
- Partial conventional DBMS
- Extended conventional DBMS
- Object-Oriented DBMS

All these approaches have tried to capitalize on the existing DBMS technology and provide an efficient solution to geographic data management. The role of the *application layer* in all four approaches is to supplement the set of capabilities offered by the underlying system architectures. In other words, the application layer either offers the set of GIS operations which are not available in the underlying DBMS or assists the interoperability when more than one systems are involved. The following Sections briefly present the alternative approaches.

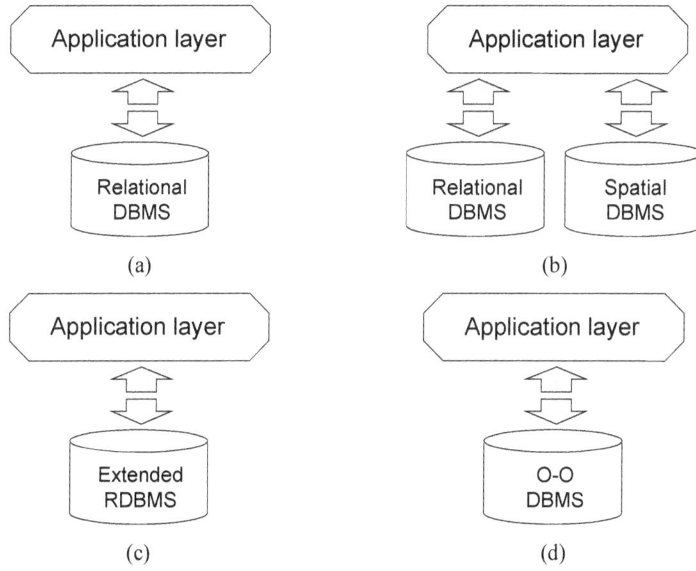


**Fig. 15.1** Architectural approaches in GIS software development.

## 15.2.1 Single Conventional DBMS

The *single conventional DBMS* architecture (Figure 15.1a) makes use of the pure relational schema to model both the spatial and non-spatial components of geographic entities. Then, a single relational DBMS is used to manage these entities. This approach fails to manage efficiently large collections of geographic entities, mainly because the representation of the spatial component (geometries) into tables results into a massive number of records; i.e., huge tables that cannot be managed efficiently. For example, if the outline of a country is represented by one million points, a table of an equal number of records will be needed to accommodate its geometric description (e.g., Figure 6.11 or 6.22b). In short, the relational model and the constraints imposed by the normal forms are not appropriate for the definition of an efficient (normalized) spatial schema.

An alternative approach commonly adopted by the single conventional DBMSs stores the geometries as binary data in a single column of a relation (Figure 15.2). This approach uses the relational database solely as a repository for the spatial component (geometry) as the DBMS can neither parse nor process this column. The application layer takes over these tasks. As the spatial processing is carried out within the application layer, this approach involves an uploading of voluminous data (spatial data are usually voluminous) on the application layer from the database. This is an extremely time-consuming task in a production environment. Hence, this architecture does not perform well for large geographic databases.

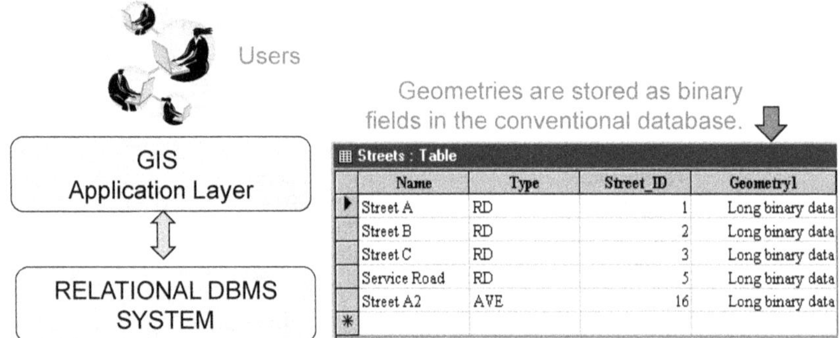


**Fig. 15.2** Geometries stored in relational tables as binary data.

## *15.2.2 Partial Conventional DBMS*

In a *partial conventional DBMS* architecture (Figure 15.1b), a relational DBMS is used to accommodate and handle the thematic information associated with the geographic entities, while a separate subsystem is used to handle the spatial data (geometries). Operators needed to define and manipulate the geographic entities are included in an application layer, which provides an interface to both subsystems and assures their interoperability. This architecture was adopted by most commercial GIS packages: e.g., Arc/Info (ESRI), MGE (Intergraph), SICAD (Siemens), and GDS (GDS); it is also referred to as *hybrid architecture*.

A relational DBMS along with a Computer Aided Design (CAD) System were usually adopted to implement this architecture. Figure 15.3 shows an example of this approach, and illustrates the reconstruction of a geographic entity from its components (geometry and theme) through common identifiers.

The main drawback of this architecture is that data describing a single geographic entity is split, stored, and handled by two separate systems. Although the two systems may efficiently support most of the management requirements, there are cases where they cannot do so and the application layer is called to process voluminous data retrieved by the two repositories. This latter scenario renders this approach inefficient in a production environment.

To clarify this, let consider a hybrid system managing the road network in New Brunswick (Figure 14.1) and three typical queries with solely spatial, solely non-spatial, and combined spatial and non-spatial predicates, respectively:

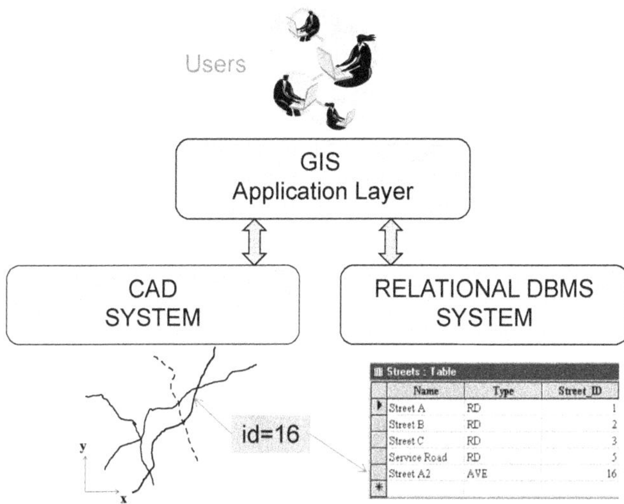



**Fig. 15.3** Example implementation of the hybrid architecture: a CAD system is coupled with a Relational DBMS. Common identifiers are used to associate the geometries with their thematic attributes.

- *Query 1:* Find all roads that are less than 5 km away from point A (spatial query; Figure 15.4a).
- *Query 2:* Find all roads that are of type avenue (AVE) (non-spatial query; Figure 15.4b).
- *Query 3:* Find all avenues that are less than 5 km away from point A (combined query; Figure 15.4c).

The user poses these queries through the user interface provided by the application layer. The first query involves a spatial predicate only. Hence, it will need to be processed by the CAD system. The application layer forwards the query to the CAD system (Figure 15.5a) and receives back the identifiers of all road segments that satisfy the spatial predicate (distance from A less than 5km). As the CAD system is very efficient in handing geometries, this query will be supported pretty well by the hybrid architecture.

The second query involves a thematic (non-spatial) predicate only. Hence, it will need to be processed by the Relational DBMS system. The application layer forwards the query to the DBMS system (Figure 15.5b) and receives back the identifiers of all road segments that satisfy the non-spatial predicate (road type: avenue). As the DBMS system is very efficient in handing non-spatial data, this query will also be supported pretty well by the hybrid architecture.

The third query is a combined one as it involves both a spatial and a thematic (non-spatial) predicate. Hence, it will need to be partially processed by both subsystems. The application layer forwards the spatial predicate to the CAD system and the non-spatial one to the DBMS system (Figure 15.5c). After it receives back the identifiers of all road segments that satisfy the spatial predicate and those that meet the non-spatial predicate, it performs an intersection to find those identifiers that show up in both sets. Obviously, the two sets may be voluminous and require long processing, while the result of the intersection may contain only a few roads. Hence, the processing may involve an uploading of voluminous data on the application layer from the database. This is an extremely time-consuming task in a production environment. Hence, this architecture does not perform well for combined queries.

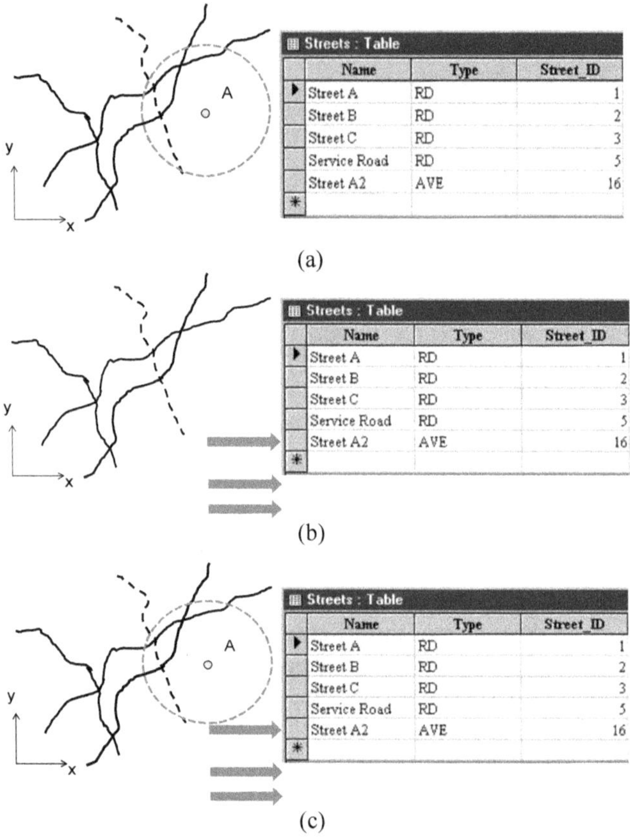

Fig. 15.4 Three typical queries in a GIS managing the road network of New Brunswick (Figure 14.1): (a) spatial predicate; (b) non-spatial predicate; and (c) both spatial and non-spatial predicate involved.

## *15.2.3 Extended Conventional DBMS*

In the *extended conventional DBMS* architecture (Figure 15.1c), a relational DBMS is modified to also support spatial data. This is accomplished by adding new constructs to the conventional DBMS so that to enhance its modeling power and provide better support for geographic applications. These new constructs include support for abstract data types (ADT), procedural fields, complex objects, composite attributes, and so on. Any additional operators needed to define and manipulate the geographic entities are included in an application layer which lies at the top of the extended DBMS.

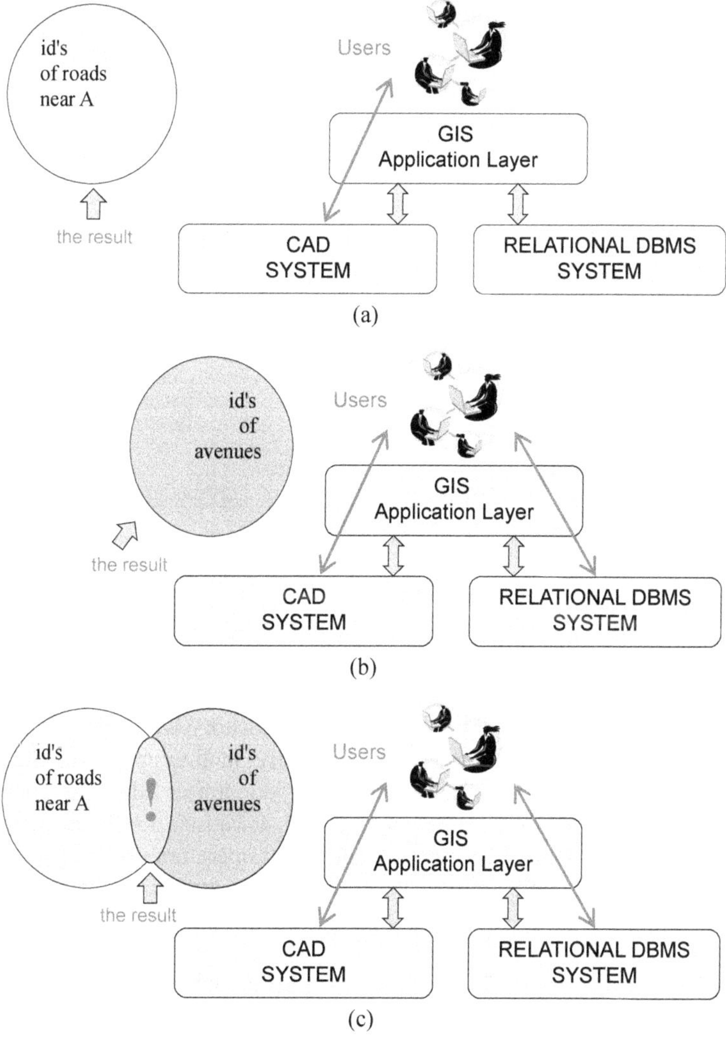


**Fig. 15.5** The processing of the three queries in Figure 15.4, respectively.

Commercial and open source relational DBMSs such as Oracle, Informix, PostgreSQL, etc., have implemented an extended relational model with object oriented concepts. This extended model is called *object-relational model* and can effectively support non-traditional applications, such as GIS applications, multimedia, etc.

Over the last couple of decades, many GIS applications were developed using the extended relational DBMSs with very satisfactory results.

The extended DBMS accommodates both the spatial and non-spatial components of geographic data into a single repository (Figure 15.6). In addition, both components can be handled internally and the need for further processing at the application layer is diminished if not eliminated. All types of queries in Figure 15.4, including the combined ones, can be executed efficiently.

Section 15.3 highlights the properties of object-relational technology. Then, Section 15.4 shows how PostgreSQL/PostGIS manages geographic data by utilizing this technology.

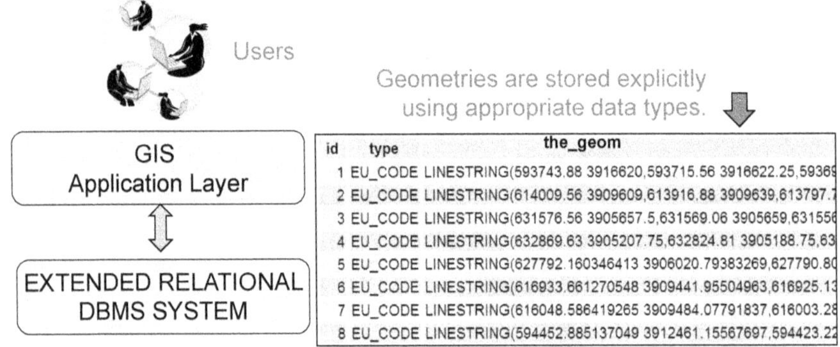


**Fig. 15.6** Geometries stored explicitly in relational tables using geometric data types.

## *15.2.4 Object-Oriented DBMS*

An alternative architecture that offers a single repository with enhanced capabilities is the *object-oriented DBMS* (Figure 15.1d). In this approach, the object-oriented model is used to model an application domain of GIS. The concepts of the underlying model (e.g., classes, inheritance, encapsulation, types, and methods; Chapter 7) are very effective. Any additional operators needed to define and manipulate the geographic entities are included in an application layer which lies on top of the object-oriented DBMS. Several prototype systems have adopted this approach, e.g., GeO2.

As described in Chapter 7, the object-oriented model has its own deficiencies; among others, the complexity of the model. Object-oriented DBMSs are difficult for non-programmers to understand and apply. Therefore, modern geographic DBMSs adopt the extended relational model that combines concepts from both the relational and object models. This model is called object-relational model and is discussed next.

## 15.3 Object-Relational Technology

*Relational technology* offers a series of *built-in data types*, such as integer, real, character, date, etc. In addition it provides a series of *functions* to handle data entries of those types, such as COUNT, AVERAGE, etc. (see Chapter 6).

*Object-relational (OR-) technology* introduces the concept of *object* (OBJECT) and offers the user the option to create *new data types* of objects (CREATE TYPE). These types are treated as objects in relational tables and can *encapsulate functions* (FUNCTIONS), materialized in an object-oriented programming language. Below is an example definition of the object type POINT in Oracle (expressed in SQL3/SQL1999).

```
CREATE TYPE POINT AS OBJECT (
     X        NUMBER,
     Y        NUMBER,
     Z        NUMBER,
     FUNCTIONS (
          DISTANCE (U:POINT, V:POINT)
          RETURNS NUMBER
);
```

This is the definition of a new type for three-dimensional points with coordinates X, Y, and Z of type NUMBER (i.e., real number). One function is encapsulated in the object type: DISTANCE. This function gets as input two points and then it calculates and returns their distance measure.

After the new type POINT is defined along with the function DISTANCE an extended SQL statement can be executed to create a table of TOWNS. Here is an example:

```
CREATE TABLE TOWNS (
     ID            NUMBER,
     NAME          VARCHAR(20),
     POPULATION    NUMBER,
     LOCATION      POINT
);
```

This statement creates a new table named TOWNS to host a set of towns in the database. The full description of those entities, i.e., identifier, spatial, and thematic dimensions (see Chapter 1) will be accommodated in a single table (Figure 15.7). The database management system will be able to handle the geometry column of type POINT according to the prior definition.

In addition, an SQL statement can compute the distance between any two towns by executing the encapsulated function DISTANCE in POINT type definition. Here is an example (for the records shown in Figure 15.7):

```
SELECT    DISTANCE(T1,T2)
FROM      TOWNS
WHERE     T1.NAME = "Fodele"
AND       T2.NAME = "Rodia";
```

The Object-Relational technology offers the tools to generate an extended schema for storing geographic entities in a single database repository and managing them efficiently through appropriate functions as defined over the geometric data types. In addition, spatial index structures can be developed and integrated into the database to accelerate the performance.

On the flip side, building a new schema along with the functions and index structures for managing of geographic entities in an object-relational database is a difficult and time consuming task. For this reason, commercial and open source DBMSs offer a spatial model to their users. Representative examples are the *Oracle Spatial and Graph* in Oracle and *PostGIS* in PostgreSQL. Both are compliant with the Open Geospatial Consortium (OGC) models and to support interoperability across systems and applications.

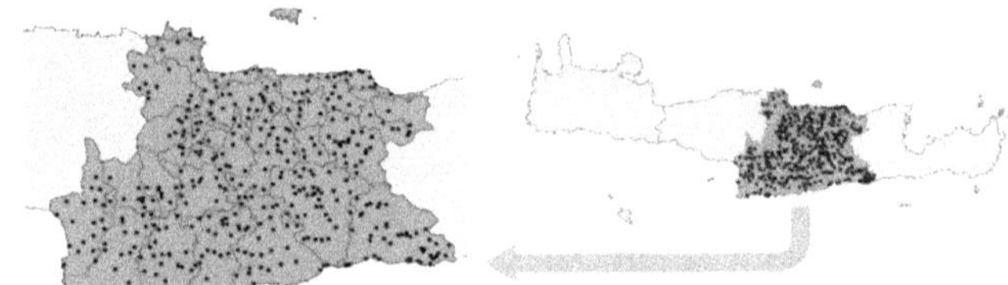

**Table: TOWNS**

| ID (NUMBER) | NAME (VARCHAR(20)) | POPULATION (NUMBER) | LOCATION (POINT) |
|---|---|---|---|
| 11492 | Agia Pelagia | 553 | POINT(592221.204941288 3918593.75) |
| 11501 | Paralia Fodele | 99 | POINT(586237.549732688 3917962.677205) |
| 5428 | Achlada | 119 | POINT(589949.637941288 3917093) |
| 5439 | Fodele | 540 | POINT(586850.762941288 3915575.5) |
| 11543 | Palaiokastro | 105 | POINT(594174.074941288 3913766) |
| 5454 | Rodia | 801 | POINT(592534.074941288 3913700) |

**Fig. 15.7** Table of towns in an Object-Relational database. ID: identifier; NAME and POPULATION: the thematic dimension; and LOCATION: the spatial dimension.

Oracle Spatial follows a hierarchical model, which is shown in Figure 15.8. In this model a *layer* accommodates a set of geometries. A *geometry* comprises a set of elements. An *element* is an instance of a basic geometry type. Figure 15.9 shows the basic *geometry types* supported by Oracle Spatial and an example complex geometry constructed of these types. Hundreds of spatial functions are defined over these geometry types, while powerful index structures based on R-tree and Quadtree variants are also available.

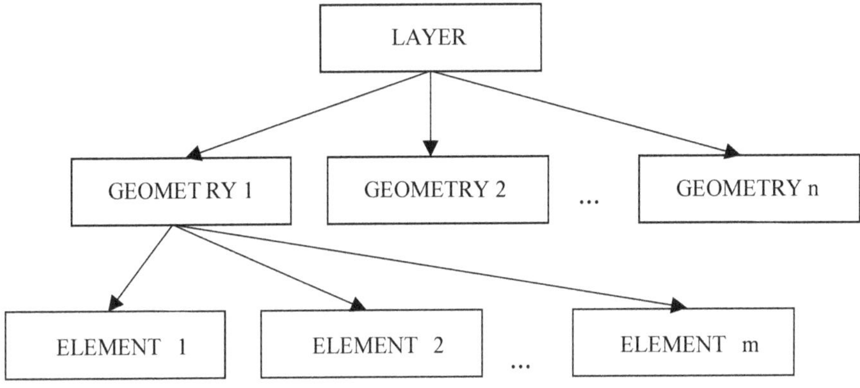

**Fig. 15.8** Oracle Spatial hierarchical data model.

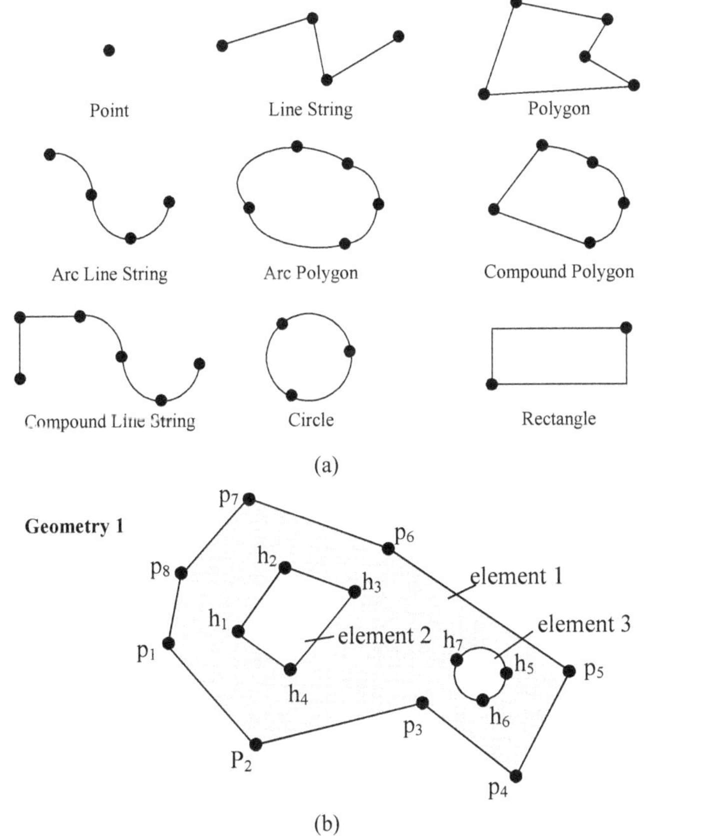



**Fig. 15.9** The basic geometry types (elements) supported by Oracle Spatial (a); and an example complex geometry consisting of three elements (b): the outer boundary of a polygon (element 1) with two holes (elements 2,3).

## 15.4 PostgreSQL and PostGIS

*PostgeSQL* is the most popular open source DBMS with a plethora of applications
and Web services (in the role of database server). PostgreSQL runs on multiple
operating systems (Windows, Linux, MacOS) and is compatible with a multitude
of programming languages and interfaces, such as Python, C/C++, ASP, Java,
ODBC, JDBC, etc. The management is predominantly via PgAdmin III software,
and other applications (e.g., PgAccess, PhpPgAdmin, WinSQL).

*PostGIS* is an extension of PostgreSQL and supports the modeling and management of geographic/spatial data in accordance with the OGC standards. It has
been developed by Refractions Research Inc. and is an open source software
(GNU General Public License). It provides special operators to compile spatial
queries, aggregation functions over spatial data, as well as spatial functions. It also
enables the assignment of projection systems to spatial data. Finally, it can be used
to visualize or share data through open applications, such as the Quantum GIS and
Map Server.

PostGIS model is a superset of the "Simple Features" model defined by the
Open Geospatial Consortium (OGC). Therefore, (part of it) is fully compatible
with the formulations to geographic data management that OGC provides.

According to OGC specifications, PostGIS follows two basic ways of express-
ing spatial objects: (a) *text form* (Well-Known Text – WKT form), and (b) *binary
form* (Well-Known Binary – WKB form). Both include information on the type of
object and its coordinates.

Some examples of the WKT type (for spatial objects) are the following:

- POINT(0 0) – point type
- LINESTRING(0 0,10 10,10 20) – polyline type
- POLYGON(0 0,10 0,20 20,0 20,0 0) – polygon type
- MULTIPOINT(0 0,10 20) – point set type
- MULTILINESTRING( (0 0,10 10,10 20), (20 30,30 30,40 30) ) – polyline set
  type
- MULTIPOLYGON( (0 0,10 0,20 20,0 20,0 0), (0 0,-10 0,-20 -20,0 0) ) – poly-
  gon set type
- GEOMETRYCOLLECTION( POINT(0 5), LINESTRING(0 5,5 10) ) – com-
  plex geometry type

Each spatial object is accompanied with a code defining the coordinate reference system (spatial referencing system identifier – SRID). The input/output formats are supported by the following functions (interfaces):

- bytea WKB = asBinary(geometry);
- text WKT = asText(geometry);
- geometry = GeomFromWKB(bytea WKB, SRID);
- geometry = GeometryFromText(text WKT, SRID);

### 15.4.1 Geographic Database Management in PostGIS

Next, some typical examples of SQL commands for the definition, construction, and manipulation of a geographic database in PostGIS are given.

The following command defines a *thematic (non-spatial) table*:

```
CREATE TABLE roads (
    ROAD_ID        int4,
    ROAD_NAME      varchar(20)
);
```

The table name is roads and has two alphanumeric attributes. The definition of a spatial table requires a column to accommodate the geometry. The following SQL command defines a spatial table:

```
CREATE TABLE parks (
    PARK_ID        int4,
    PARK_NAME      varchar(20),
    PARK_DATE      date
);
SELECT AddGeometryColumn( 'parks', 'PARK_GEOM',
    2100, 'MULTIPOLYGON', 2 );
```

The table is named parks and has three alphanumeric attributes (PARK_ID of type 4-byte integer, PARK_NAME of type text up to 20 characters, and PARK_DATE of type date; the construction date). The column hosting the geometry is called PARK_GEOM and is of type multiple polygon (MULTIPOLYGON) in two-dimensional space (2). The 2100 represents the code of the spatial (coordinate) reference system (SRID; for example, the code 4326 corresponds to WGS'84). The function AddGeometryColumn is a function that integrates PARK_GEOM column in table parks.

The syntax of function AddGeometryColumn is:

```
AddGeometryColumn(<table_name>,<column_name>, <srid>,
<type>, <dimension>)
```

and its parameters are compatible with the two tables of metadata by OGC: geometry_columns and spatial_ref_sys.

Hence, an SQL command to define the *spatial table* of roads is the following (*database definition*):

```
CREATE TABLE roads (
    ROAD_ID        int4,
    ROAD_NAME      varchar(20)
);
SELECT AddGeometryColumn( 'roads', 'ROAD_GEOM',
    -1, 'LINESTRING', 3 );
```

SRID equals to -1 means that the spatial reference system is user defined and does not follow any standard SRID. The geometry type is LINESTRING (i.e., polylines) in three-dimensional space (3).

The following command inserts a record in the table roads (*database construction*):

```
INSERT INTO roads ( ROAD_ID, ROAD_NAME, ROAD_GEOM )
VALUES (
          1,
          'Main Rd',
          GeomFromText('LINESTRING(1712 2431 12,
               1711 2432 11)',-1)
);
```

The retrieval of data from a table is done by appropriate SELECT statements (*database manipulation*). An example is:

```
SELECT ROAD_ID, AsText(ROAD_GEOM) AS GEOMETRY, ROAD_NAME
FROM roads;
```

and returs the following:

```
ROAD_ID | GEOMETRY                                    | ROAD_NAME
--------+---------------------------------------------+----------
    1   | LINESTRING(1712 2431 12,1711 2432 11) | Main Rd
    2   | LINESTRING(1991 2441 43,1992 2448 45) | Mapple St
    3   | LINESTRING(1827 2281 11,1826 2298 11) | King St
(3 rows)
```

PostGIS supports the following *operators* (see Figure 13.13):

- && checks whether the minimum bounding rectangles (MBRs) of two geometries overlap
- ~= checks whether two geometries are identical
- =: checks whether the minimum bounding rectangles (MBRs) of two geometries are identical

PostGIS supports a rich collection of *functions* for processing geospatial entities, most of which are compatible with the OGC specifications.

Table 15.1 presents a list of some representative functions available in PostGIS. These functions can be incorporated in the SELECT queries to process or analyze any data retrieved from the database. An example of such a query is the following:

```
SELECT area(PARK_GEOM)/10000 AS hectares
FROM parks
WHERE name = 'ODELL';
```

This function computes the area size of a polygonal geometry (PARK_GEOM) retrieved from the database. The query result is the following:

```
hectares
--------------
137.9
(1 row)
```

PostgreSQL/PostGIS supports the following *index structures* to accelerate the execution of operations in the database:

- *B-trees:* indexing alphanumeric attributes (see Chapter 8).
- *R-trees:* indexing geometries (see Chapter 13).
- *GiST-trees* (Generalized Search Trees): indexing geometric data based on topological and other spatial relations (in combination with the R-tree structure it achieves rapid retrieval of geometric data and relations from the database).

**Table 15.1** A list of representative functions supported by PostGIS.

| Management Functions |
| --- |
| • AddGeometryColumn(varchar, varchar, varchar, integer, varchar, integer) <br> • DropGeometryColumn(varchar, varchar, varchar)) <br> • SetSRID(geometry) |
| **Geometry Relationship Functions** |
| • Distance(geometry,geometry) <br> • Equals(geometry,geometry) <br> • Disjoint(geometry,geometry) <br> • Intersects(geometry,geometry) <br> • Touches(geometry,geometry) <br> • Crosses(geometry,geometry) <br> • Within(geometry,geometry) <br> • Overlaps(geometry,geometry) <br> • Contains(geometry,geometry) <br> • Relate(geometry,geometry) <br> • Area(geometry) <br> • Length(geometry) <br> • Buffer(geometry,double,[integer]) <br> • ConvexHull(geometry) |
| **Geometry Accessors** |
| • AsText(geometry) <br> • AsBinary(geometry) <br> • NumGeometries(geometry) <br> • NumPoints(geometry) <br> • PointN(geometry,integer) <br> • GeometryType(geometry) <br> • X(geometry) <br> • Y(geometry) <br> • Z(geometry) |
| **Geometry Constructors** |
| • GeomFromText(text,[<srid>]) <br> • PointFromText(text,[<srid>]) <br> • LineFromText(text,[<srid>]) <br> • PolyFromText(text,[<srid>]) |

Finally, PostGIS offers useful tools and interfaces to various geospatial formats, e.g., *shapefile, GML, KML*, etc.


## *15.4.2 Development of a Spatial Database in PostGIS*

Following, the development of a geographic database in PostgreSQL/PostGIS DBMS is presented. The data relates to the Prefecture of Heraklion in Crete island, Greece. Initially, the data is organized in three layers, of shapefile format. These layers are:

- Towns (towns.shp – point geometry),
- Roads (roads.shp – line geometry),
- Municipalities (municipalities.shp – polygon geometry).

These layers are visualized in Figure 15.10, and are accompanied by attributes (stored in the corresponding attribute tables), as shown in Figure 15.11.

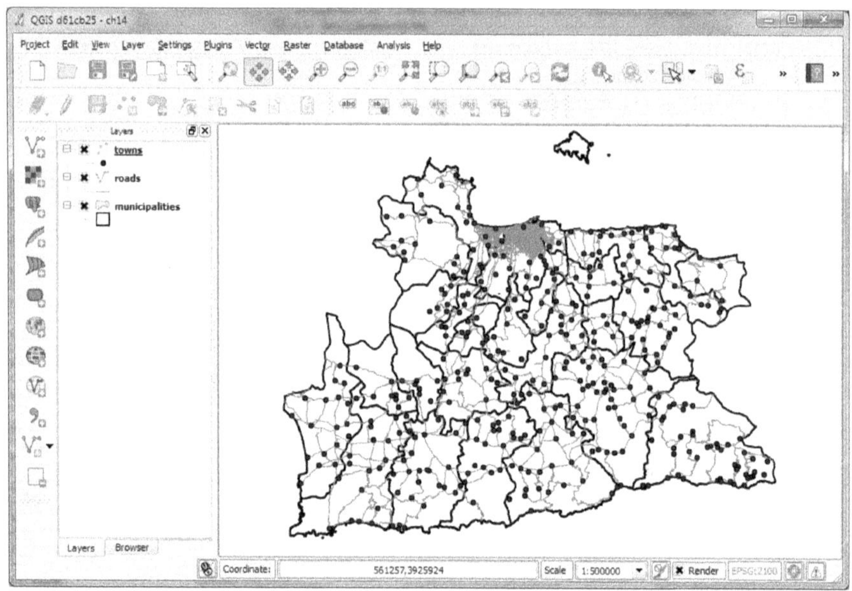

**Fig. 15.10** Visualization of the three layers for the Heraklion Prefecture in QGIS.

As the data is available in shapefile format (three layers: towns, roads, and municipalities), it will need to be transformed into a series of SQL commands, which after execution will map the content into a set of tables in PostgreSQL/PostGIS. This transformation can be done using the *shp2pgsql* (i.e., Shapefile-to-PostgreSQL) utility offered by PostGIS. This program runs in command line. Specifically, the following commands create the corresponding SQL scripts: towns.sql, roads.sql, municipalities.sql.

```
shp2pgsql -s 2100 -I towns.shp towns > towns.sql

shp2pgsql -s 2100 -I roads.shp roads > roads.sql

shp2pgsql -s 2100 -I municipalities.shp municipalities >
    municipalities.sql
```

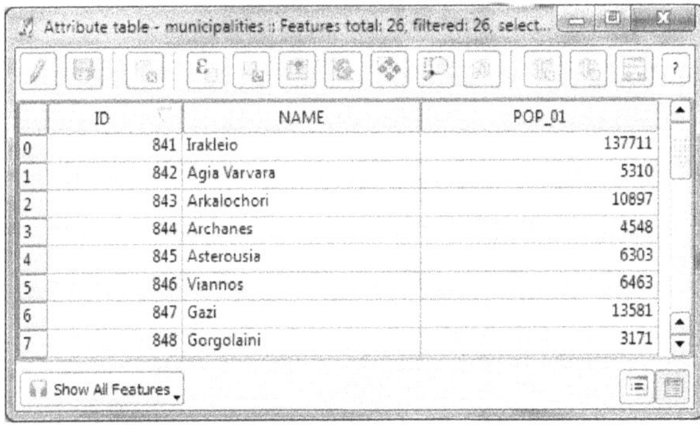

(a)

(b)

(c)

**Fig. 15.11** The attribute tables associated with the three shapefiles of Heraklion prefecture: (a) towns, (b) roads, and (c) municipalities.

The code –s 2100 corresponds to the SRID GGRS'87 (Greek Grid, EPSG:2100), while the parameter –I refers to the creation of a GiST index on the geometry column of the corresponding tables.

Following are the SQL commands (in DDL; see Chapter 6) that define the three tables (towns, roads, municipalities):

```
CREATE TABLE "towns" (gid serial PRIMARY KEY,
"id" int8,
"name" varchar(60),
"pop01" int8);
SELECT AddGeometryColumn('','towns','the_geom',
    '2100','POINT',2);
CREATE INDEX "towns_the_geom_gist" ON "towns"
    using gist ("the_geom" gist_geometry_ops);


CREATE TABLE "roads" (gid serial PRIMARY KEY,
"city_id" int4,
"dir" int2,
"n_line" varchar(1),
"speed" varchar(3),
"eu_code" varchar(10),
"length" float8,
"nameeng" varchar(35));
SELECT AddGeometryColumn('','roads','the_geom',
    '2100','MULTILINESTRING',2);
CREATE INDEX "roads_the_geom_gist" ON "roads"
    using gist ("the_geom" gist_geometry_ops);


CREATE TABLE "municipalities" (gid serial PRIMARY KEY,
"id" int8,
"name" varchar(60),
"pop01" int8);
SELECT AddGeometryColumn('','municipalities','the_geom',
    '2100','MULTIPOLYGON',2);
CREATE INDEX "municipalities_the_geom_gist"
    ON "municipalities"
    using gist ("the_geom" gist_geometry_ops);
```

Following, the SQL commands (in DML; see Chapter 6) to insert the first two records in the table of towns are given:

```
INSERT INTO "towns" ("id","name","pop01",the_geom)
VALUES ('11492','Agia Pelagia','553',
'SRID=2100;01010000008610EE68BA122241000000E080E54D41');
```

```
INSERT INTO "towns" ("id","name","pop01",the_geom)
VALUES ('11501','Paralia Fodele','99',
'SRID=2100;01010000001C907619FBE3214148A7AE5645E44D41');
```

The last piece in the above two INSERT statements relate to the coordinates of towns (point entities) encoded in hexadecimal format.

Figure 15.12 shows a snapshot of the three tables (part of), as they are stored in PostgreSQL/PostGIS. Next, some representative SQL queries to retrieve and analyze the database content are given.

---

✓ Table: Towns (geometry type: point)

```
SELECT id, name, pop01, AsText(the_geom)
FROM towns;
```

| id (identifier) | name (city/village name) | pop01 (population in 2001) | the_geom (geometry column) |
|---|---|---|---|
| 11492 | Agia Pelagia | 553 | POINT(592221 3918593) |
| 11501 | Paralia Fodele | 99 | POINT(586237 3917962) |
| 5428 | Achlada | 119 | POINT(589949 3917093) |
| ... | ... | ... | ... |

    400 rows

✓ Table: roads (geometry type: multilinestring)

```
SELECT id, speed, eu_code, AsText(the_geom)
FROM roads;
```

| id (identifier) | speed (speed limit) | eucode (E75 for national roads) | the_geom (geometry column) |
|---|---|---|---|
| 1 | 50 | E75 | MULTILINESTRING(...) |
| 2 | 90 | E75 | MULTILINESTRING(...) |
| 3 | 70 | E75 | MULTILINESTRING(...) |
| ... | ... | ... | ... |

    12228 rows

✓ Table: municipalities (geometry type: multipolygon)

```
SELECT id, name, pop01, AsText(the_geom)
FROM municipalities;
```

| id (identifier) | name (municipality name) | pop01 (population in 2001) | the_geom (geometry column) |
|---|---|---|---|
| 841 | Irakleio | 137711 | MULTIPOLYGON(...) |
| 842 | Agia Varvara | 5310 | MULTIPOLYGON(...) |
| 843 | Arkalochori | 10897 | MULTIPOLYGON(...) |
| ... | ... | ... | ... |

    26 rows

**Fig. 15.12** The three tables (part of) in PostGIS database.

The first query counts the number of towns per municipality. This is a topological query that assigns each town t in the corresponding municipality m. The search applies the topological operator *intersects*.

<u>Query 1:</u>

```
SELECT r.name as Municipality, count(t.the_geom)
    as Number_of_cities
FROM municipalities AS m, towns AS t
WHERE intersects(t.the_geom, m.the_geom)
GROUP BY m.name
ORDER BY Number_of_cities DESC;
```

Result:

```
Municipality   | Number_of_cities
---------------+-----------------
Viannos        | 46
Arkalochori    | 40
Asterousia     | 28
Gortyna        | 26
Kasteelli      | 24

...
```

The second query calculates the length (total sum) of roads that lie entirely within each municipality. In the report, the five municipalities with the largest road network are listed. The search applies the topological operator *contains*.

<u>Query 2:</u>

```
SELECT m.mane As Municipality,
    sum(length(r.the_geom))/1000 As Roads_km
FROM roads AS r, municipalities AS m
WHERE r.the_geom && m.the_geom
AND contains(m.the_geom,r.the_geom)
GROUP BY m.name
ORDER BY Roads_km DESC
LIMIT 5;
```

Result:

```
Municipality   | Roads_km
---------------+---------
Irakleio       | 595.440
Gazi           | 200.706
Arkalochori    | 147.853
Asterousia     | 146.120
Viannos        | 137.159
---------------+---------
```

Note that the first criterion in the WHERE clause refers to the filter comparison between road MBRs and municipality MBRs in pairs (operator &&; see previous Section). If they intersect, a detailed comparison between the actual geometries in the pair follows to verify the topological relationship (Figure 15.13).

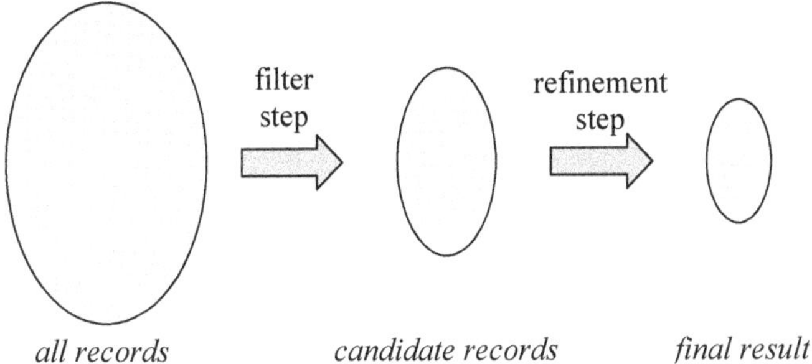


**Fig. 15.13** The two-tier query model.

The third query returns the geometries and specifically those road segments that intersect the municipality of Archalochori. The result is hosted in a new table named ark_roads, created before the retrieval of roads (first two lines).

Query 3:

```
CREATE TABLE "ark_roads" (gid serial PRIMARY KEY,
    "id" int4);
SELECT AddGeometryColumn('','ark_roads','the_geom','2100',
    'MULTILINESTRING',2);
INSERT INTO ark_roads(id, the_geom)
SELECT r.gid, r.the_geom
FROM roads AS r, municipalities AS m
WHERE r.the_geom && m.the_geom
AND intersects(m.the_geom,r.the_geom)
AND m.name = 'Arkalochori'
```

The result of this query is visualized in Figure 15.14a. To select only the segments of the road network falling fully within Archalochori, the topological function *intersects* must be replaced by the function *contains*.

The next query searches for the neighboring municipalities of Arkalochori. It is implemented by applying the topological function *touches*. As the result is also geometric (the geometries of neighboring municipalities), it is hosted in a new table named ark_neigh and visualized in Figure 15.14b.

Query 4:

```
CREATE TABLE "ark_neigh" (gid serial PRIMARY KEY,
    "id" int4);
SELECT AddGeometryColumn('','ark_neigh','the_geom','2100',
    'MULTIPOLYGON',2);
```

```
INSERT INTO ark_neigh (id, the_geom)
SELECT n.gid, n.the_geom
FROM municipalities as m, municipalites as n
WHERE m.name = 'Arkalochori'
AND touches(m.the_geom, n.the_geom);
```

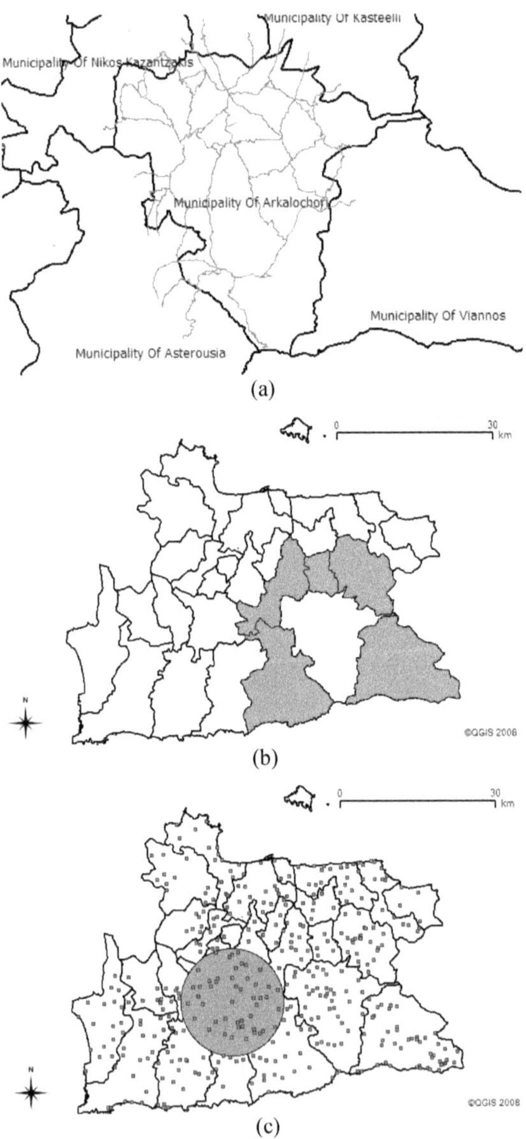



**Fig. 15.14** Visualization of queries that return geometries.

The last query creates a buffer zone of radius 10km around the city of Larani. The buffer zone is hosted in a new table named buffer, and visualized in Figure 15.14c. The number 16 in the definition of the buffer refers to the number of vertices used to represent the circle as polygon geometry.

<u>Query 5:</u>

```
CREATE TABLE "buffer" (gid serial PRIMARY KEY,"id" int4);
SELECT AddGeometryColumn('','buffer','the_geom','2100',
    'POLYGON',2);
INSERT INTO buffer (id, the_geom)
SELECT id, Buffer(the_geom, 10000, 16)
FROM towns
WHERE name = 'Larani';
```

# References and Further Reading


Aronoff, S., 1989. *Geographic Information Systems: A Management Perspective*. WDL Publications.

Egenhofer, M., 1989. A formal definition of binary topological relationships. In the *Proceedings of the International Conference on Foundations of Data Organization and Algorithms*, pp. 457-472.

Garcia Molina, H., Ullman, J.D., and Widom, J., 2000. *Database Systems Implementation*. Prentice Hall.

Melton, J., and Mattos, N., 1996. *An Overview of SQL 3: The Emerging New Generation of the SQL Standard*. Tutorial No. T5, VLDB Conference, Bombay.

Mitchell, T., 2005. *Web Mapping*. O'Reilly:USA.

OGC – *Open Geospatial Consortium*, http://www.opengeospatial.org/

Oracle Corp. 2003. http://www.oracle.com/

Oracle Spatial, 2003. *Oracle Spatial – Reference Manual*. Oracle Corp.

OSGeo, 2009. http://www.osgeo.org

PostgeSQL, 2009. http://www.postgresql.org/

PostGIS, 2009. http://postgis.refractions.net

Rigaux, P., Scholl, M., and Voisard, A., 2002. *Spatial Databases with Applications to GIS*. Morgan-Kaufmann.

Stefanakis, E., and Prastacos, P. 2008. Development of an Open Source Based Spatial Data Infrastructure. *AppliedGIS Journal*, Vol 4(4).

Stonebraker, M., and Moore, D., 1996. *Object-Relational DBMSs: The Next Great Wave*. Morgan-Kaufmann.

Shashi, S., and Chawla, S., 2003. *Spatial Databases – A Tour*. Prentice Hall.

Worboys, M.F., 1995. *GIS – A Computing Perspective*. Taylor-Francis.

W3C, 2003. *World Wide Web Consortium*. http://www.w3.org/

Yeung, A.K.W, amd Hall, G.B., 2007. *Spatial Database Systems: Design, Implementation and Project Management*. Springer.

# Index

CPSIA information can be obtained
at www.ICGtesting.com
Printed in the USA
FSHW020948140319
56369FS